AF895831

Biology

Bassim Hamadeh, CEO and Publisher
John Remington, Senior Field Acquisitions Editor
Michelle Piehl, Senior Project Editor
Alia Bales, Production Editor
Jess Estrella, Senior Graphic Designer
Stephanie Kohl, Licensing Coordinator
Don Kesner, Interior Designer
Natalie Piccotti, Director of Marketing
Kassie Graves, Vice President of Editorial
Jamie Giganti, Director of Academic Publishing

Copyright © 2019 by Cognella, Inc. All rights reserved. No part of this publication may be reprinted, reproduced, transmitted, or utilized in any form or by any electronic, mechanical, or other means, now known or hereafter invented, including photocopying, microfilming, and recording, or in any information retrieval system without the written permission of Cognella, Inc. For inquiries regarding permissions, translations, foreign rights, audio rights, and any other forms of reproduction, please contact the Cognella Licensing Department at rights@cognella.com.

Trademark Notice: Product or corporate names may be trademarks or registered trademarks, and are used only for identification and explanation without intent to infringe.

Cover images:
Copyright © Depositphotos/A-R-T-U-R.
Copyright © Depositphotos/lightsource.
Copyright © Depositphotos/zhuzhu.
Copyright © Depositphotos/ridofranz.
Copyright © Depositphotos/Yaruta.
Copyright © Depositphotos/panuruangjan.
Copyright © Depositphotos/Khakimullin.
Copyright © Depositphotos/IgorTishenko.
Copyright © Depositphotos/VERSUSstudio.

Interior image:
Copyright © Mokkie (CC BY-SA 4.0) at https://commons.wikimedia.org/wiki/File:Ammonite_1.JPG

Printed in the United States of America.

ISBN: 978-1-5165-0895-2(pbk) / 978-1-5165-0896-9 (br)

cognella® | ACADEMIC PUBLISHING

Biology

TEXTBOOK AND ACTIVITIES FOR THE FLIPPED CLASSROOM

First Edition

Kevin M. Bonney and Lori Nicholas

cognella® | ACADEMIC PUBLISHING

Contents

Scientific Inquiry

1.1 INTRODUCTION

Scientific inquiry is an important tool for gaining understanding of the natural world and the universe in which we live. The methods of science are numerous and diverse and can be applied to a wide range of situations. These methods are not static but are continually being modified, expanded, and improved, as is our understanding of natural processes. However, all science is based upon a certain set of rules and conventions. This chapter discusses the fundamentals of scientific inquiry and provides a foundation from which to approach the information provided in the rest of this textbook.

Fundamentals of Science

Science can be defined as the intellectual and practical process of systematically studying the natural world through observation and experimentation. In other words, science is a method of generating knowledge about humans and the world in which we live. In fact, the word science is derived from the Latin word *scientia*, meaning knowledge. Science is a very useful way of gaining knowledge, but it is not the only way. There are many ways to learn about humans and the world other than science, including astrology, sociology, psychology, and theology. None of these means of understanding are necessarily *the* right way or *the* wrong way. Rather, each of these approaches serves a particular purpose for which it is best suited. There are many questions best answered by science, such as how chemical signals are transmitted between neurons. There are also many questions best answered by approaches other than science, such as whether humans have a soul. Science generally works best for answering questions that involve making observations and testing explanations for those observations. Although the following discussion focuses on science alone, it is important to remember that, in the real world, science is often thoroughly interconnected with and reliant upon skills and knowledge derived from other disciplines.

On a fundamental level, science can be categorized as either descriptive or hypothesis-driven. Descriptive science involves observing, describing, and classifying natural phenomena, then using that information to form generalizations. This type of science often relies on **inductive reasoning**, which is a type of logical thinking that generates conclusions by forming

LEARNING OBJECTIVES

- Define the terms hypothesis, law, science, and theory
- Compare deductive reasoning with inductive reasoning
- Discuss the main steps and cyclical nature of the scientific method
- Categorize different types of data and variables
- Identify the hypothesis, experimental procedure, and conclusions of a published scientific paper
- Apply the scientific method to an experimental activity
- Describe the pros and cons of the peer review process
- Discuss ways to evaluate whether a publication is an appropriate source of information to cite in a research study or paper

FIGURE 1.1: **Examples of inductive and deductive reasoning**—In inductive reasoning, you look for patterns and then make a conclusion. In deductive reasoning, you use known information to make a conclusion.

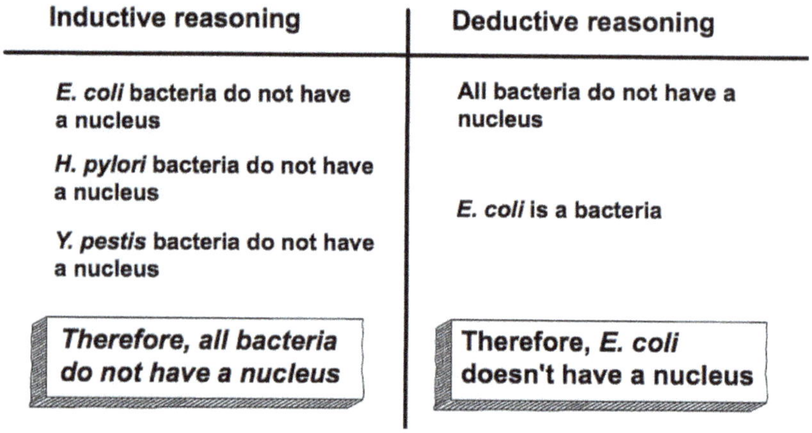

connections between multiple related observations. Hypothesis-driven science focuses on testing hypotheses through experimentation. A **hypothesis** is a proposed explanation for an observation that can be tested. Hypotheses are based on **deductive reasoning**, which is a type of logical thinking in which specific results are predicted based on known information and general principles. Sometimes a hypothesis is referred to as an educated guess; however, many scientists disagree with this comparison because hypotheses are based on evidence and are testable, whereas guesses do not necessarily meet either of those criteria. Because a hypothesis can be tested, it is said to be falsifiable, meaning that it can be shown to be false. In fact, the goal of the scientific method is to determine whether the data collected either supports or refutes the explanation proposed by the hypothesis. If the data align with the hypothesis, the hypothesis is accepted; if the data do not fit within the expectations proposed by the hypothesis, the hypothesis is rejected. See Figure 1.1 for examples of these two different kinds of reasoning.

Hypotheses are often (but not necessarily) written as "If/then" statements. One example of a hypothesis is "If reading textbooks increases knowledge retention, then students who read this textbook will score higher on the exam in this class than students who do not read this book." However, the statement "If ghosts help a student, that student will pass the test" is not a hypothesis because it cannot be observed or experimentally tested and falsified.

In the first example, if data collected during the experiment showed that students who read this book did score higher on the exam than students who do not read this book, that data would support the hypothesis, and the hypothesis would be accepted. However, if the data collected showed there was no difference in test scores between students who read this book and those who did not, or if students who did not read the book earned higher scores on the test, then the hypothesis would be falsified and rejected. A new hypothesis could then be tested. Oftentimes, in the practice of science, many hypotheses are experimentally tested and falsified before a hypothesis is accepted, or meaningful conclusions can otherwise be drawn from the collected data.

After a number of repeated observations and experiments produce a large collection of evidence that supports a broad conclusion, a scientific theory may be generated. A **theory** is an explanation for why some aspect of the natural world occurs the way it does that is well substantiated by a large amount of evidence. Scientific theories differ from hypotheses because they are broader in scope, are supported by a greater amount and variety of evidence, and are general enough to support the generation of new hypotheses. A scientific theory should also not be confused with a **law** of science, which is a statement that summarizes a set of observations about *what* occurs in the natural world but does not explain *why* it occurs.

The meaning of the word *theory* in science is widely misunderstood because of confusion with colloquial usage of the word. It is sometimes suggested that scientific theories do not warrant acceptance, are inherently flawed or incomplete, or are akin to opinions or guesses because they are *just* theories. On the contrary, rather than being a guess or an opinion, scientific theories are so strongly supported by evidence that they are capable of withstanding scrutiny from other scientists and are widely accepted by many experts. Furthermore, if any convincing evidence existed that invalidated or brought into question what a theory states, the theory would no longer exist (it would either have been changed or rejected as a theory). That is not to say that once a theory is proposed it is unquestionable and unchangeable, or that it can completely explain every detail about the topic it addresses. As our technological capabilities and ability to understand the world continually advance, even the most widely accepted theories might be adjusted and revised to reflect new knowledge. This illustrates the inquiry-driven nature and testability of science.

The Scientific Method

Science is a way of learning about the world by applying a type of methodology known as the scientific method. The scientific method is often illustrated as a linear progression of five or six steps with a clear beginning and end. Although distilling the process of science down to this simple formula can be helpful for gaining a rudimentary understanding of how science works, it is important to understand that science is much more complex than this conceptualization of the scientific method indicates. For one, there is not only one method by which science is always conducted. There are many methods used in science, and many ways that they can be combined or adapted to accommodate a particular purpose. Second, science is not linear. Whether speaking of an individual experiment or the entire history of science, science does not travel in one direction from start to finish. Science is an iterative process. Because of this, the **scientific method** is best thought of as an evolving process that consists of making observations, asking questions, proposing hypotheses, conducting experiments, analyzing data, and refining this approach until conclusions can be drawn and shared with others (Figure 1.2).

Each new line of scientific inquiry begins with making an observation. In hypothesis-driven

FIGURE 1.2: The cyclical nature of the scientific method—The scientific method is not a linear process but is more cyclical in nature.

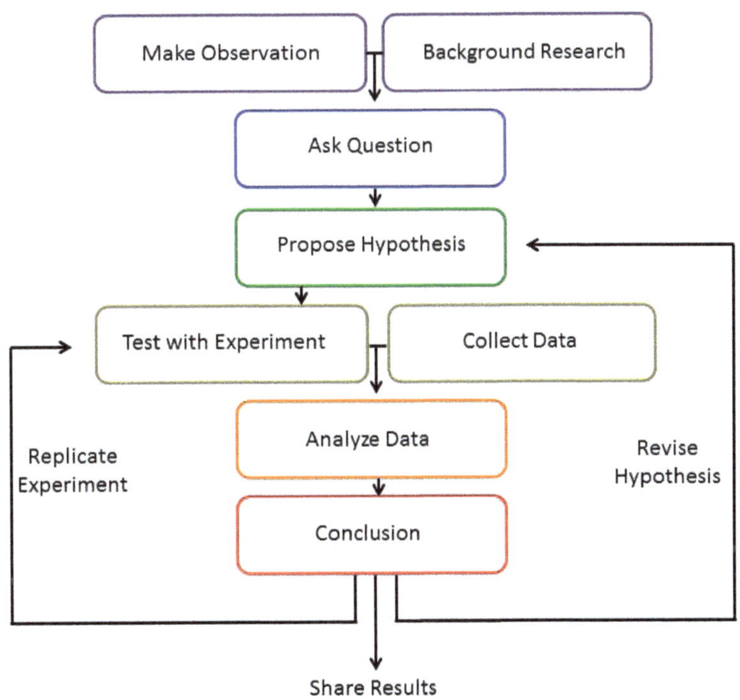

science, scientists make observations about the natural world not for the purpose of describing the way things are but rather to identify why or how things are the way they are. Biologists may make observations about living things or about the interactions between living things and their environment. These observations can be based upon personal experience and thoughts or result from learning about observations made by others.

Observations lead scientists to ask questions and make predictions that can be tested and potentially answered by conducting experiments. Before commencing an experiment, scientists also conduct background research to learn what is already known about the topic. Completing thoughtful and thorough background research helps ensure the scientific method will yield meaningful results that build upon the current body of knowledge and answer questions that have not already been sufficiently answered. Background research also helps scientists design appropriate and effective experiments by anticipating the need for materials, equipment, training, and outside expertise, as well as to account for the amount of time and funding needed and to troubleshoot potential problems.

The next step is to propose a hypothesis, which is a testable explanation for a question about why or how something in the natural world exists the way it does. Observations may generate many questions and hypotheses, but only one hypothesis can be tested at a time with a single experiment. Once a hypothesis is formulated, predications are made about the results that may be obtained during the experiment and what the implications of those results may be. It is important to develop a hypothesis and predictions before conducting an experiment and collecting data, rather than retrospectively trying to propose a hypothesis that explains data already collected. This helps ensure experimental work is purposeful and objective. After predictions are made, experiments are conducted. Data are collected during experiments and potentially before and after the experiment is conducted as well. There are many types of data that could be collected, some of which are described later in this chapter. The next step is for data to be analyzed and used to draw conclusions. However, experiments must often be repeated before conclusions can be drawn to obtain sufficient and reliable data. Initial data analysis may lead a scientist to revise or refine their initial hypothesis. The process of science often consists of many iterations of these steps that sometimes lead to the next step but often lead back to the previous step or through a cycle of revising and refining the approach to the current step. Results obtained at any point along this path may also lead to the abandonment of a particular line of investigation in favor of a different line of inquiry that is more interesting, more important, more time sensitive, or more feasible. The ultimate goal of science is to share findings with others. Learning about what previous experiments have shown often leads to new observations that initiate new rounds of scientific inquiry. Because of this, it may be best to conceive of the scientific method as a cycle rather than a line pointing in one direction as shown in Figure 1.2.

Experimental Variables and Groups

One of the first steps in designing an experiment is identifying the variables. A **variable** is any condition or item that can be changed or controlled during an experiment. There are three main types of variables to be aware of when planning an experiment. Each experiment may involve more than one of each type of variable. The first type of variable is the **independent variable**, which is the condition or item deliberately changed or manipulated during the experiment. The second type of variable is the **dependent variable**, which is what the scientist observes and measures during the experiment. The dependent variable gets its name because what happens to it depends on what the independent variable does. The third type of variable is most often referred to as a control variable but can also be called a constant variable, an extraneous variable, or a standardized variable. **Control variables** are factors that are kept equal and left unchanged through the experiment. Although control variables are not what the scientist is trying to test or measure, they must be closely monitored throughout the experiment to ensure they remain constant. This is important because uncontrolled changes in these variables could affect the dependent variable and alter or obscure the results of the experiment.

For example, imagine a scientist is conducting an experiment to test the hypothesis that plants will grow more under red-colored light than under green-colored light. In this experiment, the color of the light (red or green) is the independent variable because it is what the scientist deliberately changes (Figure 1.3). The growth of the plant is what will be observed and measured and is expected to change in response to changes in the independent variable, so it is the dependent variable (Figure 1.3). Any other factor that is not one of the independent or dependent variables, yet may affect the outcome of the experiment, could be a considered a control variable. In this example, control variables include the amount of water, oxygen, and nutrients each plant receives (Figure 1.3). It is important to carefully monitor these variables and make sure they remain equal between plants, otherwise there is no way to know whether differences observed in the dependent variable (plant growth) are a result of the independent

FIGURE 1.3: Three types of variables in an experiment examining how exposure to different types of light affects the growth of plants—In an experiment with plants grown in sunlight and red and green light, there are an independent variable, a dependent variable, and controlled variables.

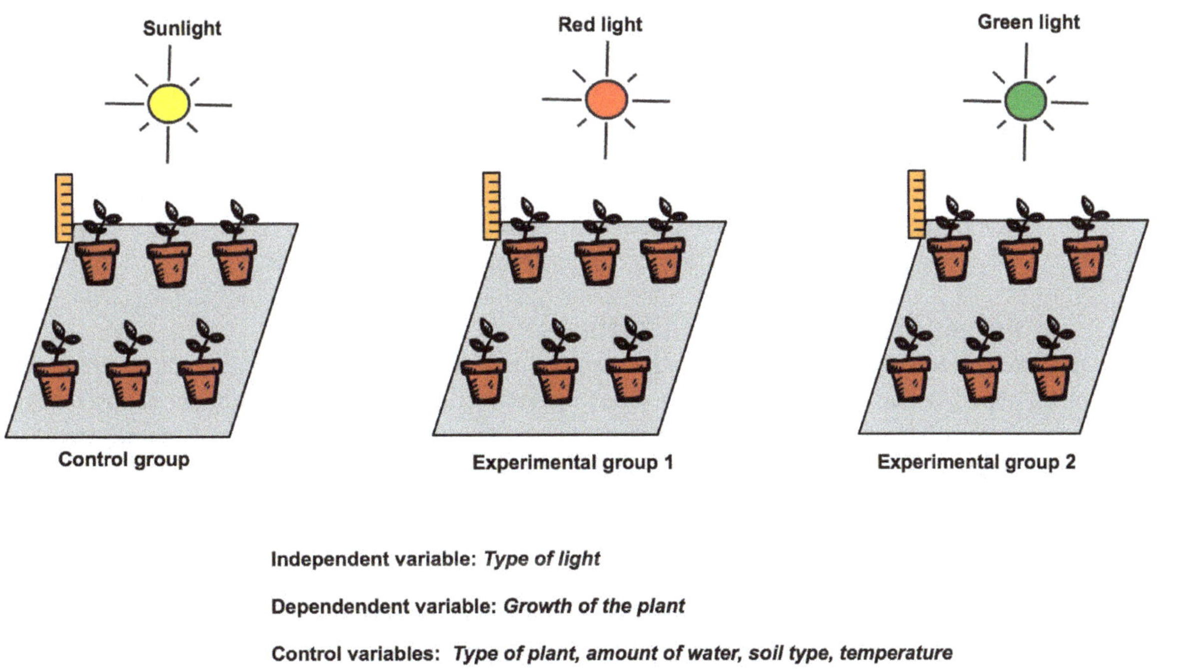

variable (light color) or another factor. It is worth mentioning that control variables cannot simply be eliminated from an experiment so the scientist does not have to worry about them. In this case, the experiment could not be successful if the plants were not provided water, oxygen, and nutrients.

It is important not to confuse control variables with control groups. In scientific experiments, observations and hypotheses usually apply to multiple objects or individuals, rather than a single object or individual. These objects and individuals are grouped based on how the independent variable is applied to them. Groups in which changes to the independent variable are tested are called **experimental groups**, whereas groups in which the independent variable is not applied or changed are **control groups**. In the hypothetical experiment described earlier, plant growth observed in the experimental groups should be compared with growth observed in a control group to determine whether the independent variable induced an increase in growth, a decrease in growth, or no change in growth compared with the average amount of growth that would normally occur in the absence of the independent variable. Plants grown in red light would comprise one experimental group, and plants grown in green light would comprise a second experimental group. Plants grown in white light (unfiltered sunlight) and plants grown without any light could form control groups. The white light group would be considered a positive control group, because the independent variable was applied to that group, whereas plants grown in no light would be referred to as a negative control group, because they did not receive the independent variable.

As with control variables, it is important that the control group does not vary from the experimental group in any way other than application of the independent variable. Otherwise, it cannot be known for certain whether any difference observed is attributable to the independent variable or is a result of the effect of other factors. This is why clinical trials for medical treatments often include a placebo group, rather than a group of people not given anything. A placebo is a treatment (e.g., pill, injection, exercise, etc.) that seems identical to the real treatment being tested in every way except that it does not have the active therapeutic ingredient. It is well documented that the act of participating in a perceived therapeutic intervention, such as taking a pill, may exert a real psychological and/or physiologic effect. This is known as the placebo effect. By giving the control group in a medical experiment a placebo that mimics the independent variable, the placebo effect can be ruled out as a potential cause for changes in dependent variables observed during the experiment.

Sample Size and Other Considerations

Many factors in addition to those already mentioned must be considered when forming experimental and control groups. One important consideration is the number of observations or individuals to include in each group. This is called the **sample size**. When determining the appropriate sample size, scientists must consider a number of factors including the effect size (the magnitude of the difference expected to be observed between groups), the inherent variability in each group (often measured as a standard deviation), and the desired power or significance level of the results (how mathematically certain one wants to be that the results observed are caused by the independent variable rather than random chance or other factors). To conserve resources and limit costs, scientists often attempt to work with the smallest sample size necessary to make accurate and reliable conclusions. Another factor related to sample size is replication, or the repetition of an experimental condition. The replication number can be increased by increasing the sample size when conducting the experiment or by repeating the same experiment with the same experimental conditions. To prevent results from being influenced by bias or random variation, the sorting of objects or individuals into groups should also be randomized. This not a haphazard process but is achieved through use of a computer program or other deliberate means of generating true randomization. Finally, to apply conclusions from an experiment to make generalizations or inferences about a larger population, it is important to ensure representativeness. Representative groups accurately reflect the variation of relevant characteristics observed in the population of interest.

Types of Data

Data can be classified into a variety of different types. First, data can be categorized as either qualitative or quantitative. **Qualitative data** approximates and describes characteristics, whereas **quantitative data** defines specific measurements. Qualitative data includes observable attributes, such as appearance, color, profession, religion, smell, and taste, as well as personal reports about feelings, opinions, and behaviors. Qualitative data that can neither be measured numerically nor ordered along a scale, such as colors or professions, is also known as **nominal data**. Scientific reports often focus on quantitative data, but each type of data is useful for specific purposes.

Examples of quantitative data include age, area, cost, height, length, score, temperature, time, volume, weight, and other numerical values that can be measured. Quantitative data can be further classified as either continuous or discrete. **Continuous data** can be represented by any numerical value within an infinite or defined interval. **Discrete data**, on the other hand, can only be represented by a finite set of values. In other words, continuous data can be always represented on a finer and finer level using smaller units of measurement and more decimal places, but discrete data cannot be further divided beyond a certain point. Examples of continuous data include temperature, time, and volume. The number of students in a classroom, the number of marbles in a jar, and the age of a person in years are all examples of discrete data. Another way to visualize the difference between continuous data and discrete data is to compare an analog clock, such as a wall clock with hands that rotate around a circular face, with a digital clock that displays numbers on a computerized screen. On an analog clock, time can be measured in hours, minutes, seconds, and even fractions of a second. However, the accuracy of measuring time with a digital clock is limited by which discrete numbers the clock displays. If the digital clock only displays hours and minutes, there is no way to measure the time in seconds or fractions of a second using that clock.

Sometimes the line between qualitative data and quantitative data may be unclear, for example, if students rate how good this textbook is on a scale from 1 to 5 where 1 is the worst and 5 is the best. This type of data is known as ordinal data. **Ordinal data** is a type of data where variables have a natural or implied order, but the distances between values cannot be known or reliably measured. In the example of the textbook rating scale, it is clear that 2 is better than 1, and 3 is better than 2, but the difference between 2 and 3 cannot be measured or defined. Furthermore, 2 cannot be defined as twice as good as 1, and the difference between 1 and 2 does not necessarily mean the same as the difference between 2 and 3. This is in stark contrast to the measurements on a ruler where the length of 2 inches can be reliably defined as twice the length of 1 inch, and the distance between the 1-inch mark on a ruler and the 2-inch mark is the same as the distance between the 2-inch mark and the 3-inch mark. Although ordinal data may involve numbers, it cannot be analyzed and interpreted in the same way as other quantitative data. One common misconception is that ordinal data, such as that obtained from a satisfaction survey, can be averaged mathematically, and the averages from two different measurements can be meaningfully compared. This type of inappropriate data analysis is both incorrect and misleading. There are, however, many ways to accurately report and compare sets of qualitative data. For example, the frequency of each response (i.e., the number of 1s, 2s, 3s, 4s, and 5s) or the percentage of total responses recorded for each of those categories can be displayed in a table and compared with other similar data sets. Another type of data that blurs the line between qualitative and quantitative is binary data. **Binary data** results from categorizing items into one of two mutually exclusive groups, such as "true or false" and "right or wrong." Although binary data can easily be converted into numbers, it is actually a type of qualitative data.

Sharing Results

Science is both an iterative process and a collaborative discipline. Communication of scientific processes and findings among scientists, as well as between scientists and nonscientists, is extremely important for advancing the field and using science to benefit society. Scientists share their results in many ways. These include informal methods of sharing work, such as discussions with peers, blog posts, and publication of opinion pieces and popular science articles, as well as more formal methods of sharing work, such as presenting findings at a scientific conference or through peer-reviewed publications. The peer-review process can be thought of as the backbone of modern scientific research. **Peer review** involves the critical evaluation of academic or scientific work by knowledgeable professionals in the same field. The purpose of peer review is to evaluate work submitted for publication for accuracy, validity, originality, use of appropriate methods, integrity and signs of bias, and overall quality of the scientific work and its written description. Because experts vet the claims presented in peer-reviewed publications, these articles are often considered to be of greater integrity and significance than work that has not been peer reviewed. Peer review is often employed to screen publications submitted to competitive scientific journals, thereby influencing the prestige of the work and that of its authors.

Different types of peer review are used by different publications. Double-blind peer review is often considered to provide the highest level of integrity. In double-blind peer review, neither the authors nor the reviewers are aware of each others' identities, which is presumed to reduce the influence of personal or institutional bias. In a single-blind peer review, the names of the reviewers are unknown to the author, but the identities of authors are provided to the reviewer. This system allows reviewers to include judgments about the authors' credentials in their assessment of the work under review. The potential benefits and risks of this approach are debated, and its effectiveness may vary in different contexts. In the third type of peer review, known as open peer review, the identities of all authors and reviewers are made available to each other. For each type of peer review, the staff of the publication to which the work has been submitted often selects the reviewers to prevent collusion among authors and reviewers. However, many publications allow authors to choose or recommend specific reviewers.

When a manuscript is submitted for peer review, there are three possible outcomes. Rarely, the manuscript will be accepted for publication without a request for any revisions. More commonly, reviewers will request the author to make changes to the work, which may involve additional experimentation, data analysis, or editing of the written report. The authors can choose to revise the work and resubmit to the same journal, which will then ask its editorial staff or peer reviewers to reconsider the work for publication. Alternatively, authors may choose to not address the changes suggested by reviewers and instead submit their manuscript to a different publication, which may be more appropriate, less competitive, or simply happen to view the work more favorably. The third outcome for work submitted to a peer-reviewed publication is that it is rejected. A manuscript may be rejected for many reasons. These include problems in the experimental methods, procedures, data analysis, interpretation, or writing perceived to be too serious or systemic to be adequately fixed. Other factors that may lead to rejection are lack of sufficient originality and significance of the work to warrant publication or lack of relevance to the particular journal. The majority of reputable scientific journals are specialized, publishing only articles related to specific topics deemed of interest to the journal's readership by its editorial board. Often, manuscripts without serious technical flaws are rejected by a journal due simply to lack of relevance. Because of the amount of work invested in producing a research manuscript, authors often choose to resubmit rejected manuscripts to other journals. Oftentimes, authors of scientific articles are able to find another journal that will accept a rejected manuscript for publication, with or without significant editing. However, a number of scientific manuscripts will never be accepted by any reputable publication.

When scientists, other professionals, or members of the public read peer-reviewed articles, they are provided a certain amount of confidence that the work is accurate and reliable. This implied integrity is not offered by publications that do not undergo a similar process of evaluation. However, both scientists and nonscientists have criticized the peer-review process. One criticism is that the quality of peer review is highly variable and, in some cases, does not serve its purported purpose. Peer view is also costly, in terms of both time and money. The peer-review process often delays publication of research by months to a year or more, which, in extreme cases, could prevent access to potentially life-saving information from people who are dying. Some people also argue that the peer-review process prevents a certain amount of legitimate work from ever been shared publicly and may exert a bias toward certain types of work or results. Publications in the natural sciences and other fields are often cited as having a positivity bias. This means they preferentially report studies that confirm a hypothesis or show an effect, yet obscure through elimination the vast majority of scientific work that shows rejected hypotheses or no observed effect from a certain trial. This may result in other researchers needlessly investing significant amounts of effort and money to repeat experimental approaches others have already found to be ineffective or undesirable. Further, journals must employ staff to facilitate the peer-review process, which adds a cost that is passed off to either the readers of the journal or the scientists publishing their work. Yet, peer reviewers are rarely paid for their work, which amounts to the donation of many hours of labor over the course of a career. Finally, the careers of scientists are often strongly tied to the number and prestige of their publications. Some question whether peer reviewers and editorial boards of scientific journals should wield so much power over the careers of researchers. Based on these factors, a growing number of journals do not use peer review. When reading any scientific report or any written piece on any topic, it is important to be mindful of the entire process that led to the sharing of that work.

Evaluating Sources of Information

Properly conducted scientific research is inspired and guided by knowledge gained from reviewing the research findings published by other scientists studying related topics. Whether writing an article for a peer-reviewed journal or composing a blog post about science, scientists rely on scholarly sources of information to provide rationale and context for their experiments and conclusions. A **scholarly source** is an article written by scientists or other trained professionals who are experts in their fields. In the sciences, scholars are usually people who have earned a graduate degree, such as a Ph.D. or MD, in a scientific or medical field, and who conduct specialized research at an academic center, governmental organization, or other research organization that is well regulated and recognized as legitimate by other scholars. Although scholars write many types of articles, including blog posts, opinion articles for newspapers and magazines, and both fiction and nonfiction books, not all writing published by scholars is necessarily considered scholarly sources. Strictly speaking, scholarly sources are only those published in peer-reviewed journals. Scholarly sources on scientific topics often, but not always, contain data obtained from experiments conducted by the author or authors. However, some scholarly sources are written to summarize, evaluate, and provide original insight about the experimental findings of other scientists but do not contain new experimental data collected by the author.

To determine whether a publication is a scholarly source, the first thing to consider is whether it is published in a legitimate, peer-reviewed, academic journal. Journals that use peer review usually prominently display a statement indicating articles are "peer-reviewed" or "refereed" on their website and in the front of their printed publications, though individual articles usually do not indicate whether or not they were peer-reviewed. Because anyone can claim to produce a peer-reviewed publication, whether or not they adhere to an appropriate process of peer review, the next step is to evaluate whether the journal can be trusted to provide accurate information, and whether other scholars in the field generally consider that information significant and relevant. Making this determination can be difficult, even for well-trained professionals, because thousands of scientific journals of widely varying quality and integrity are published around the world, and many illegitimate publications are designed to look like scholarly sources. Some clues to look for when evaluating whether a publication is a scholarly journal include whether an esteemed professional scientific society or well-known university endorses the journal, and whether the authors' names and professional affiliations are clearly indicated for each article. Scholarly articles always cite their sources as either endnotes or footnotes, and often include a long list of references to other peer-reviewed publications that can be consulted if a reader has questions about claims presented in the article. Scientific journal articles often follow a similar format, starting with a title that states the main subject of the article, followed by an abstract summarizing the main points of the article, then several sections of introductory or background information, and a description of the methods and materials used to generate the article, as well as any data and conclusions produced by the author(s).

Scholarly articles are made available in several ways, including as part of printed journals and books, and as electronic files. Although many scholarly articles are freely available to the public, a number of peer-review journals charge subscription or access fees to read the articles they publish. However, universities and other academic and scientific organizations usually cover the cost of these fees for many relevant publications to provide students, faculty, and

TABLE 1.1: Search Engines and Databases for Accessing Scientific Publications

Search Engine/Database	Internet URL
Google Scholar	scholar.google.com
JSTOR	jstor.org
Library of Congress	loc.gov
Microsoft Academic	academic.microsoft.com
National Center for Biotechnology Information (NCBI) MeSH Database PubMed	 ncbi.nlm.nih.gov/mesh ncbi.nlm.nih.gov/pubmed/
ProQuest	proquest.com
Semantic Scholar	semanticscholar.org
Web of Science	webofknowledge.com

employees with free access. One of the possible ways to access peer-reviewed publications is to obtain a printed or electronic copy of the journal in which the article is published. This process can be facilitated by the use of standard Internet search engines and library card catalogs. However, given that articles may appear in hundreds or even thousands of different publications for each field of science, using specialized search engines and databases that archive scientific articles is much more efficient and effective. Several database engines useful for accessing scientific publications are listed in Table 1.1.

When planning a scientific study, writing a paper about a scientific topic, or simply reading about a scientific topic to learn more information, use of scholarly sources provides some level of confidence in the accuracy of the information presented. However, not every claim made in a scholarly source will necessarily be accurate and complete, and many sources not strictly considered scholarly sources contain a wealth of truthful, correct, and important information. For example, many scholars and journalists publish accurate descriptions of science in popular newspapers, magazines, and in books; yet these publications often may not contain an abstract, a materials and methods section, or citations of other scholarly sources. Some indications that a source should not be trusted include use of sensational titles, overly subjective, hyperbolic, or inflammatory language, testimonials about products for sale or links to commercial websites, and the lack of information about who wrote and published the article. When in doubt, the best approach is to try to corroborate a questionable claim with information from another independent source. As mentioned earlier, even with many years of practice, it can be difficult to distinguish reliable sources from unreliable ones, but relying on multiple independent sources reduces the chance of being misled by a deceptive author or publisher.

In addition to evaluating the integrity of the source of scientific information, it is also important to evaluate the veracity of individual claims within each source. To start, it is a good idea to evaluate claims that seem scientific to determine whether they reflect true science or pseudoscience. The term **pseudoscience** refers to claims purported to be scientific and factual but are, in fact, not based on evidence collected using the scientific method. Unlike true science, pseudoscientific claims do not result from testable and falsifiable hypotheses about observable phenomenon, or from well-constructed experiments that produce results that would withstand peer review. Clues that claims may be the result of pseudoscience include vague and overstated predictions or conclusions, unclear or unstated research methods, reliance on anecdotal evidence, and lack of citations to peer-reviewed sources supporting conclusions, as well as logical fallacies and statements about unobservable or unquantifiable phenomenon.

Even scientific claims should be carefully assessed to determine their true meaning, significance, and implications. For example, to interpret the meaning of conclusions provided in a scientific article, it is important to know whether each claim represents causation or correlation. Scientific studies are often focused on demonstrating **causation**, which is when one factor, action, or event directly produces an observable change in another factor, action, or event. Causation should not be conflated with **correlation**, which refers to a relationship between two observable phenomena that can be used to make predications about how those phenomena behave in relation to each other. Sometimes recognizing a correlation provides valuable insight about how natural phenomena behave and may even be predictive of a causation that has not yet been established because enough data has not been collected. However, correlations often represent coincidences and unrelated trends in natural phenomenon that do not reliably predict how those phenomena will behave in the future. Therefore, correlations are often incorrectly interpreted and should be carefully evaluated before being used to justify conclusions or future actions.

1.2 SUMMARY

- Science generates knowledge through systematic study of the natural world
- A hypothesis is a proposed explanation for an observation that can be tested, whereas a theory is an explanation for a natural phenomenon that is well substantiated by a large amount of evidence
- The scientific method is an iterative process of making observations, asking questions, proposing hypotheses, conducting experiments, and analyzing data
- In an experiment, an independent variable is deliberately changed or manipulated to determine its effect; a dependent variable is what is observed and measured; and control variables are kept equal and unchanged to avoid extraneous variation
- To ensure results are meaningful, scientific experiment must include appropriate control groups, experimental groups, and sample sizes
- Many types of data can be collected during experiments, including binary, continuous, discrete, nominal, ordinal, qualitative, and quantitative
- Scientific research is supported by references to scholarly sources written by scientists or other trained experts

1.3 KNOWLEDGE CHECK

1. Which of the following best defines the word *hypothesis*?
   a. a broad explanation substantiated by a large amount of evidence
   b. a condition or item that can be changed during an experiment
   c. an educated guess about how a natural phenomenon occurs
   d. a proposed explanation for an observation that can be tested
   e. a set of observations about what occurs in the natural world

2. The type of thinking that forms a general conclusion by relating specific observations is:
   a. deductive reasoning
   b. experimental grouping
   c. hypothesis-driven science
   d. inductive reasoning
   e. scientific method

3. A change or effect that a scientist measures during an experiment is:
   a. a control variable
   b. a dependent variable
   c. an extraneous variable
   d. an independent variable
   e. an invariable

4. The number of words in this chapter is an example of both:
   a. continuous data and discrete data
   b. discrete data and quantitative data
   c. quantitative data and ordinal data
   d. ordinal data and qualitative data
   e. qualitative data and nominal data

5. In an experiment that tests whether plants grow more under red or green light, which of the following would be an example of a dependent variable?
   a. amount of water
   b. color of light
   c. height of plant
   d. humidity level
   e. sample size

6. List the steps of the scientific method and discuss why the process of science is not necessarily linear.

7. Explain how to analyze and evaluate ordinal data.

8. Discuss the purpose of control variables.

1.4 ACTIVITY

Plan an Experiment

Scientific articles are often cited for the results and conclusions they provide but also valuable is the information provided about the scientific process the authors went through to obtain their data and conclusions. The purpose of this activity is to practice identifying important aspects of the scientific method revealed in a published scientific journal article.

Read the following article and answer the accompanying questions for Activity 1.4 about experimental design and methods.

Campbell, M. W., Carter, J. D., Proctor, D., Eisenberg, M. L., de Waal, F. B. M. (2009). Computer animations stimulate contagious yawning in chimpanzees. *Proceedings of the Royal Society B, 276*, 4255–4259. doi:10.1098/rspb.2009.1087.

1.5 ACTIVITY

Black Box Experiment

The goal of this activity is to apply the scientific method to determine the contents of a "black box." Each scientific team will be given a "black box," which is a sealed container with one or more unknown items inside, as well as a list of items that may be inside the container and an identical empty container. A ruler, scale, and other tools may also be provided. Without opening the container, the team must use observations, predictions, data, and reasoning to determine what is in the box. First, generate a hypothesis as to what is inside your box based on your observations. Next, plan an experiment to determine what is in your box. Then, collect, record, and analyze data including measurements, drawings, and other types of observations. Finally, determine your conclusions including whether to accept or reject the hypothesis. After this is complete, check to see if your hypothesis and conclusion were correct by opening the box.

ACTIVITY QUESTIONS

1.4

1. What observation(s) did the authors make about a natural phenomenon?
2. According to the article, what pertinent information was already known (from background research and other observations) about what the authors observed?
3. What was the authors' hypothesis?
4. Was the authors' hypothesis testable and falsifiable?
5. Describe the results predicted by the hypothesis.
6. What findings would cause the authors to accept their hypothesis?
7. What findings would cause the authors to reject their hypothesis?
8. List the control, dependent, and independent variable(s) in the experiment.
9. Identify the control and experimental group(s) in the experiment.
10. What types of quantitative and qualitative data were collected?
11. Identify the sample size.
12. List the materials and equipment used for the experiment.
13. List the steps of the experimental procedure used for the experiment.
14. Identify the main data/findings of the experiment.
15. Identify the conclusion(s) of the experiment.
16. Did the authors accept or reject their hypothesis?

1.5

1. List your observations and describe how they were made.
2. Identify your hypothesis of what is inside the box.
3. Describe your predictions.
4. Describe your experiment on how to determine what is in the box.
5. Record your data, including any drawings or measurements that you made.
6. Discuss your conclusions.
7. Open your box. State whether your hypothesis and/or conclusion were correct.
8. Reflect upon which parts of the scientific method were most important in determining the contents of the "black box."

1.6 REFERENCES

Creswell, J. W. (2013). *Research design: Qualitative, quantitative, and mixed methods approaches.* Thousand Oaks, CA: Sage Publications.

Hyde, K. F. (2000). Recognising deductive processes in qualitative research. *Qualitative Market Research: An International Journal*, *3*(2), 82–90.

Tang, et al. (2010). The scientific method and scientific inquiry: Tensions in teaching and learning. *Science Education, 94*(1), 29–47.

White, E. (2014). Science, pseudoscience, and the frontline practitioner: The vaccination/autism debate. *Journal of Evidence-Based Social Work, 11*(3), 269–74.

Introduction to Cell Biology

LEARNING OBJECTIVES

- List the characteristics of living things
- Describe the level of organization used to classify living things
- Compare and contrast prokaryotic and eukaryotic cells
- Identify examples of autotrophs and heterotrophs
- Describe the function of the main types of organelles
- Discuss the meaning and significance of the Endosymbiont Theory
- Demonstrate knowledge of the terms hydrophilic and hydrophobic
- Compare and contrast the processes of active transport, facilitated diffusion, osmosis, and passive transport

2.1 INTRODUCTION

Biology is the field of science that focuses on the study of living things. Although just one of many fields of science, biology itself can be subdivided into many specializations including anatomy and physiology, biochemistry, ecology and environmental science, genetics, marine biology, microbiology and immunology, and quantitative biology and bioinformatics. These highly varied specializations are all united by the shared goal of answering several basic questions. The first question biologists attempt to answer is: What characteristics determine whether something is alive? The second question studied by many biologists is: How can living things be ordered and classified based on their shared characteristics? The third major focus of biologists is: How do living things function? And the fourth, but certainly not last, question that is a main focus of biology is: How do living things change over time? This chapter, as well as the remainder of this textbook, will provide an overview of the ways biologists have answered these questions.

Characteristics of Living Things

One of the most fundamental questions in biology is: How can living things be classified to distinguish them from each other and from nonliving things? Although this seems like a straightforward question, so much remains to be known about the universe that current attempts to answer this question are undoubtedly incomplete and will require revision in the future. Recent estimates suggest that 86% of the species on Earth have not even been discovered yet! That is not to mention that there is much about known species, and even our own bodies, that we cannot explain. Based on what we do know, living things can be differentiated from nonliving things based on a set of eight shared characteristics. Although objects widely considered to be nonliving may exhibit some of these characteristics, living things manifest all of them in one form or another. The eight abilities or characteristics that define life are adaptation, cellular organization, energy processing, growth and development, heredity, homeostasis, reproduction, and responsiveness (Table 2.1). In addition to these characteristics, it is also essential that living things are

TABLE 2.1: Characteristics of Living Things

Characteristic	Description
Adaptation	Modification of the structure or function of any part or behavior of an organism that results in increased ability to survive and reproduce in its environment
Cellular Organization	Containing a hierarchy of ordered sub-cellular and supra-cellular structures
Energy Processing	Use of energy from the sun or other source to carryout chemical reactions and other cellular processes
Growth and Development	Changes in the number, size, and/or type of cellular and body structures throughout lifetime
Heredity	Passage of genetic traits of one individual to another
Homeostasis	Regulation and maintenance of internal conditions in a range that supports life despite changes in the environment
Reproduction	Making of a new organism containing the genetic material of the original (parent) organism(s)
Responsiveness	Changes in behavior in response to favorable or unfavorable environmental stimuli

capable of actively maintaining stable internal conditions to support the processes of life. This condition, known as **homeostasis,** involves precisely balancing levels of salts, water, oxygen, and other molecules at appropriate levels, as well as maintaining a steady internal temperature and other vital conditions.

Types of Cellular Organisms

All living organisms are made of cells. **Cells** are defined as the fundamental unit of life capable of carrying out all of the functions necessary for a living thing to survive. All cells are surrounded by a membrane that controls movement of molecules into and out of the cell, and all cellular organisms rely on DNA as their genetic material. Beyond this, cells differ widely in terms of structure and function.

Cellular organisms can be divided into two main types: eukaryotic and prokaryotic (Table 2.2). The most fundamental distinction between eukaryotic cells and prokaryotic cells is that eukaryotic cells contain a nucleus and prokaryotic cells do not. The name **eukaryote** is derived from Greek for "true kernel" or "true nucleus," because the nucleus was thought to look like a kernel of corn when viewed under early microscopes. Eukaryotic cells are generally much larger and more complex than prokaryotic cells. In addition to a nucleus, eukaryotic cells contain many other specialized structures called **organelles**, which can be thought of as "the organs of the cell." Most organelles are surrounded by a lipid membrane, which provides structure and organization, and regulates the passage of molecules into and out of the organelle.

One notable exception to this rule is the ribosome, which is an organelle composed entirely out of a molecule called ribosomal RNA and is not surrounded by a membrane.

Organisms can also be categorized based on whether they are multi-cellular or composed of only one cell (Table 2.2). All multi-celled organisms including animals, plants, and fungi are eukaryotes. Some eukaryotes are also composed entirely of a single cell (unicellular). Examples of unicellular eukaryotes include a type of fungus called yeast and most types of algae, which are often thought of as plants but are quite different, as well as a type of animal-like microorganism called protozoa. Collectively, these unicellular eukaryotes, along with a few related multicellular organisms such as kelp, are known as protists. Most eukaryotes contain the same types of organelles, although they may differ in size and number from cell to cell. There are a few notable distinctions between certain types of cells. For example, plant cells contain chloroplasts for producing energy through photosynthesis and a large central vacuole for storing water and nutrients (Figure 2.1). These are not found in other types of cells. Unlike animal cells, the cells of plants and fungi are surrounded by a rigid cell wall outside of their plasma membrane to provide structural support. This is important because these organisms lack internal or external skeletons. Conversely, animal cells contain two types of organelles called centrioles and lysosomes, but the cells of plants and fungi do not (Figure 2.2).

Prokaryotes are distinguished from eukaryotes because their cells do not contain a nucleus or any other membrane-bound organelles. The word *prokaryotic* means "before nucleus," reference the idea that the first prokaryotes evolved before the first eukaryotes. Prokaryotic cells are generally smaller and simpler than eukaryotic cells, and are more similar to the cells that formed the earliest life forms on earth than are eukaryotes. Prokaryotic cells are generally less than 1% of the size of eukaryotic cells. An advantage of being this small is that nutrients and waste products can easily move into and out of the entire cell. Therefore, they do not require some of the specialized structures necessary for relatively large eukaryotic cells to carry out the processes of life. Each prokaryote does contain a cell wall, plasma membrane, ribosomes, and a concentration of DNA known as a nucleoid (Figure 2.3). The two main types of prokaryotes are **bacteria** and **archaea**. Each of these groups is highly diverse, containing many different species found throughout a wide range of environments. There are thousands of different species of bacteria, some of which cause disease in humans and other organisms, and many that live in the environment and do not cause disease. Archaea were

FIGURE 2.1: **Plant cell**—All plant cells share a number of common organelles, including chloroplasts and cell walls.

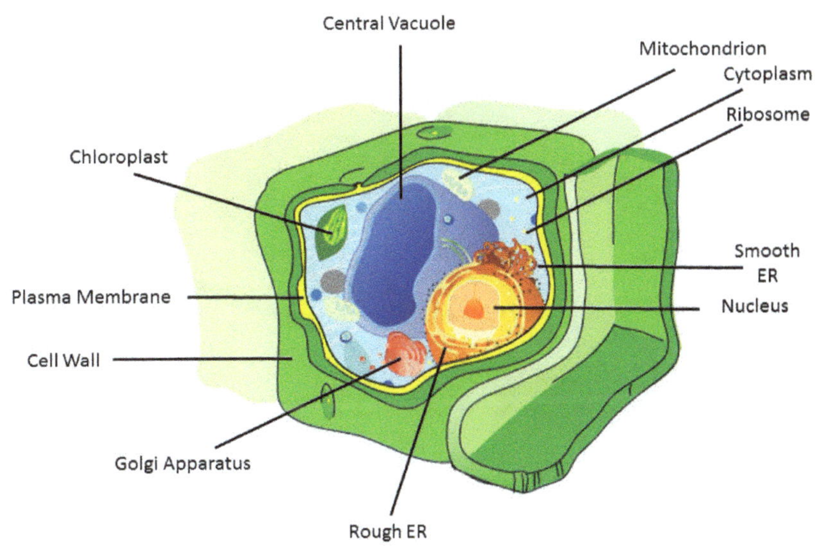

FIGURE 2.2: **Animal cell**—Animal cells contain a diverse set of organelles with a range of functions including protein synthesis and poison detoxification.

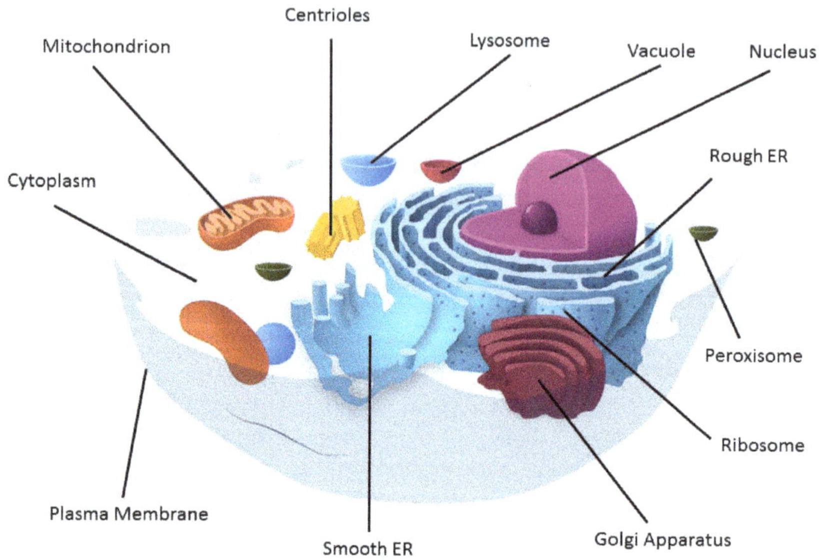

FIGURE 2.3: **Prokaryotic cell**—Prokaryotic cells are small and simple, yet still contain all of the components necessary to sustain life.

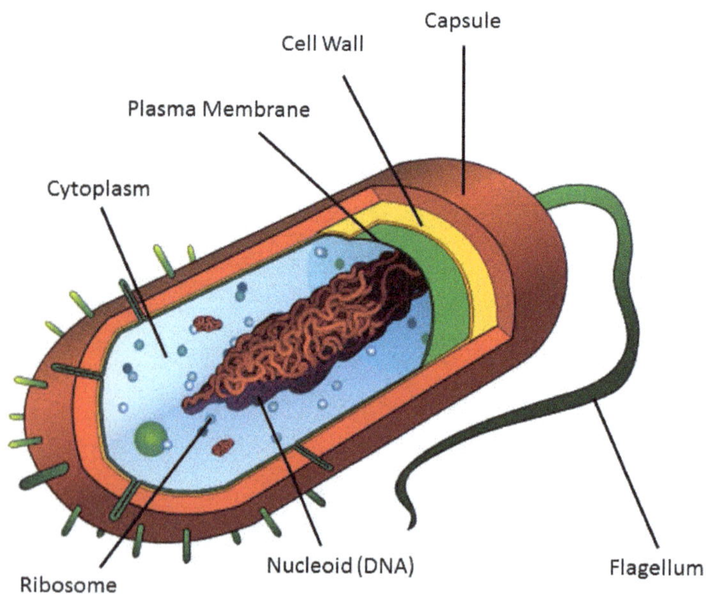

once thought to also be bacteria, but detailed analysis has revealed the two types of organisms are significantly different genetically and structurally. Genetically, archaea are now known to be just as closely related to eukaryotes as they are to bacteria.

Ribosomes are also the only organelle present in prokaryotes. Prokaryotes must still contain all of the cellular components necessary for life. In addition to possessing ribosomes, all prokaryotes include a plasma membrane, a cell wall, and DNA. Because prokaryotes are free-living organisms, they are often surrounded by a sticky capsule that helps them thrive in their environment. Many prokaryotes also possess flagella, which are whip-like tails that propel them through their environment.

Both eukaryotic and prokaryotic organisms can be further characterized by whether they are autotrophs or heterotrophs (Table 2.2). An **autotroph** is an organism that obtains food and energy without eating other organisms or consuming organic molecules produced by other organisms. Almost all plants are photoautotrophs that use photosynthesis to capture energy from the sun. Some autotrophs do not undergo photosynthesis and instead are able to use other forms of chemical or heat energy to power their cellular processes. These are known as chemoautotrophs. A **heterotroph** is an organism that obtains food and energy from eating other organisms or by

TABLE 2.2: Characteristics of Different Cell Types

	Eukaryotes				Prokaryotes
Characteristic	Animals	Fungi	Plants	Protists	Bacteria and Archaea
Nucleus	Yes	Yes	Yes	Yes	No
Ribosomes	Yes	Yes	Yes	Yes	Yes
Cell Wall	No	Yes	Yes	Yes/No	Yes
Chloroplasts	No	No	Yes	Yes/No	No
Lysosomes	Yes	No	No	No	No
Cellularity	Multicellular	Multicellular	Multicellular	Unicellular	Unicellular
Energy	Heterotroph	Heterotroph	Autotroph	Autotroph/Heterotroph	Autotroph/Heterotroph

consuming organic molecules produced by other organisms. Animals, fungi, and some protists and prokaryotes are heterotrophs. Heterotrophs can be divided into four groups based on how they obtain their food: (i) holozoic heterotrophs, such as humans, digest their food inside their body; (ii) mutualistic heterotrophs engage in a food production system with another organism that benefits both organisms; (iii) parasitic heterotrophs obtain food from living inside a host organism, which does not benefit and often harms the host organism; and (iv) saprobiontic heterotrophs externally digest dead and decaying organic matter.

Viruses, Viroids, and Prions

Although many things can be easily classified as either living or nonliving, others seem to occupy a grey area. This includes three types of infectious particles known as viruses, viroids, and prions. Although each of these exhibit several of the eight characteristics of living things described earlier, most scientists agree they fail to fulfill all of the criteria required to be defined as alive. Viruses are often lumped together with bacteria because both are very small and can cause illness; however, viruses are much more rudimentary than their bacterial counterparts. Each virus particle contains its own genetic material (DNA or RNA), a protective protein coat known as a **capsid**, and sometimes a membrane called an envelope. Yet viruses lack cells, which is one of the main criteria of life. Viruses also cannot reproduce on their own without infecting a host cell, do not grow, and do not use their own energy. Viruses are generally much smaller than eukaryotes or prokaryotes; however, the largest known virus is larger than the smallest known bacterium. Viroids and prions are even smaller and less complex than viruses. Viroids are infectious RNA particles that cause some diseases in plants, but do not affect animals. Prions are infectious proteins linked to several serious neurological diseases in animals including mad cow disease and its human equivalent, Creutzfeldt-Jakob disease.

Cell Structures and Their Functions

Each cellular structure and organelle is responsible for the same general function regardless of the type of cell in which it is found. Like the organs of our body, organelles and other cellular structures do not work in isolation but are rather continually engaged in a complex orchestra of biochemical processes. Still, to understand the role of each cellular structure, it helps to examine them one by one (Table 2.3).

Capsules are one cellular structure found in prokaryotes but not eukaryotes. Capsules are located on the outside of the cell wall and are made primarily of carbohydrates and protein. Sometimes capsules are referred to as slime layers if they are not as dense and structurally organized as a true capsule. Capsules are sticky and allow prokaryotes to adhere to each other and to surfaces, such as the enamel of a tooth or the surface of a rock. More importantly, capsules provide protection from dehydration and from other cells and viruses. Bacteria that infect animals may use their capsule to avoid being killed or eaten by the host immune system.

The **cell wall** is a structure found outside the plasma membrane of a variety of different cells, including those of bacteria, fungi, and plants. Animal cells do not contain cell walls. Cell walls vary considerably among different organisms in terms of size and composition, but all cell walls are mostly made of carbohydrates. Plant cell walls are composed mostly of the carbohydrate cellulose, whereas bacterial cell walls are made of a combination of protein and sugar called peptidoglycan. The main functions of cell walls are to provide structural support and to prevent dehydration and loss of shape. The cell walls of bacteria are more complex than those of plants because they are also used to protect against the host immune systems during infection. Many common antibiotics, including penicillin, fight bacterial infections by inhibiting production of bacterial cell walls, preventing bacterial survival and reproduction.

The **plasma membrane**, or cell membrane, is a structure that separates a cell from its external environment. All cells are surrounded by a plasma membrane comprised primarily of two layers of phospholipids called a phospholipid bilayer. Plasma membranes also contain a number of other lipids, as well as carbohydrates, proteins, and other molecules (Figure 2.4). The **fluid mosaic model** is used to describe observations that the plasma membrane is a movable and changeable collection of different types of molecules drifting in a fluid phospholipid bilayer. The plasma membrane is attached to the cytoskeleton, which allows the cell to change shape, move, and undergo processes, such as phagocytosis that involve repositioning of the plasma membrane.

Cholesterol is one molecule abundant in plasma membranes. Cholesterol provides stability to the plasma membrane, keeping it from being too flexible and from allowing too many molecules to pass through. Cells vary in the amount of cholesterol found in their plasma membranes. The more cholesterol contained in a plasma membrane, the less fluid and permeable it is. Despite this, plasma membranes are much more fluid and dynamic than cell walls, which is important for allowing cells to change shape when necessary and to restructure parts of the plasma membrane to carry out certain functions. For example, one function of the plasma membrane is to bind hormones and other signaling molecules with receptors that respond to chemical signals from other cells. When hormone signals are present, a cell may concentrate all of the receptors for that hormone in the area of the cell surface receiving the signal. Each receptor on a plasma membrane recognizes and responds to only one or a small number of molecules called ligands. However, receptors sometimes malfunction and respond inappropriately. Receptors may also be taken over by viruses attempting to gain entry into the cell. Viruses such as human immunodeficiency virus (HIV) use receptors already present on host cells to gain entry and cause infection.

TABLE 2.3: Cell Structures and Their Functions

Structure	Function
Capsule	Provides protection for prokaryotic cells
Cell wall	Provides protection and structural support
Centriole	Involved in cell division of animal cells
Chloroplast	Produces energy for plants cells through photosynthesis
Cytoplasm	Holds contents of cell; site of some important chemical reactions
Endoplasmic reticulum (ER)	Rough ER: Modifies proteins and membranous vesicles Smooth ER: Synthesizes lipids; detoxifies poisons
Flagellum	Helps move cell through exterior environment
Golgi apparatus	Modifies and sorts molecules made by the ER
Lysosome	Digests molecules, bacteria, and worn-out organelles
Mitochondrion	Produces energy for the cell through cellular respiration
Nucleoid	Location of DNA in prokaryotes
Nucleus	Location of DNA in eukaryotes
Peroxisome	Breaks down components of lipids and proteins
Plasma Membrane	Selective barrier that controls passage of substances into and out of the cell
Ribosome	Site of protein synthesis
Vacuole	Stores water and other substances

In addition to enclosing the cell, providing structure, and mediating binding of external molecules to receptors, the plasma membrane also regulates the movement of substances into and out of the cell. A few small molecules, such as carbon dioxide and some lipids, are even capable of diffusing directly through the plasma membrane. However, many can only cross the plasma membrane when directed through a special channel in a trans-membrane protein. Because the plasma membrane only allows certain molecules to pass through its border at certain times, it is said to be selectively permeable. Much of this selectivity is a result of one very important property of the phospholipid bilayer that makes up the majority of the plasma membrane, which is that one side (the phosphate group) is polar, and the other side (the lipid tail) is nonpolar. Polar molecules are those that have an electrical charge on part of the molecule different from the electrical charge on another part of the molecule; nonpolar molecules have an evenly distributed electrical charge throughout. Polar molecules interact well with water, so they are said to be **hydrophilic**, which literally means "water loving." Nonpolar molecules do not mix well with water and are referred to as **hydrophobic**. These properties can be observed by placing sugar and oil into water. Sugar is polar and hydrophilic, so it easily dissolves in water. On the other hand, oil is nonpolar and hydrophobic, so it will stay separated from the water, usually in a distinct layer on top of the water.

The **cytoplasm** is what fills the interior of the plasma membrane. It consists of a jelly-like substance known as cytosol and contains all of the cell's organelles, as well as a complex system of microtubules and other movable fibers known as the cytoskeleton. The cytoskeleton helps support the structure of the cell, aids in positioning organelles, and is actively involved in

FIGURE 2.4: Plasma membrane structure

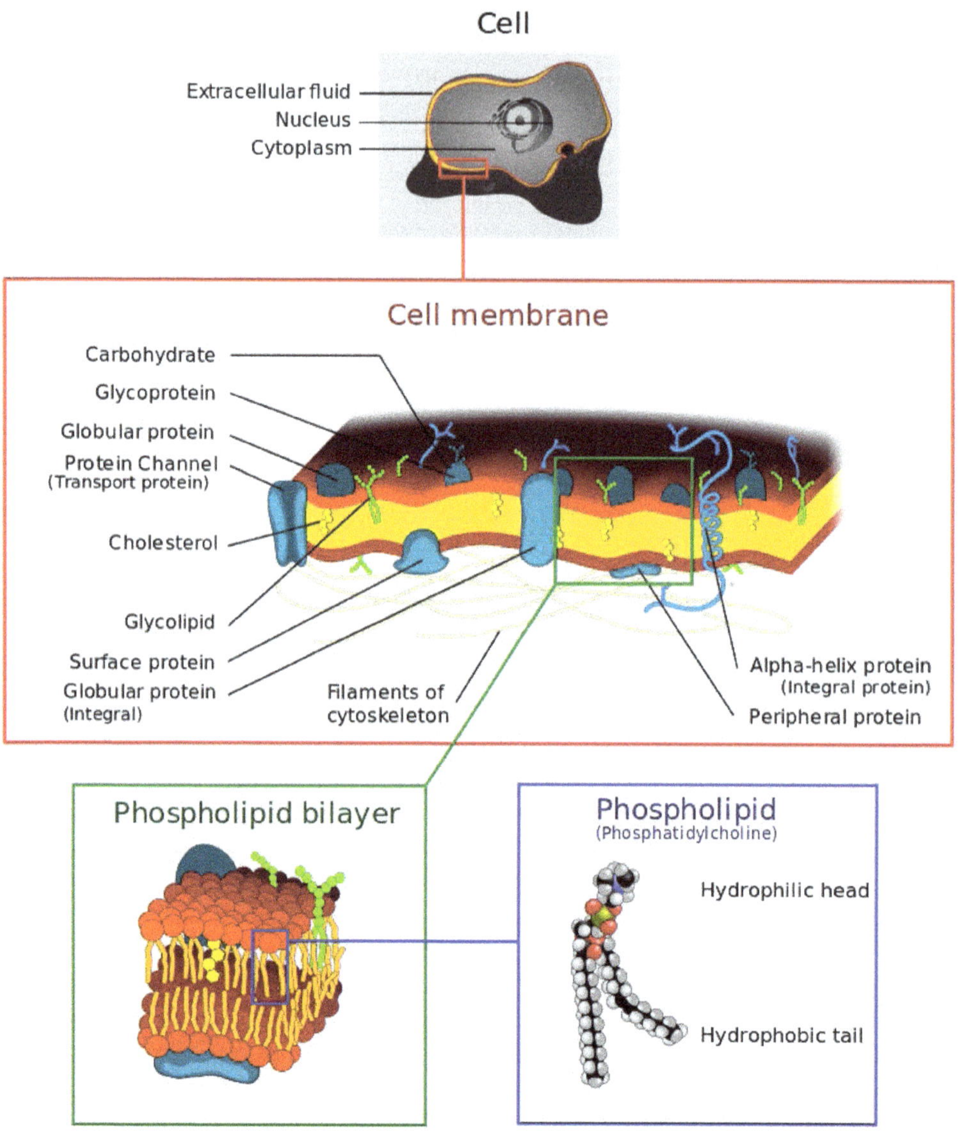

moving cellular components during cell division. In single-celled organisms, the cytoskeleton also helps the organism achieve motility. Also in the cytoplasm are numerous dissolved molecules involved in various chemical reactions. All cell types contain cytoplasm, although it is sometimes referred to as protoplasm in prokaryotes.

Centrioles are structures found in animal cells and some lower plants like ferns. The function of centrioles is not completely understood; however, it is clear that centrioles form the center from which fibers called microtubules are organized. Microtubules are involved in a number of cellular processes that involve movement of the cell or organelles. It is thought that centrioles are particularly important for cell division in animal cells, perhaps by helping direct the movement of chromosomes during mitosis. Interestingly, most other eukaryotic cells do not require centrioles for this process.

Some eukaryotic and prokaryotic cells have a single **flagellum** or multiple flagella, which are long, tail-like structures often used to propel the cell through the environment. In humans, flagella are most notable on male reproductive cells called sperm. Some cells have smaller hair-like projections called **cilia** that serve a similar function. For cells held in place within a layer of cells, cilia and flagella can be used to move water or other extracellular liquids across their outer surface to facilitate the capture of nutrients or to move undesired particles away from the cell.

The DNA of prokaryotes is concentrated in a region of the cell called the **nucleoid**, which may also contain RNA, enzymes, and other proteins. The nucleoid is not bound by a membrane and is thus not considered a true nucleus.

In eukaryotes, DNA is contained in a membrane-bound organelle called a **nucleus**. The nucleus controls the synthesis of ribosomes and proteins, and is the site of transcription (the making of mRNA from a DNA template) in eukaryotes. See Figure 2.5 to see the nucleus and its structure. The nucleus is usually the most easily visualized organelle when a cell is examined with a microscope. Historically, one of the first classification systems of cells developed involved identifying cells as either eukaryotes and prokaryotes because the nucleus was one of the few cellular structures that could be observed with the early microscopes.

Ribosomes are the cellular components responsible for protein synthesis in all eukaryotic and prokaryotic cells. Unlike most eukaryotic organelles, ribosomes are not surrounded by a membrane and instead are made of ribosomal RNA. All ribosomes are composed of two pieces known as the large subunit and the small subunit. Some ribosomes are attached to other structures, such as the membrane of the nucleus or the rough endoplasmic reticulum, whereas others, known as free ribosomes, are suspended in the cytoplasm. The function of both attached and free ribosomes is to direct the synthesis proteins from the genetic code contained in messenger RNA through a process known as translation. However, most proteins synthesized by attached ribosomes are destined for export outside of the cytoplasm, whereas proteins produced by free ribosomes usually remain and function within the cytoplasm.

The endoplasmic reticulum consists of an extensive series of

FIGURE 2.5: Nucleus—The nucleus is surrounded by a double membrane and contains the nucleolus, where ribosomes are made, and the genetic material in the form of chromatin.

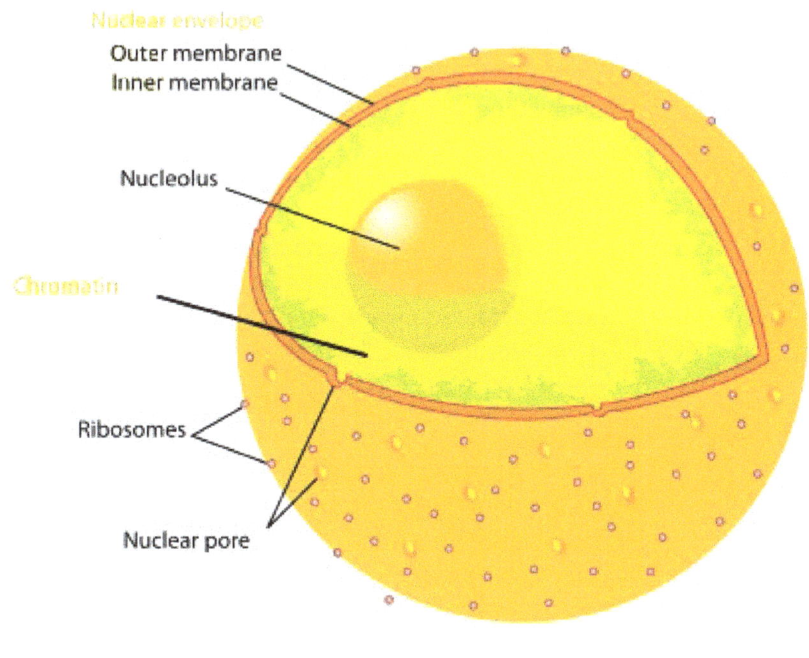

membrane-bound compartments known as cisternae. The endoplasmic reticulum is connected to the membrane of the nucleus and, along with the nuclear membrane, Golgi apparatus, lysosomes, plasma membrane, and transport vesicles, forms an interconnected membranous network within the cell known as the endomembrane system. The region of the endoplasmic reticulum closest to the nucleus is known as the **rough endoplasmic reticulum** because its surface appears rough and bumpy when viewed under a microscope. This is because the presence of attached ribosomes, which are smaller than the endoplasmic reticulum (Figure 2.6). Part of the endoplasmic reticulum is not covered with ribosomes and is called the smooth endoplasmic reticulum. Each region of the endoplasmic reticulum carries out a distinct function. The main function of the rough endoplasmic reticulum is to modify proteins synthesized by ribosomes to prepare them for secretion outside of the cell. The modifications made to proteins by the rough endoplasmic reticulum include attachment of carbohydrates that aid their function and encapsulation by a lipid membrane to form a transport vesicle that can move through the cytoplasm toward the plasma membrane. The rough endoplasmic reticulum also produces the phospholipids that comprise the plasma membrane. The **smooth endoplasmic reticulum** is involved in many cellular processes, including production of lipids, detoxification of drugs and poisons, and storage of calcium ions. The smooth endoplasmic reticulum is especially important for the function of cells in the testes and ovaries because it the part of the cell that produces the sex hormones testosterone and estrogen. The liver is also reliant on the function of the smooth endoplasmic reticulum because the role of the liver in the body is to help filter drugs and poisons out of the blood and convert them into less harmful substances. The cells of all eukaryotic organisms contain endoplasmic reticulum (Figure 2.6).

All eukaryotic cells contain a set of flattened membranous sacs called the **Golgi apparatus,** which is part of the endomembrane system along with the endoplasmic reticulum (Figure 2.6). The role of the Golgi apparatus is to package and sort molecules produced by the endoplasmic reticulum. Transport vesicles released by the endoplasmic reticulum fuse with one side of the Golgi apparatus, known as the "cis face" or receiving side, and empty their contents into the Golgi apparatus where they are modified and repackaged into vesicles to be released from the opposite side, known as the "trans face" or shipping side. In a way, the Golgi apparatus can be thought of as the post office of the cell. The amount of Golgi apparatus varies among different types of cells. It is most abundant in cells that secrete a large amount of molecules into the extracellular space, such as the secretory cells of glands, the digestive tract, and the immune system.

Lysosomes are organelles that act much like the stomach of the cell. Lysosomes contain an acidic solution of digestive enzymes capable of breaking down carbohydrates, lipids, proteins, nucleic acids, and worn-out organelles. When cells take in large molecules from the exterior environment, those molecules often end up in a lysosome. The process of cellular eating in which large particles are taken into a cell is known as **phagocytosis**. When cells take in food molecules through the process of phagocytosis, the molecule is first surrounded by the plasma membrane, which pinches off around the molecule to form a sac known as a food vacuole. The food vacuole then moves toward and fuses with a lysosome to mix the digestive enzymes with the food and break it into smaller molecules the cell can use for nutrients and energy production. Another important function of lysosomes is the destruction of bacteria and other invaders or unwanted particles. Cells of the immune system are capable of engulfing entire bacteria in the same way that food particles are taken up. Some unicellular organisms, such

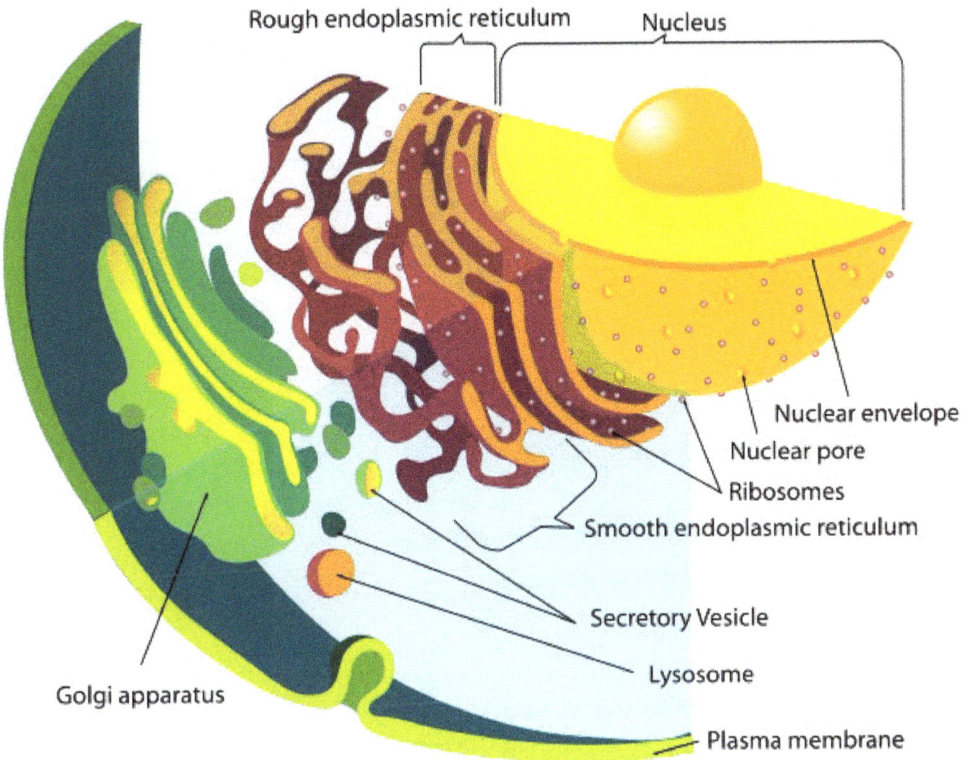

FIGURE 2.6: **Endomembrane system**—The endomembrane system includes the nucleus, smooth and rough endoplasmic reticulum, and the Golgi apparatus.

as amoeba, phagocytize bacteria as a source of food (Figure 2.7). When bacteria or other nonfood molecules are involved in phagocytosis, the vacuole formed is called a phagocytic vacuole rather than a food vacuole. When a phagocytic vacuole fuses with a lysosome, the combined structure is known as a phagolysosome. Only certain cells undergo phagocytosis. For example, animal-like protists called amoebas rely on phagocytosis for obtaining nutrients and energy, but plant-like protists undergo photosynthesis instead. Cells may also take in small particles dissolved in extracellular fluid through a process called **pinocytosis**, which literally translates to "cellular drinking." When cells expel molecules to the exterior environment through a reverse of this process, it is called **exocytosis**. Lysosomes are also known as the "recycling center" of the cell, because they are involved in not only breaking down and removing molecules from the cell, but they also help to convert components of some of these molecules into substances the cell can use.

Vacuoles are membrane-bound containers derived from the endomembrane system. In addition to food vacuoles that contain molecules taken in by the cell for the purpose of obtaining nutrients and energy, there are many other types of vacuoles that serve a variety of purposes. Some vacuoles store digestive enzymes or proteins for later use by the cell. Other vacuoles, especially in plants, contain poisonous or bad-tasting chemicals to make the plant unpalatable to a potential predator. Other vacuoles are used for storing water. The largest organelle inside plant cells is usually a water storage vacuole known as a central vacuole. Because plants generally cannot move around to find new sources of water when there is a shortage, central vacuoles play an important role in preventing plants from dying from dehydration.

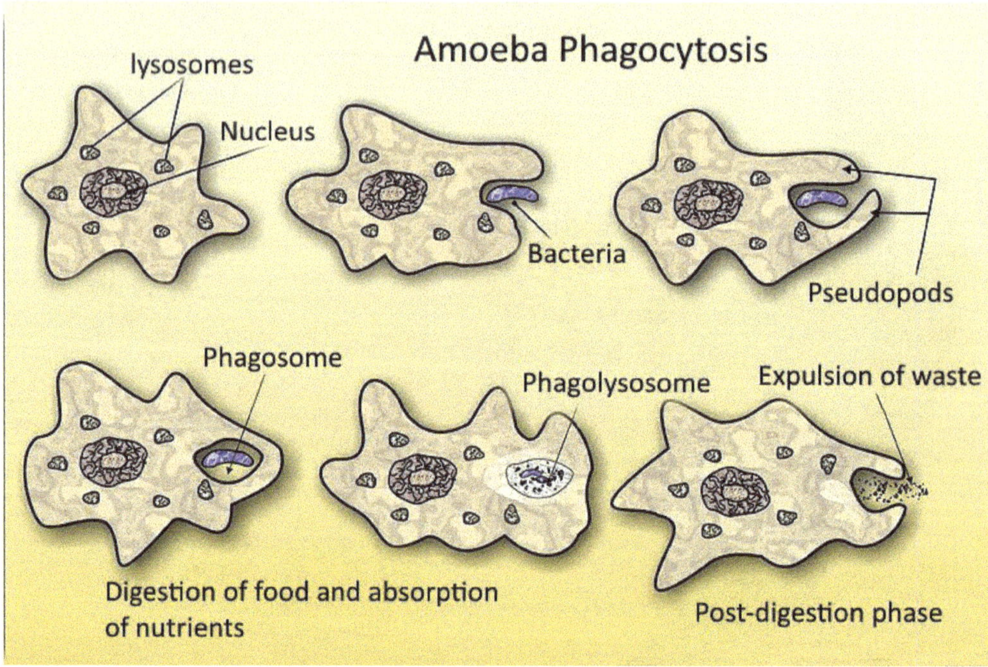

FIGURE 2.7: **Phagocytosis**—Amoebas are single-celled organisms that obtain food by eating bacteria through a process called phagocytosis.

When plants are exposed to large amounts of water, the central vacuoles of their cells fill up to capacity; when water is scarce, the central vacuoles shrink as they become empty. The pressure on the cell wall created by a full central vacuole filling up the interior of the cell help plants maintain their rigid structure. When dehydrated, the vacuoles exert less pressure on the cell wall, causing the plant to become flaccid and wilt.

Mitochondria are one of two organelles that contain their own DNA and ribosomes and are enclosed by two membranes instead of one (Figure 2.8). Mitochondrial DNA contains relatively few genes, but it does contain most of the genes required for its own functioning. A number of hereditary diseases involving problems with cellular metabolism are caused by mutations in mitochondrial DNA rather than mutations in the chromosomal DNA found in the nucleus. This is because mitochondria are the organelles responsible for producing the majority of all energy consumed by many eukaryotic cells, especially those of animals (Figure 2.8). All eukaryotic organisms, including those of fungi and plants, use mitochondria to produce energy. Plants actually use the products of photosynthesis (glucose and oxygen) to drive cellular respiration and can then recycle the byproducts of cellular respiration (carbon dioxide and water) during photosynthesis.

Chloroplasts are organelles found in plants as well as some protists and prokaryotes. Similar to mitochondria, chloroplasts contain their own DNA and ribosomes, and are enclosed by two membranes instead of one (Figure 2.9). The job of chloroplasts is to capture energy from the sun to produce organic molecules through a process called photosynthesis. During **photosynthesis,** atoms from carbon dioxide and water are bonded together using energy harnessed from the sun to form the energy-storing carbohydrate glucose. Individual glucose monomers can then be linked together to form more complex carbohydrates for long-term energy storage or for other uses, such as providing structural support as part of the cell wall. Oxygen is also produced as a byproduct of photosynthesis and is released into the atmosphere where it may be breathed in by other organisms. Energy from sunlight is captured by a green pigment known as chlorophyll contained in disc-like structures within the chloroplast known as thylakoids. Thylakoids are found in stacks called grana, located inside a fluid called stroma that fills the inside of chloroplasts.

Peroxisomes are small organelles found in most eukaryotes. The function of peroxisomes is to detoxify drugs and poisons, such as alcohol, and to break down lipids, proteins, and other molecules. The name of this organelle is derived from the fact that hydrogen peroxide is produced as a byproduct from the breakdown of these molecules. Hydrogen peroxide can be damaging to other parts of the cell, so it is retained in peroxisomes until it can be safely broken down into oxygen gas and water.

FIGURE 2.8: Mitochondrion—The mitochondrion has a double membrane, its own DNA and ribosomes, and is the site of cellular respiration.

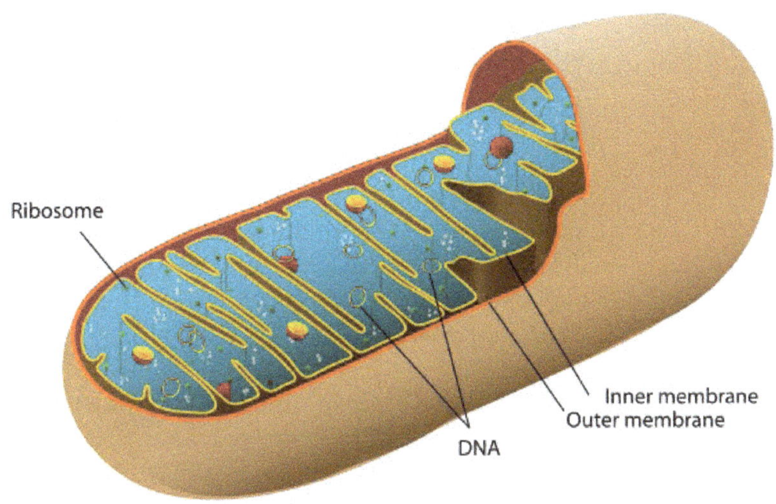

FIGURE 2.9: Chloroplast—Chloroplasts are sites of photosynthesis. Each chloroplast contains its own DNA and ribosomes, and has two membranes.

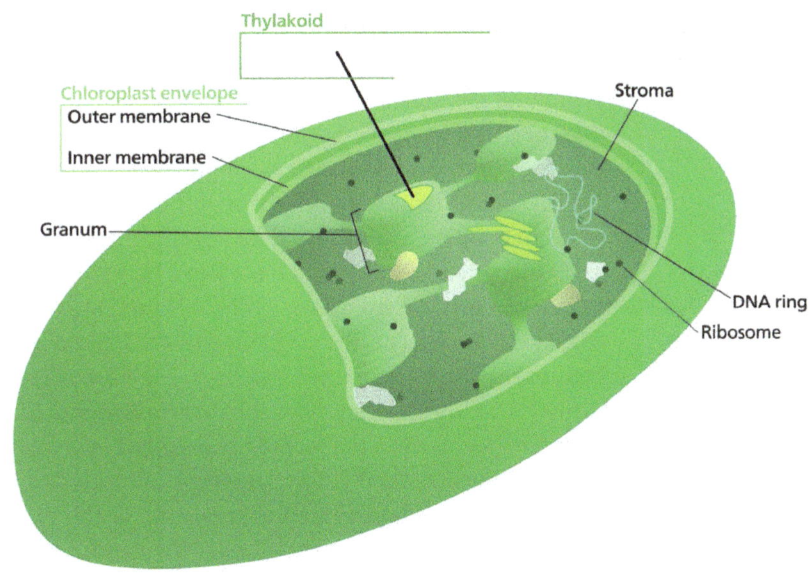

Cellular Respiration, Glycolysis, and Fermentation

In eukaryotes, the process of breaking down carbohydrates, lipids, and other molecular fuels into a form of energy that cells can use is called **cellular respiration**. The type of cellular respiration that occurs in mitochondria is called "aerobic" cellular respiration because oxygen is required to obtain energy from the breakdown of biological molecules. It is because of aerobic cellular respiration that humans and many other organisms must continually take in fresh oxygen to survive; without oxygen, our cells cannot harvest enough energy from food molecules to carry out life-sustaining chemical reactions. The type of energy produced by mitochondria, **adenosine triphosphate (ATP),** is known as the "energy currency" of the cell because it is used to power most other chemical reactions in the cell.

In addition to the aerobic cellular respiration that occurs in mitochondria, eukaryotic cells are also capable of producing energy through two other processes known as glycolysis and fermentation. **Glycolysis** is the breakdown of the carbohydrate glucose into two smaller molecules called pyruvate, releasing two molecules of ATP. In cells that undergo aerobic cellular respiration, glycolysis precedes the aerobic portion of the process, and it is the pyruvate produced by glycolysis that enters the mitochondria, where it is broken down further to release even more ATP. **Fermentation** is the breakdown of glucose or other small carbohydrates to release ATP as well as byproducts including carbon dioxide and various acids or alcohols, depending on the type of fermentation. Both glycolysis and fermentation occur in the cytoplasm rather than in the mitochondria and do not involve the use of oxygen but produce less ATP than aerobic cellular respiration from the same amount of carbohydrate. Many types of fungi, such as the yeast used to produce bread, beer, and wine, rely primarily on fermentation to produce energy because they live in environments with little or no oxygen.

Evolutionary Origin of Chloroplasts and Mitochondria

The **Endosymbiont Theory** explains how complex eukaryotic cells are thought to have evolved from simpler prokaryotic cells (Figure 2.10). The theory states that, at one time, the cells of some prokaryotes became enlarged and began to develop a series of membrane-bound organelles, including a nucleus, effectively becoming primitive eukaryotes. Later, smaller prokaryotes known as protobacterium entered these cells as either prey or as a parasite, and managed to avoid being destroyed or ejected, thus becoming an endosymbiont, or a cell living inside another cell. Because the endosymbiont was capable of producing its own energy, this was beneficial for the larger host cell. This relationship was also beneficial to the endosymbiont because it provided protection and nutrients. Eventually the two cells became dependent upon this endosymbiotic relationship and established a permanent arrangement of living and reproducing together. This scenario is thought to have occurred multiple times, involving protobacteria capable of producing energy through aerobic respiration, as well as primitive cyanobacteria capable of producing energy through photosynthesis. Descendants of the former are known as mitochondria, and descendants of the latter are known as chloroplasts.

Evidence supporting this theory can be found by examining the structures and behaviors of modern mitochondria and chloroplasts (Figure 2.10). Unlike other organelles, mitochondria and chloroplasts are enveloped by two distinct membranes. The presence of the second membrane can be explained by observing bacteria taken in to immune cells through phagocytosis. During phagocytosis, the membrane of the phagocytic cell engulfs a bacterium and then pinches off from the rest of the membrane to form a membranous sac called a phagocytic vacuole. Because the bacterium is also surrounded by its own membrane, it appears to be surrounded by a double membrane while in the phagocytic vacuole. Second, mitochondria and chloroplasts contain their own ribosomes and DNA distinct from host cell ribosomes and DNA. This is a trait shared by free-living bacteria but not by other organelles. The third main piece of evidence that supports the Endosymbiont Theory is that mitochondria and chloroplasts are capable of growing and reproducing somewhat independently inside their host cells in a manner similar to the way that bacteria replicate. This can sometimes be observed in host cells adapting to increased energy needs. For example, cells of a growing plant may increase the number of chloroplasts inside each cell to increase the amount of energy produced through photosynthesis. When humans undergo strenuous aerobic exercise, such as training

FIGURE 2.10: Endosymbiont Theory—The Endosymbiont Theory explains the unique evolutionary origin of chloroplasts and mitochondria.

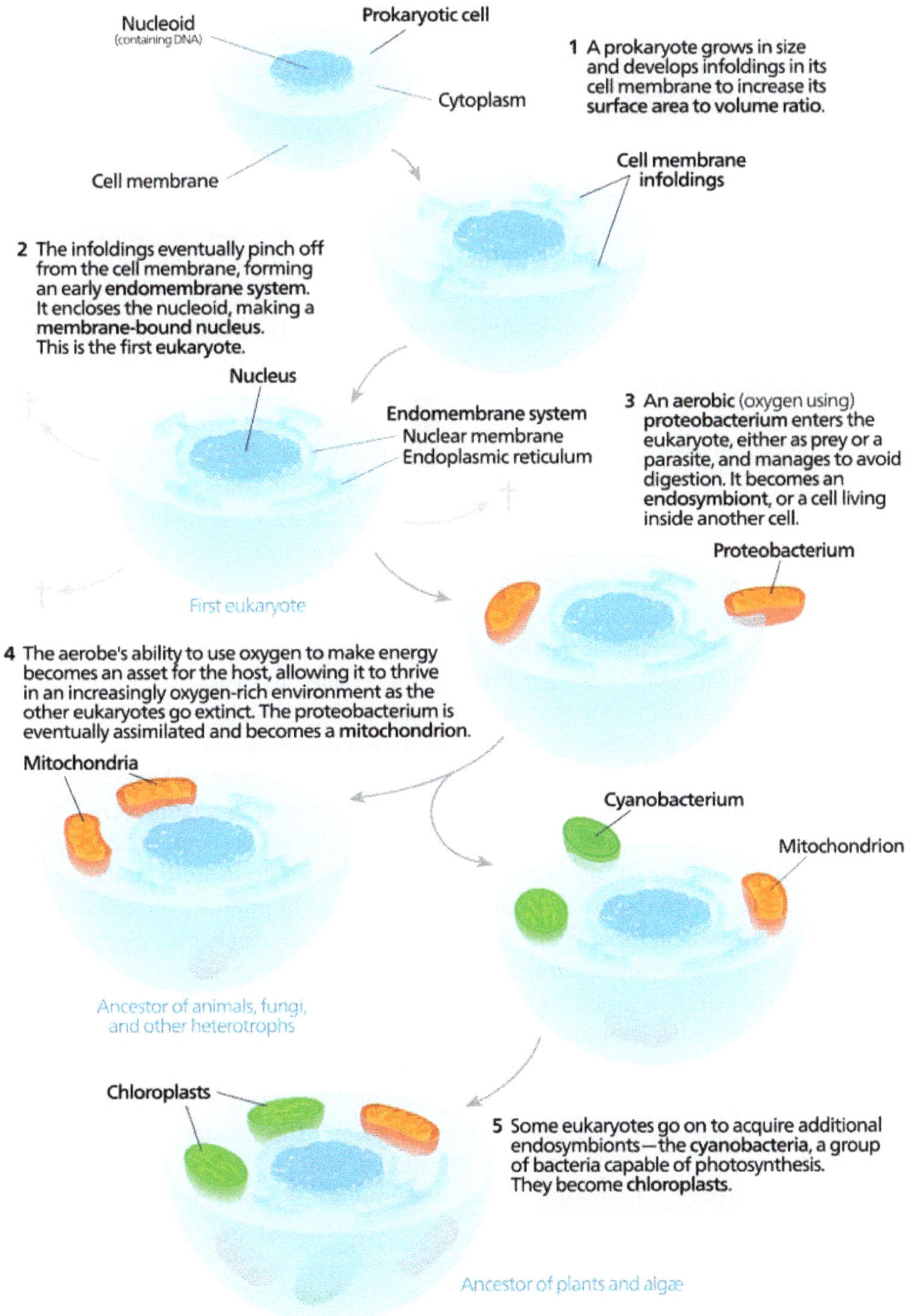

for a marathon, the number of mitochondria with the cells of their muscles may increase to provide more energy for movement through increased cellular respiration.

Osmosis and Diffusion

Diffusion is the tendency for molecules to spread out evenly into the available space. Substances diffuse by following a concentration gradient, moving from regions where the concentration of that substance is higher to regions where the concentration of that substance is lower. This process is driven by the random movement of molecules and does not require input of additional energy. Each different substance will diffuse along its own concentration gradient. For example, if chloride ions, sodium ions, and water are all present in the same fluid but at different concentrations, they will each diffuse at a different rate. In addition to the strength of the concentration gradient, other factors that influence the rate of diffusion include mass and temperature; lower mass molecules diffuse more quickly than higher mass molecules, and diffusion rates for all molecules generally increase as temperature increases.

Living things rely on diffusion to move water, nutrients, and other molecules into and out of their cells. For example, if the concentration of oxygen is higher outside of the cell than inside, oxygen will move into the cell. At the same time, if the concentration of carbon dioxide is higher inside the cell than outside, carbon dioxide will diffuse out of the cell. This is the process that constantly occurs in the lungs of animals, exchanging the waste product of cellular respiration (carbon dioxide) with the fresh oxygen necessary to fuel additional cellular respiration and release energy from food molecules inside cells.

Diffusion of some small molecules such as carbon dioxide can occur directly through the plasma membrane. Sometimes particles diffuse through a biological membrane directly without necessitating the use of energy. This process is called passive transport because no energy input is required in order for it to occur. When the molecules that are passively diffusing across a selectively permeable membrane are water molecules, the term used for the process is **osmosis**. Osmosis and diffusion can both move molecules into and out of the cell, depending on the direction of the concentration gradient (Figure 2.11). Maintaining the correct balance of water and other molecules in cells is crucial for survival. When the concentration of water is higher outside of a cell than inside, water will flow into the cell via osmosis, and the cell will expand. The type of environment where this occurs is called **hypotonic**. If a cell is exposed to a hypotonic environment for too long, it may burst open and die after taking in too much water. Conversely, a **hypertonic** environment is one in which the concentration of water is lower outside of the cell than inside. Cells placed in a hypertonic environment will shrink as water moves down its concentration gradient to the exterior of the cell. If the concentration of water is equal on both sides of a plasma membrane, no net movement of water across the membrane will occur, and the cell will neither expand nor shrink. This is known as an **isotonic** environment. What determines the concentration of water is the amount of other molecules dissolved in the water. For example, the concentration of water is higher in fresh water than in salt water because the salt molecules are counted amongst the total number of molecules.

Facilitated Diffusion

Although molecules have a tendency to move down their concentration gradient, not all molecules can easily move through the lipid bilayer of a plasma membrane. Large molecules, polar molecules, and electrically charged atoms (ions) often require assistance to move across the nonpolar lipid interior of the plasma membrane. This assistance comes in the form of tunnel-like proteins that span the entire width of the membrane, known as transport proteins. The passive movement of molecules across the plasma membrane through a transport protein and down their concentration gradient is known as **facilitated diffusion**. Cells are also capable of moving molecules across the plasma membrane against their concentration. This process, known as active transport, requires both a transport protein and the input of energy to combat the force of the concentration gradient. Active transport is crucial for many cellular processes, including the function of nerve cells, which requires establishment of an electrical gradient by concentrating positive and negative ions on one side of the membrane.

FIGURE 2.11: **Osmosis**—During osmosis, water molecules move across a semipermeable membrane, down their concentration gradient, from an area where water molecules are highly concentrated to an area where they are less concentrated.

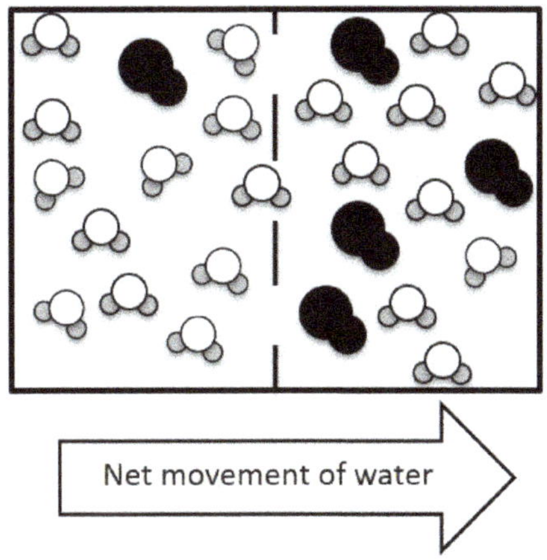

2.2 SUMMARY

- Characteristics shared by all living things include adaptation, cellular organization, energy processing, growth and development, heredity, homeostasis, reproduction, and responsiveness
- Living cells can be divided into two main categories, eukaryotic and prokaryotic, based on whether they contain a nucleus and membrane-bound organelles
- Autotrophs produce their own organic food molecules, whereas heterotrophs must consume other organisms or their products to obtain food and energy
- Organelles are distinct structures in cells that carry out specialized functions including digestion of foreign particles, membrane production, photosynthesis, and protein synthesis
- Nonpolar hydrophobic molecules interact well with other nonpolar molecules but not with polar molecules like water; polar molecules are hydrophilic and interact well with water but not with nonpolar molecules
- The Endosymbiont Theory proposes that eukaryotic cells evolved from a beneficial relationship between primitive host cells and energy-producing bacteria that were the ancestors of chloroplasts and mitochondria
- Osmosis and diffusion move water and other molecules from areas of high concentration to areas of low concentration
- Transport proteins help polar molecules and molecules moving against their concentration move through the plasma membrane via facilitated diffusion (passive and active transport)

2.3 KNOWLEDGE CHECK

1. Which one of the following statements is true?
   a. Eukaryotes are generally smaller than prokaryotes
   b. Eukaryotes can be free-living organisms but prokaryotes cannot
   c. Eukaryotes have a nucleus but prokaryotes do not
   d. Eukaryotes have a plasma membrane but prokaryotes do not
   e. Eukaryotes have ribosomes but prokaryotes do not

2. Organisms that obtain food and energy without eating other organisms or consuming organic molecules are known as:
   a. autotrophs
   b. endosymbionts
   c. heterotrophs
   d. hydrophilic
   e. organelles

3. What is the function of the smooth endoplasmic reticulum?
   a. break down lipids and proteins
   b. control passage of molecules into and out of the cell
   c. produce energy through cellular respiration
   d. synthesize proteins
   e. synthesize lipids and detoxify poisons

4. A cell that secretes a large amount of molecules into the extracellular space is expected to have a higher amount of which of the following organelles than an otherwise identical cell that does not secrete molecules?
   a. centrioles
   b. Golgi apparatus
   c. lysosomes
   d. peroxisomes
   e. vacuoles

5. The origin of which two organelles can be explained by the Endosymbiont Theory?
   a. central vacuole and chloroplast
   b. chloroplast and mitochondrion
   c. mitochondrion and nucleus
   d. nucleus and ribosome
   e. plasma membrane and cell wall

6. List the structures found in eukaryotic cells but not prokaryotic cells.

7. Compare and contrast the processes of active transport, diffusion, osmosis, and passive transport.

8. The term "amphipathic" means "having both hydrophilic and hydrophobic parts." Identify a molecule that is amphipathic and describe its role in the cell.

2.4 ACTIVITY
Visualizing Cell Structures

The purpose of this activity is to visualize the structure of different cell types and identify each type of organelle present. Using a full page of paper for each cell type, draw an animal cell, a plant cell, and a prokaryotic cell. Label each organelle in your drawing. If possible, use different colors. Use the following list of organelles in your drawing (label all organelles present in each cell type):

Capsule	Nucleoid
Cell wall	Nucleus
Centriole	Peroxisome
Chloroplast	Plasma membrane
Cytoplasm	Ribosome
Flagellum	Rough endoplasmic reticulum
Golgi apparatus	Smooth endoplasmic reticulum
Lysosome	Ribosome
Mitochondrion	Vacuole

2.5 ACTIVITY

Reviewing Cell Structures

The purpose of this activity is to review the function of each organelle while simultaneously creating a study aid for future use. Make a set of flash cards that can be used to review the function of each organelle. Start by collecting index cards or cutting blank pieces of paper into quarters until you have a total of 18 pieces. On one side of each card or paper, write the name of each organelle; summarize the function of that organelle on the opposite side.Include each of the following organelles:

Capsule
Cell wall
Centriole
Chloroplast
Cytoplasm
Flagellum
Golgi apparatus
Lysosome
Mitochondrion
Nucleoid
Nucleus
Peroxisome
Plasma membrane
Ribosome
Rough endoplasmic reticulum
Smooth endoplasmic reticulum
Ribosome
Vacuole

2.6 ACTIVITY

Osmosis and Diffusion

This activity provides a hands-on opportunity to demonstrate and analyze osmosis and diffusion. Follow the directions for completing the following case study and submit answers to all of the accompanying questions:

Bonney, K. M. (2014). Case study: From gummy bears to celery stalks: Diffusion and osmosis. *Journal of College Science Teaching, 43*(6), 72–76.

2.4

1. Submit drawings of an animal cell, a plant cell, and a prokaryotic cell with the structure and function of each of the following organelles labeled, as appropriate. Alternatively, you may be instructed to submit a photograph of your work and keep the original work to use as a study tool for a future test or quiz.

Capsule	Nucleoid
Cell wall	Nucleus
Centriole	Peroxisome
Chloroplast	Plasma membrane
Cytoplasm	Ribosome
Flagellum	Rough endoplasmic reticulum
Golgi apparatus	Smooth endoplasmic reticulum
Lysosome	Ribosome
Mitochondrion	Vacuole

2.5

1. Submit a set of flashcards with the name and function of each of the following organelles labeled. Alternatively, you may be instructed to submit a photograph of your work and keep the original work to use as a study tool for a future test or quiz.

Capsule	Nucleoid
Cell wall	Nucleus
Centriole	Peroxisome
Chloroplast	Plasma membrane
Cytoplasm	Ribosome
Flagellum	Rough endoplasmic reticulum
Golgi apparatus	Smooth endoplasmic reticulum
Lysosome	Ribosome
Mitochondrion	Vacuole

2.6

1. Submit answers to the questions that correspond to this case study, as directed by your instructor.

2.7 REFERENCES

Archibald, J. N. (2015). Endosymbiosis and eukaryotic cell evolution. *Current Biology, 25*(19), R911:R921.

Bonney, K. M. (2014). Case study: From gummy bears to celery stalks: Diffusion and osmosis. *Journal of College Science Teaching, 43*(6), 72–76.

Diekmann, Y. and Pereira-Leal, J. B. (2013). Evolution of intracellular compartmentalization. *Biochemical Journal, 449*, 319–331.

Diener, T. O., McKinley, M. P., and Prusiner, S. B. (1982). Viroids and prions. *Proceedings of the National Academy of Sciences of the United States of America, 79*(17), 5220–5224.

Pan, P. and Umbreit, W. W. (1972). Growth of mixed cultures of autotrophic and heterotrophic organisms. *Canadian Journal of Microbiology, 18*(2), 153–156.

Songer, C. J. and Mintzes, J. J. (1994). Understanding cellular respiration: An analysis of conceptual change in college biology. *Journal of Research in Science Teaching, 31*(6), 621–637.

FIGURE CREDITS

FIG. 2.1: Adapted from "Plant Cell," https://commons.wikimedia.org/wiki/File:Plant_cell_structure_no_text-2.svg. Copyright in the Public Domain.

FIG. 2.2: Adapted from "Animal Cell," https://commons.wikimedia.org/wiki/File:Animal_Cell_Unannotated.svg. Copyright in the Public Domain.

FIG. 2.3: Adapted from "Prokaryotic Cell," https://commons.wikimedia.org/wiki/File:Average_prokaryote_cell-_unlabled.svg. Copyright in the Public Domain.

FIG. 2.4: Copyright © Dhatfield (CC BY-SA 3.0) at https://commons.wikimedia.org/wiki/File:Cell_membrane_detailed_diagram_4.svg.

FIG. 2.5: Mariana Ruiz, "Nucleus," https://commons.wikimedia.org/wiki/File%3ADiagram_human_cell_nucleus.svg. Copyright in the Public Domain.

FIG. 2.6: Mariana Ruiz, "Endomembrane system," https://commons.wikimedia.org/wiki/File%3AEndomembrane_system_diagram_en.svg. Copyright in the Public Domain.

FIG. 2.7: Miklos, "Phagocytosis," https://commons.wikimedia.org/wiki/File:Amoeba_phagocytosis.jpg. Copyright in the Public Domain.

FIG. 2.8: Mariana Ruiz Villarreal, "Mitochondrion," https://commons.wikimedia.org/wiki/File%3AAnimal_mitochondrion_diagram_en.svg. Copyright in the Public Domain.

FIG. 2.9: Adapted from "Chloroplast," https://commons.wikimedia.org/wiki/File%3AChloroplast_(borderless_version)-en.svg. Copyright in the Public Domain.

FIG. 2.10: Copyright © Kelvinsong (CC BY-SA 3.0) at https://commons.wikimedia.org/wiki/File:Serial_endosymbiosis.svg.

Atoms and Molecules

3.1 INTRODUCTION

Although understanding cells and other basic cell structures is a helpful starting point for building a foundation of knowledge about biology, living things also function on a subcellular level. Despite lacking the tools necessary to explore the microscopic and molecular worlds for all but the last couple hundred years, scientists have collected a vast library of information about the building blocks of life. One of the most interesting findings in all of biology is that the enormous diversity of different forms of life that exist on Earth are all composed of the same few components. This chapter explains how these tiny building blocks come together to form the molecular components from which all living organisms are built.

Building Blocks of Life

All living things, as well as the environment that surrounds them, are made of elements. **Elements** are defined as distinct substances that cannot be chemically converted or broken down into other substances. For example, water is not an element because it can be broken down into oxygen and hydrogen; however, oxygen and hydrogen are elements because they cannot be broken down into substances distinct from other elements. Elements are made of small particles called atoms that all share the same chemical properties of that element. **Atoms** are composed of three types of smaller particles called protons, neutrons, and electrons. Protons and neutrons are found in the core (or nucleus) of the atom, whereas electrons orbit the core of the atom in a three-dimensional sphere known as an electron cloud (Figure 3.1). **Protons** are positively charge particles, **electrons** have a negative electric charge, and **neutrons** are uncharged. The distinct physical and chemical properties of each element are determined through a combination of factors including the total number of protons contained in the atom, the number of protons relative to the numbers of electrons and neutrons, and the way the electrons are structured around the atom's core.

The structure of an atom's electron cloud is most important in determining the behavior of the atom in chemical reactions. Electrons are

LEARNING OBJECTIVES

- Describe the structure of atoms
- Identify what distinguishes different types of elements from each other
- Describe how the order of elements in the periodic table was determined
- List the four types of macromolecules and examples of each
- Identify the building blocks and chemical bonds that comprise each type of macromolecule
- Compare and contrast the functions and structures of each type of macromolecule
- Compare and contrast the structures of properties of saturated fats and unsaturated fats
- Describe the importance of macromolecules to everyday life
- Define the term calorie and identify factors that determine the amount of calories obtained from certain foods
- Identify the four emergent properties of water that help support life on Earth
- Define the terms acid and base, and explain how the strength of each is indicated by the pH scale

FIGURE 3.1: Atom structure—Atoms consist of positively charged protons, uncharged neutrons, and negatively charged electrons.

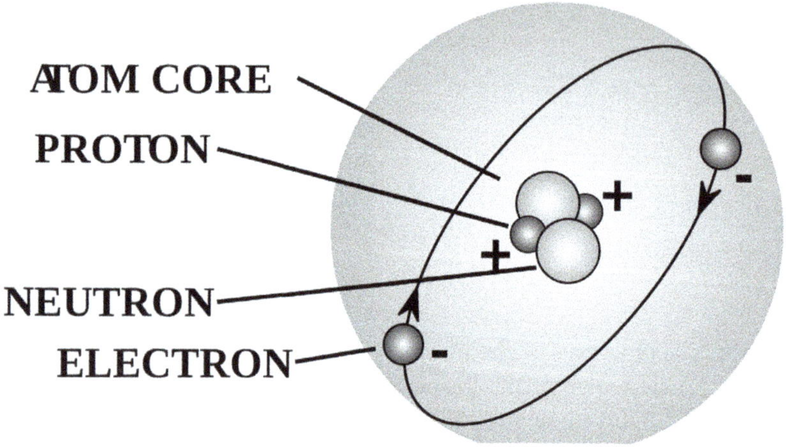

sometimes depicted as orbiting the core of an atom along a fixed path, similar to how the moon orbits the earth. However, it is more accurate to depict electron orbits as part of a cloud surrounding the atom's core. The electron cloud of an atom could be thought of as similar to the atmosphere of the Earth in that it is a three-dimensional space surrounding a sphere with distinct layers in which particles may move. Using that analogy, electrons orbiting an atom's core are like airplanes and other flying bodies moving through the atmosphere. Similar to how the Earth's atmosphere is divided into distinct layers (troposphere, stratosphere, mesosphere, thermosphere, exosphere), the electron clouds of most atoms are composed of multiple concentric layers, known as electron shells. Each **electron shell** is located at a different distance from the atom's core and contains electrons with different amounts of energy. Individual electrons orbit the atom's core within one specific electron shell, and each electron shell can only contain a certain number of electrons (the first shell can contain two electrons; the second and third shells can each contain eight electrons; and the outer shells, if present, can contain even more electrons). Therefore, the more electrons an atom has, the more electron shells it must have to accommodate them. For example, carbon atoms (Figure 3.2A) have six electrons that fit into two electron shells (two electrons are in the first shell, and four electrons are in the second shell). Calcium atoms have 20 electrons, so they must have 4 electron shells to contain them (Figure 3.2B); 2 electrons are in the first shell, 8 in the second, 8 in the third, and the remaining 2 are in the fourth shell.

Regardless of the total number of electrons an atom has, it is only the number of electrons contained in the outer electron shell that determines many of the chemical properties of that atom, including which other atoms it will form chemical bonds with. In fact, only the electrons in an atom's outer electron shell, which are called **valence electrons**, will interact with other atoms. The interactions between valence electrons of adjacent atoms may result in the formation of a chemical bond that holds the atoms together. Atoms with full outer electron shells are inert, meaning they will not form chemical bonds with other atoms. Atoms that do not have full outer electron shells have a tendency to fill their outer shells by sharing electrons with other atoms in the form of chemical bonds, or by taking or giving away the number of electrons necessary to achieve a full outer shell. When two or more atoms share pairs of electrons, the type of bond formed is called a **covalent bond**. The term *covalent bond* is a general term for chemical bonds that involve the sharing of electrons between atoms. Covalent bonds are further divided into two categories based on how electrons are shared between the atoms. When atoms share electrons equally, the bond that forms is known as a **nonpolar**

FIGURE 3.2A: Electron shell diagram of carbon—Chemical properties of atoms are based, in part, on the number of electrons in their outermost (valence) electron shell.

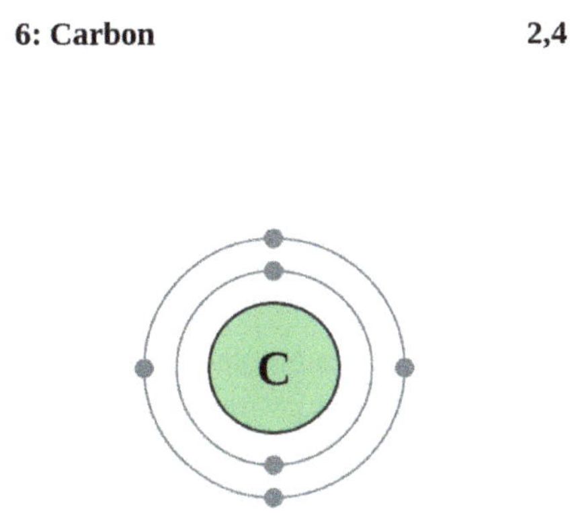

FIGURE 3.2B: Electron shell diagram of calcium

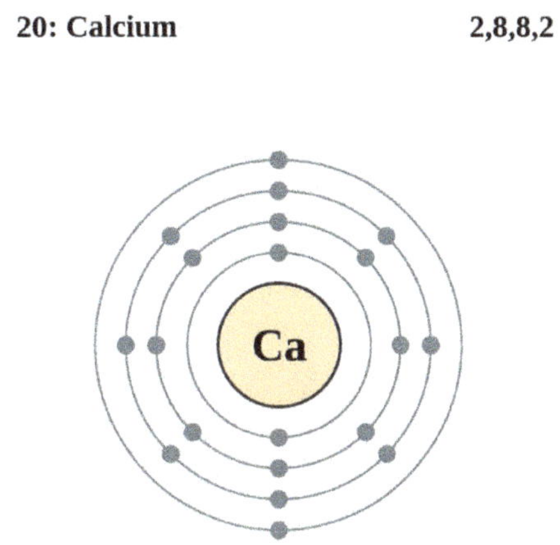

covalent bond (Figure 3.3). When atoms share electrons unequally because one atom attracts the electron more strongly than the other, the bond that forms is known as a **polar covalent bond**. Specific examples of covalent bonds are sometimes given other names to denote their function or location. Covalent bonds that form between the subunits that proteins are made from are called **peptide bonds**, and covalent bonds between subunits of nucleic acids are called **phosphodiester bonds.**

If one atom gives or takes electrons from another atom, the bond that forms is called an **ionic bond** (Figure 3.4). Ionic bonds tend to not be as strong as covalent bonds, and atoms joined by ionic bonds are more likely to separate from each other than those connected by covalent bonds. When an ionic bond is broken, the atoms that were part of that bond will no longer have the same number of electrons as protons and thus will no longer be neutrally charged; the atom that gave away one or more electrons will take on a positive charge, and the atom that accepted one or more electrons would have a negative charge. This is because electrons are negatively charged. Atoms that have a charge as a result of exiting an ionic bond are called **ions.** If the overall charge of an ion is negative, it is referred to as an **anion**, whereas an ion with a positive net charge is called a **cation.**

For example, elemental sodium atoms have 11 protons and 11 electrons, and are neutrally charged, yet they have only 1 electron in their third (valence) electron shell, so they have a tendency to give 1 electron to another atom to

FIGURE 3.3: Examples of covalent bonds—Covalent bonds are strong chemical bonds found in a diverse array of molecules.

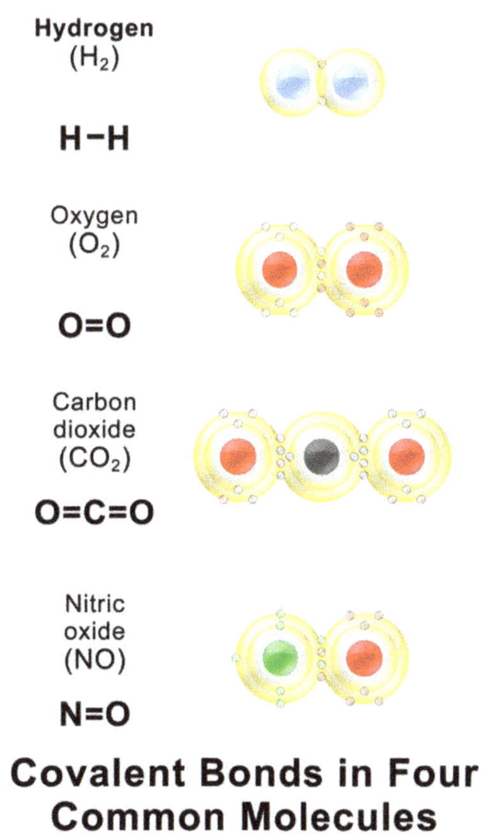

have a full valence shell (the second one). Chlorine atoms are also neutrally charged in their elemental form, with 17 protons balancing out 17 electrons. In this state, chlorine atoms are one electron short of filling their third electron shell, so they have a tendency to take one electron from another atom. Because of this, sodium and chlorine atoms often form ionic bonds with each other, resulting in a compound called sodium chloride. When sodium chloride (NaCl) is dissolved in water, the ionic bonds will break, releasing positively charged sodium ions (Na+) and negatively charged chlorine ions (Cl-). Although the ionic compound sodium chloride is commonly referred to as table salt, sea salt, or sometimes simply salt, all compounds formed by ionic bonds are technically considered **salts**. Other examples of salts include ammonium nitrate, copper sulfate, and potassium chloride.

One other type of chemical bond that commonly involves biological molecules is the hydrogen bond. Hydrogen bonds are the bonds that form between individual water molecules within liquid water and ice (Figure 3.5). This type of bond is even weaker than ionic bonds, which makes it relatively to easy for hydrogen bonds to form and be broken. **Hydrogen bonds** do not involve sharing or exchanging of electrons but instead result from an attraction between a slightly positively charged hydrogen atom in one molecule and a slightly negatively charged atom in a nearby molecule. This charge separation is a result of the fact that some atoms have a greater affinity for negatively charged electrons than others and will unequally share them with

FIGURE 3.4: Formation of an ionic bond—Ionic bonds are formed when two oppositely charged atoms exchange electrons to complete their valence shells.

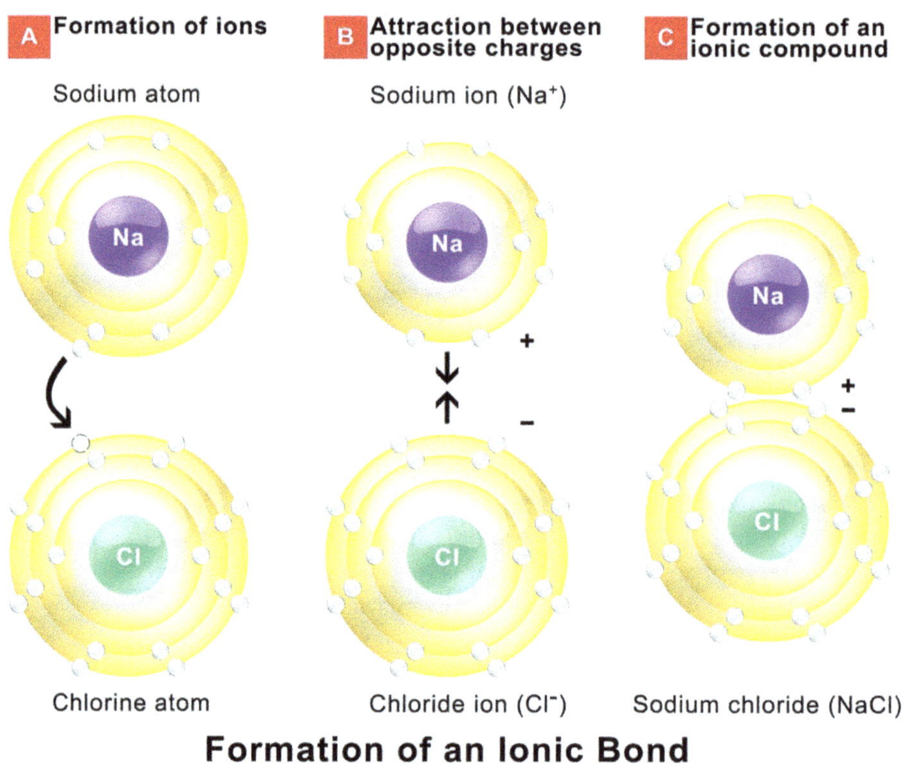

other atoms in a covalent bond. In the case of water molecules, oxygen atoms hold onto more of the negative charge of shared electrons than do the hydrogen atoms they are covalently bonded to. This creates a partial negative charge on the portion of the water molecule with the oxygen atom and a partial positive charge on the portion of the molecule where the hydrogen atoms are. The strength of an atom's tendency to attract electrons is a property called **electronegativity**. Oxygen atoms have a higher electronegativity than most other atoms found in biological molecules.

FIGURE 3.5: Water molecules are attracted to each other by hydrogen bonds—The atoms in a water molecule are held together by covalent bonds, but water molecules are attracted to each other by hydrogen bonds.

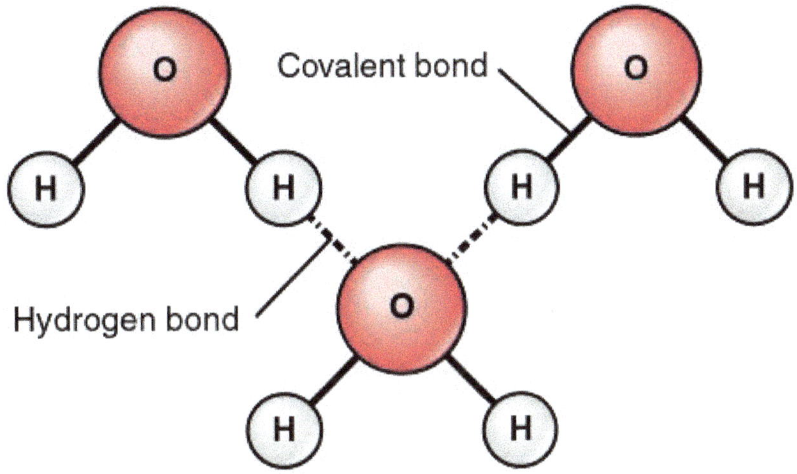

Despite the importance of electrons in determining how an atom will react with other atoms, what distinguishes the atoms of one element from the atoms of a different element is not the number of electrons but rather the number of protons. All of the atoms of an element always have the same number of protons, even though the number of electrons and neutrons may vary among atoms of the same element. The number of protons an atom has is different for each element and is known as the element's **atomic number**. For example, hydrogen has only one proton and has an atomic number of 1, whereas carbon has an atomic number of 8 because carbon atoms have eight protons. The atom with the highest known number of protons is ununoctium. Its atomic number is 118.

Protons and neutrons have much more mass than electrons, so the mass and weight of an element is mainly determined by the total number of protons and neutrons contained within the atoms of that element. The **atomic mass** of an atom, expressed in atomic mass units, is approximately equivalent to the combined number of protons and neutrons contained in its core. Protons are positively charged, neutrons are neutral or uncharged, and electrons are negatively charged. Therefore, the relative number of protons and electrons determines whether an atom has an electric charge. Atoms with more protons than electrons are positively charged, atoms with the same number of protons and electrons are neutrally charged, and atoms with more electrons than protons are negatively charged.

It is common for atoms of an element to have the same number of neutrons as protons, but this is not always the case. For some elements, such as carbon, nitrogen, and uranium, the number of neutrons in each atom may vary. Atoms of an element with a different number of neutrons than protons are called **isotopes**. Isotopes with extra neutrons have a higher mass and are referred to as "heavier" than standard atoms of the same element. Isotopes may possess additional chemical properties, such as radioactivity, which results from atoms releasing energy-containing particles from their core. Elements that may exist as isotopes are often categorized by the average number of neutrons that atoms of that element possess, unless a particular isotope has been isolated.

FIGURE 3.6: Periodic table of elements—The period table of elements is organized based on atomic number, atomic mass, and shared physical and chemical properties of the elements.

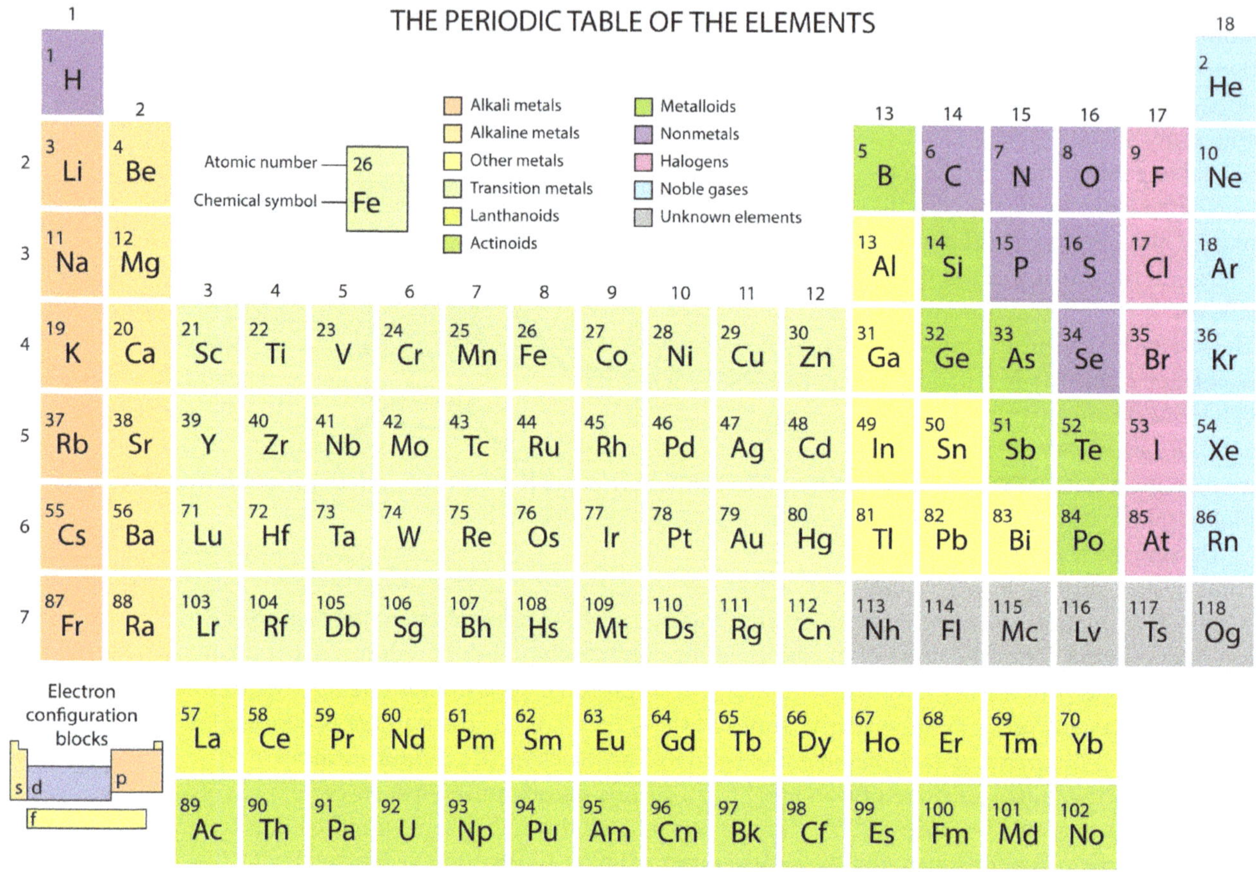

TABLE 3.1: Elements Abundant in Human Body

Element	Symbol	Percentage of atoms in human body
Oxygen	O	65.0
Carbon	C	18.5
Hydrogen	H	9.5
Nitrogen	N	3.2
Calcium	Ca	1.5
Phosphorus	P	1.0
Sulfur	S	0.3

In the periodic table of elements, atoms are ordered into rows based on their atomic number and are ordered into columns based on their atomic mass (Figure 3.6). The periodic table can be read from left to right and top to bottom, starting with the lowest atomic number and atomic mass in the upper left corner. The shape of the periodic table is further structured by grouping elements together based on shared physical and chemical properties. Although 118 distinct types of atoms, or elements, have been discovered, just 7 different elements account for almost 99% of the mass of the human body: calcium, carbon, hydrogen, nitrogen, oxygen, phosphorus, and sulfur (Table 3.1). With the exception of calcium, which is important for maintaining strong bones and sending electrical signals throughout the body, these elements are most often found in one of four types of large biological molecules, or **macromolecules**. The four types of biological macromolecules important for the structure and function of living things are carbohydrates, proteins, nucleic acids, and lipids. All of these molecules have carbon and hydrogen atom backbones, which makes them **organic** molecules. Small groups of atoms called functional groups can be attached to the carbon-hydrogen backbone to build many different kinds of molecules. Carbohydrates, proteins, and nucleic acids are all types of polymers, which are chains made from repeating, structurally similar subunits called monomers. Lipids are a structurally diverse group of molecules that are usually not composed of repeating subunits and are thus not considered polymers.

Carbohydrates

Carbohydrates are a group of molecules, including sugars and starches, that are most often made by plants and serve as a substantial food source for many animals. Carbohydrate structures can be recognized because they usually have the same number of carbon atoms and oxygen atoms, and approximately twice as many hydrogen atoms. This is often represented by the formula $(CH_2O)n$, where *n* is the number of carbon atoms in the molecule. For example, glucose contains six carbon atoms; using the formula $(CH_2O) \times 6$, it can be determined that glucose contains 6 carbon atoms, 12 hydrogen atoms, and 6 oxygen atoms. The chemical formula of glucose would customarily be written as $C_6H_{12}O_6$. Many carbohydrates can also be recognized by their name, which often ends with the suffix "-ose," as in "glucose," "sucrose," and "cellulose."

The atoms within carbohydrates are linked together by covalent bonds. Each covalent bond involves the sharing of one pair of electrons between two atoms and is often represented as a line connecting the letters that represent the atoms (Figure 3.7). Chemical structures may be drawn using a letter to represent each atom or by using use a type of chemical shorthand that omits the "C" used to indicate carbon atoms that are part of a chain or ring structure (Figure 3.7). When this type of shorthand is used, one can assume that there is a carbon atom whenever there is a chemical bond not connected to a letter, indicating an atom other than carbon is present.

Carbohydrates are made of building blocks called **monosaccharides**. Some carbohydrates, known as simple sugars, exist as single monosaccharides, whereas others are disaccharides, or two monosaccharides bonded together. The chemical bonds that bind individual monosaccharides together within disaccharides are called glycosidic bonds or **glycosidic linkages**. An example of a monosaccharide is glucose (Figure 3.7). An example of a disaccharide is sucrose (Figure 3.8). Notice that sucrose is made of two monosaccharides, glucose and fructose, joined by a glycosidic bond. The function of monosaccharides and disaccharides

FIGURE 3.7: **Glucose structure**—Glucose is a monosaccharide composed of a six-carbon ring structure.

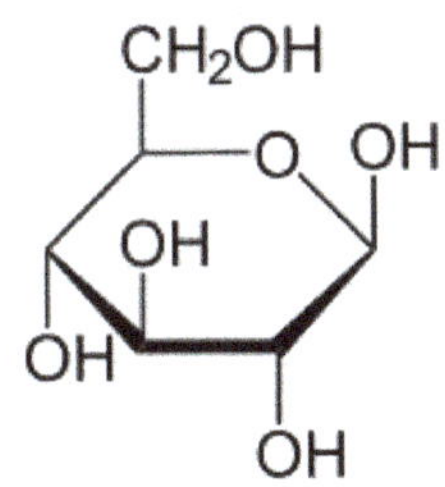

FIGURE 3.8: **Sucrose structure**—Sucrose is a disaccharide composed of two monosaccharides: fructose and glucose.

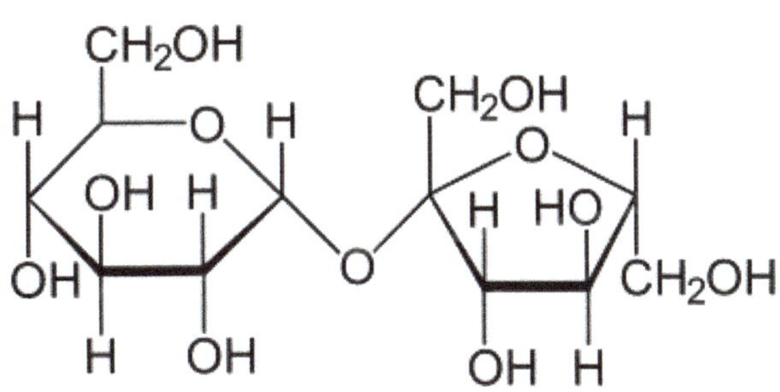

like glucose and sucrose is energy storage. Plants produce these sweet-tasting molecules to store energy captured during photosynthesis, and this energy is transferred to your body when you eat the fruit of those plants.

Complex carbohydrates exist as long chains of monosaccharides called polysaccharides. As in disaccharides, glycosidic linkages connect individual monosaccharides to form polysaccharides. Two examples of polysaccharides are cellulose, which provides structural support to the stems and leaves of plants (Figure 3.9), and glycogen (Figure 3.10), which is used to store energy in the cells of animals. Although cellulose and glycogen are both polymers of glucose, the glucose monomers are connected to each other in different ways, forming distinct three-dimensional structures. This illustrates the important point that the diverse functions and physical properties of carbohydrates are determined by not only which type of monomer they are composed of but by how those monomers are linked together (Table 3.2) The structure of the polysaccharides also affects how they can be used as a source of food and energy. For example, humans do not produce enzymes capable of efficiently breaking apart the bonds that hold the glucose subunits of cellulose together, so we cannot use cellulose as a source of energy. However, humans do produce enzymes capable of breaking down starch, so starch is an abundant source of plant-based energy in our diets.

TABLE 3.2: Examples of Carbohydrates and Their Functions

Example	Function
cellulose	provides structural support in plant cells; polymer of glucose
chitin	provides structural support in insect exoskeletons; polymer of glucose
glucose	found in red blood cells; transports oxygen throughout body
glycogen	stores energy in liver and muscle cells; polymer of glucose
lactose	source of energy found in milk; disaccharide of glucose and galactose
starch	stores energy in plants; polymer of glucose
sucrose	natural sugar from plants such as sugar cane; disaccharide of glucose and fructose

FIGURE 3.9: Polysaccharide structure (cellulose)—Cellulose is a polysaccharide that provides structural support in plants.

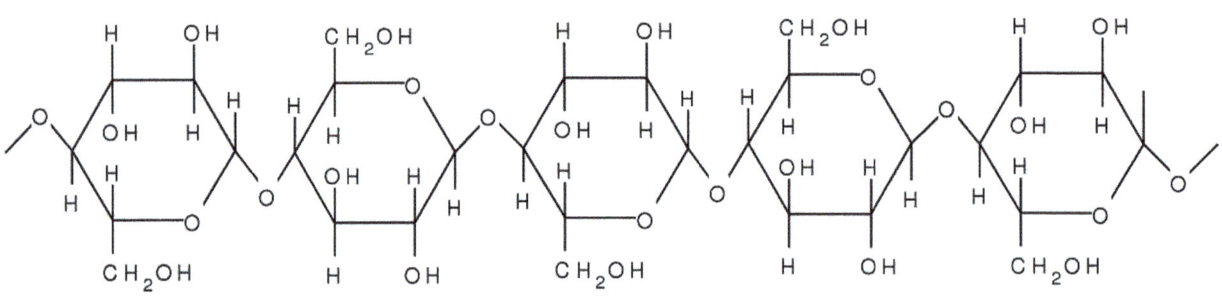

FIGURE 3.10: **Polysaccharide structure (glycogen)**—Glycogen is a polysaccharide used for long-term energy storage.

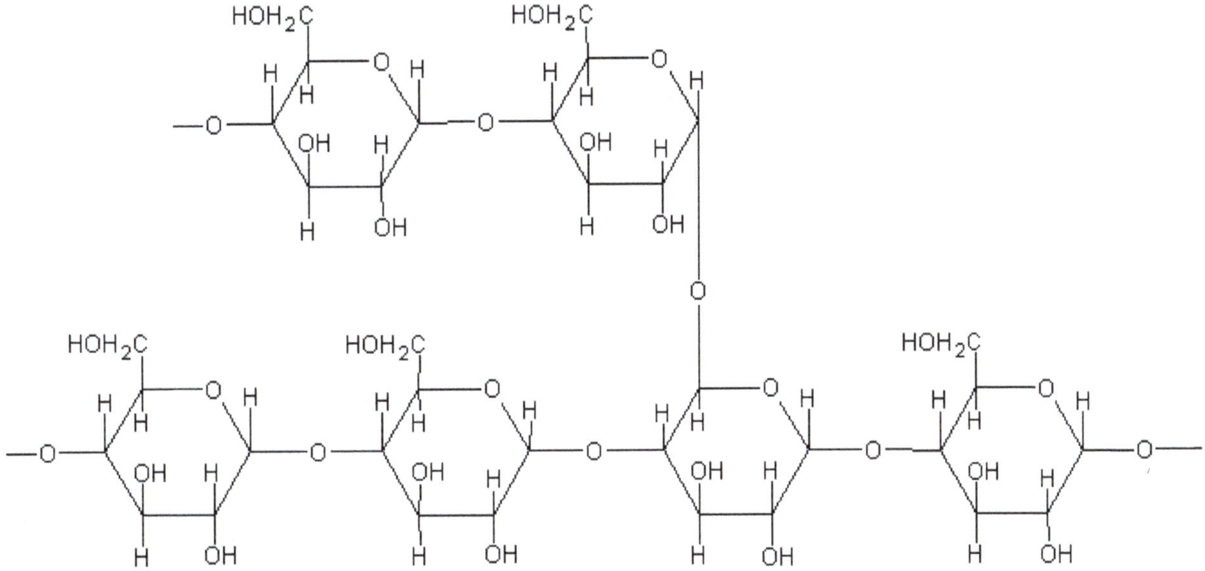

Proteins

Proteins are made of building blocks called **amino acids** connected by a type of chemical bond called a peptide bond. **Peptide bonds** are covalent bonds that form between the nitrogen-containing "amino" group of one amino acid and the carbon-based "carboxyl" group of an adjoining amino acid. Long chains of amino acids are called polypeptides until they are folded up into a specific three-dimensional structure called a protein. In addition to carbon, hydrogen, and oxygen, each amino acid contains nitrogen atoms, and some also contain sulfur. Each amino acid is made of a carbon-hydrogen backbone, which has an amino group, a carboxylic acid group, and an R group attached to it (Figure 3.11).

FIGURE 3.11: **Amino acid**—The basic structure of all amino acids contains an amino group, a carboxyl group, a hydrogen atom, and a variable side chain called an R-group.

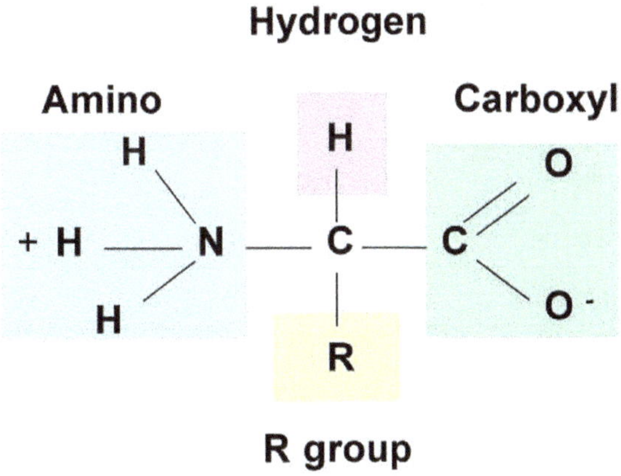

As the building blocks of proteins, amino acids are an important dietary component. Nine amino acids (histidine, isoleucine, leucine, lysine, methionine, phenylalanine, threonine, tryptophan, and valine) are known as **essential amino acids** for humans because they cannot be synthesized from other molecules, and thus it is essential that sufficient amounts of these molecules be consumed as part of our diet. There are a total of 20 different amino acids that have each been assigned a standardized 3-letter abbreviation and a 1-letter abbreviation (Table 3.3). What makes each amino acid different is the side chain, or R group, that the amino acid contains. In chemical shorthand, "R" is sometimes used to represent a part of a molecule that is variable. It is the R groups that determine the chemical properties of each amino acid. The R groups also determine how polypeptides will fold up into three-dimensional proteins.

TABLE 3.3: Amino Acid Abbreviations and Chemical Properties

Amino acid	Three-letter abbreviation	One-letter abbreviation
Alanine	Ala	A
Arginine	Arg	R
Asparagine	Asn	N
Aspartate	Asp	D
Cysteine	Cys	C
Glutamate	Glu	E
Glutamine	Gln	Q
Glycine	Gly	G
Histidine	His	H
Isoleucine	Ile	I
Leucine	Leu	L
Lysine	Lys	K
Methionine	Met	M
Phenylalanine	Phe	F
Proline	Pro	P
Serine	Ser	S
Threonine	Thr	T
Tryptophan	Trp	W
Tyrosine	Tyr	Y
Valine	Val	V

Of the four types of macromolecules, proteins are the most varied in structure and function. Some are made of only one amino acid and some contain several polypeptides, each made of thousands of amino acids. Proteins take on a seemingly limitless variety of shapes across a wide range of sizes. The function of proteins is highly dependent upon the three-dimensional shape that the polypeptide folds into, and conditions or genetic mutations that change the shape of proteins often disrupt their function. Some examples and functions of proteins are shown in Table 3.4. The many functions of proteins include helping to provide immunity against infections, transporting oxygen and other important molecules throughout the body, facilitating the completion of chemical reactions, and providing structural support and mobility to cells and tissues. One of the most important types of proteins are **enzymes**, which act as catalysts to accelerate the rate of chemical reactions. Another important group of proteins are **hormones**, which are chemical messengers produced by specialized glands that help control a wide range of bodily functions, such as digestion and growth. Some of the most important hormones in the human body, such as insulin, are proteins; however, a number of other crucial hormones, including the sex hormones estrogen and testosterone, belong to a different class of molecules called lipids.

Lipids

Lipids are a group of biological molecules that are insoluble in water. Unlike the other three types of macromolecules, lipids are not polymers. That is to say, lipids are not made of chains of similar subunits. Because of this, the structures of various lipids may appear quite different from each other, but most often they include long chains or ring structures (Figure 3.12). One way to recognize the structure of a lipid is that it is made of mostly carbon and hydrogen

TABLE 3.4: Examples of Proteins and Their Functions

Example	Function
antibodies	part of the immune system that defends the body against infections
hemoglobin	found in red blood cells; transports oxygen throughout body
lactase	Enzyme that catalyzes (speeds up) breakdown of the sugar lactose
insulin	hormone (chemical signal) that control cell's intake of the glucose
keratin	structural protein that gives strength and durability to hair and skin
myosin	motor protein in muscles that produce movement

atoms, with relatively few atoms of other types. The other atoms most commonly found in lipids are oxygen and phosphorous.

Two types of lipids that have important roles in human diet and health are saturated fats and unsaturated fats. Most animal and human fats exist in a structure known as a triglyceride, which is composed of a small molecule known as glycerol joined to three carbon-hydrogen chains, which may be either saturated or unsaturated, depending on the fat (Figure 3.12). When these carbon-hydrogen chains are not joined to other molecules, they are known as fatty acids (Figure 3.12). **Saturated fats** are so named because they contain the greatest number of hydrogen atoms possible given the number of carbon atoms they contain. Therefore, the carbon chains of saturated fats are *saturated* with hydrogen atoms. **Unsaturated fats** have one or more double bonds between some carbon atoms. Because each carbon atom can and must form exactly four chemical bonds, the carbon atoms involved in a double bond are able to bond to fewer hydrogen atoms than would be possible if there was no double bond. Fats that contain one double bond in their hydrogen-carbon chain are called monounsaturated fats, and fats with more than one double bond are called polyunsaturated fats. Saturated fats are found in animal fat, butter, cheese, and other dairy products. They tend to be solids at room temperature because the saturated fat molecules are able to pack together tightly. Saturated fats are less healthy dietary components than unsaturated fats because they increase levels of LDL (bad) cholesterol circulating in the blood, decrease levels of HDL (good) cholesterol, and subsequently increase the risk of heart disease. Unsaturated fats are found in avocados, nuts, olive oil, and vegetable oils. These tend to be liquid at room temperature because their bent structures cannot pack together as tightly as the more uniform and relatively flat structures of saturated fat molecules.

Perhaps the unhealthiest fats of all are a type of artificially produced unsaturated fat called trans fats. **Trans fats** contain double bonds and are thus unsaturated fats, yet they have three-dimensional structures that resemble those of saturated fats. Trans fats occur naturally, in small amounts, in some meat and dairy products, but the majority of trans fats consumed by humans are artificially produced through a chemical process that adds hydrogen atoms to unsaturated vegetable oils. Another name for this type of fat often seen on food labels is hydrogenated oil. The addition of hydrogenated oils helps to increase the shelf life of packaged foods, such as baked goods, corn and tortilla chips, and fried foods. This helps to reduce the overall cost of some foods. However, consumption of trans fats has been linked to increase

FIGURE 3.12: **Lipid structures**—Lipids structures are highly variable, yet all lipids contain mostly carbon and hydrogen atoms.

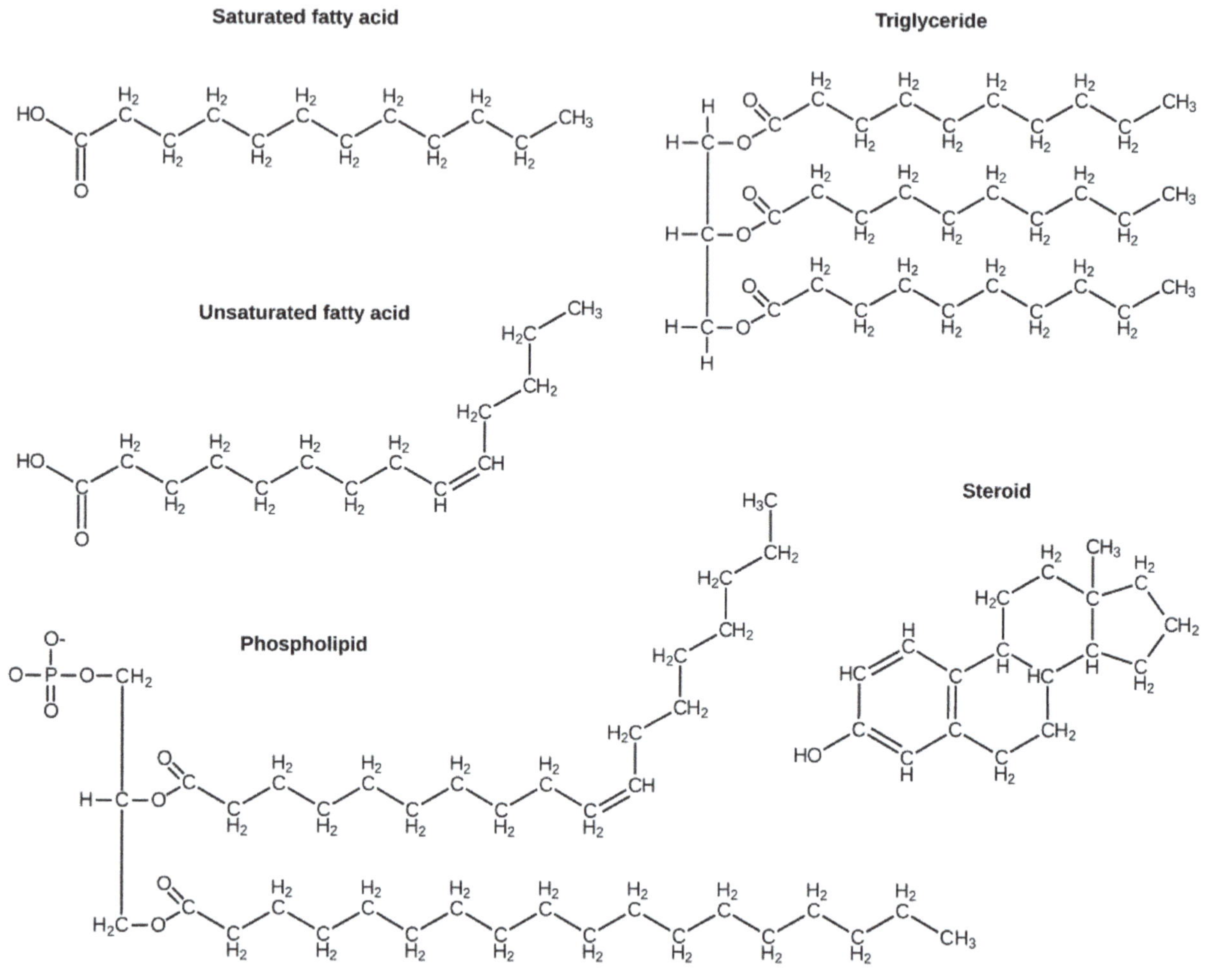

risk of heart disease. This is due, in part, to the fact that consumption of trans fat leads to increased levels of LDL cholesterol as well as decreased levels of HDL cholesterol.

Phospholipids are chains of fatty acids that contain a phosphate group and are a main structural component of cellular membranes. Lipids are generally hydrophobic, meaning they do not mix well with or are insoluble in, water. This is because the nonpolar chains of carbon and hydrogen in lipids do not interact well with water molecules, which are highly polar. The polar phosphate groups of phospholipids are hydrophilic, allowing one end of the phospholipid to interact with water, whereas the other end does not. In cellular membranes, two layers of phospholipids line up next to each other to conceal their hydrophobic tail regions on the inside of the membrane, away from water, while pointing their hydrophilic phosphate groups toward the water on the inside and outside of the cell. In addition to their roles inside the body and as food sources, lipids are also the main component of many widely used materials such as detergents, gasoline, propane, and wax.

TABLE 3.5: Examples of Lipids and Their Functions

Example	Function
cholesterol	steroid involved in formation of cell membranes and hormones
cortisol	hormone that directs the function of cells and tissues
oils	long-term energy storage molecule
phospholipids	main structural component of cellular membranes
triglyceride (fat)	long-term energy storage molecule long-term energy storage molecule
waxes	protective coating on plant and animals cells that prevents moisture loss

Nucleic Acids

Nucleic acids are polymers made of building blocks called nucleotides, which are each composed of a sugar molecule, a nitrogen-containing base, and a phosphate group. The phosphate groups of individual nucleotides are connected to each other by a type of chemical bond called a **phosphodiester bond** to form long chains of nucleic acid. One type of nucleic acid, deoxyribonucleic acid (DNA), is important for storing and transmitting hereditary genetic information. The other type of nucleic acid, ribonucleic acid (RNA), is important for the synthesis of proteins from genetic instructions contained in DNA and is involved in facilitating certain chemical reactions. Each nucleotide contains carbon, hydrogen, oxygen, and nitrogen, as well as phosphorous, which is not found in carbohydrates, proteins, or in most types of lipids. Unlike the other three types of macromolecules discussed in this chapter, nucleic acids are not a significant source of nutrients or calories in the human diet and will therefore not be discussed in detail in this chapter. However, DNA and RNA will be discussed in detail throughout several of the later chapters in this book.

Macromolecules and Calories

The structure of macromolecules is directly related to the amount of energy provided by eating food. Energy is stored in the chemical bonds that hold macromolecules together. When enzymes in the body break down the macromolecules in food, energy is released from the chemical bonds that held the atoms in those molecules together. The energy contained in chemical bonds and macromolecules is measured in units called calories. The term **calorie** is defined as the amount of energy needed to raise the temperature of one gram of water by one degree Celsius. However, when we refer to the number of calories in food, we are technically referring to Calories with a capital "C." One *Calorie* is actually equal to 1,000 *calories* and is also known as a kilocalorie (*kilo* means 1,000). Despite this important distinction, it is conventional to refer to the Calories in food as calories.

Water

Although carbohydrates, lipids, nucleic acids, proteins, and other carbon-based molecules are all important components of living things, the one molecule that really makes life as we know it possible is water. Water molecules contain an oxygen atom linked to two hydrogen atoms through polar covalent bonds. Because electrons are shared unequally within water molecules, the overall charge of each water molecule is unevenly distributed across its atoms. A molecule with an uneven charge distribution is called a **polar molecule**. The properties of water molecules that contribute to the ability of water to sustain life on Earth are derived from its strong polarity. The four emergent properties of water that help support life on Earth are cohesion, versatility as a solvent, temperature moderation, and expansion upon freezing.

In liquid water, individual water molecules are linked together by hydrogen bonds. The close association of molecules by linkages, such as hydrogen bonds, is called **cohesion**. The cohesiveness of water molecules can be observed when sucking water through a straw or by examining the surface tension of liquid water exposed to air. Cohesion is also necessary for plants to be able to transport water from the ground, through their roots and stems, and on to their leaves without having a circulatory system capable of pumping water though the plant. During much of the day, water continually evaporates into the air through pores in the leaves of plants, causing adjacent water molecules to be pulled further into the leaf to take their place. Chains of water molecules are pulled upward through the entire plant because of this force, but if not for a strong cohesiveness between water molecules, this chain would break and water could not be transported very far.

In addition to being highly cohesive, water is also a highly versatile solvent. A **solvent** is a substance in which other molecules dissolve. The substances that dissolve in a solvent are called **solutes**. When one or more solutes dissolve in a solvent, forming a homogenous mixture, a **solution** is formed. When water is the solvent, the solution is referred to as an aqueous solution. The solvent properties of water are essential for transporting oxygen, nutrients, and other important molecules within living organisms and throughout the environment. Because water is a polar molecule, polar molecules such as oxygen and sodium ions dissolve readily in water, whereas nonpolar molecules, such as lipids, do not dissolve in water.

Two properties of water extremely important for sustaining life on Earth are its efficiency at moderating temperature and that it expands at cold temperatures as it freezes. Water is capable of both absorbing a large amount of heat from warmer air and releasing a large amount of heat to cooler air without experiencing a significant change in its own temperature. On an environmental level, this helps regulate temperature extremes in oceans and in ecosystems near water, and on an individual level, this modulates extreme changes in body temperature during temporary exposure to hot or cold air. Additionally, water is one of few known substances that expands as it freezes, causing it to become less dense. As a result, when air temperatures drop below the freezing point of water, water molecules near the surface freeze and float, rather than sink, becoming a protective layer that prevents deeper water from also freezing. This allows organisms living in the water to survive through the winter.

Acids and Bases

Water is a dynamic substance, and the chemical bonds between molecules within aqueous solutions may frequently break and reform. In some instances, a hydrogen atom will detach from one water molecule and bind to a different water molecule with which it shares a hydrogen bond. When this occurs, the electron from that hydrogen atom remains with the original water molecule, and only the proton from the hydrogen atom shifts to the other water molecule. This single proton is called a hydrogen ion and has a positive charge. Hydrogen ions are commonly referred to as $H+$. However, despite being depicted this way, hydrogen ions do not remain as independent protons in aqueous solutions. Instead they are attached to another water molecule, forming what is known as a hydronium ion (H_3O^+). The other hydrogen atom and the oxygen atom left behind from the original water molecule are referred to as a hydroxide ion, which is commonly written as OH-. Hydroxide ions have a negative charge due to the loss of a positively charged proton. This type of dissociation is rare among water molecules, but is nonetheless extremely important to living things. This is because hydrogen ions and hydroxide ions are highly reactive and are involved in many essential chemical reactions.

The concentration of H+ and OH- are equal in pure water, but the addition of certain substances to aqueous solutions can cause a shift in this balance. A substance that causes an increase in the H+ concentration of a solution is called an **acid**. Conversely, any substance that reduces the H+ concentration of a solution is called a **base**. When the concentration of H+ and OH- in a solution are equal, the solution is referred to as neutral; when the concentration of H+ is higher than the concentration of OH- the solution is referred to as acidic, and when the concentration of H+ is lower than the concentration of OH- the solution is referred to as basic. The degree to which a solution is acidic or basic can be measured using the pH scale. The pH scale starts at 0 and extends as high as 14; lower numbers indicate stronger acids, while higher numbers indicate stronger bases. A pH of 7 is neutral, meaning the concentration of H+ and OH- is equal. It is important to know that the pH scale is logarithmic, meaning that for each whole number that separates two solutions on the pH scale, the H+ and OH- concentrations of those solutions differ by a factor of ten. For example, a pH of 6 indicates the H+ concentration is ten times higher than a pH of 7. A solution with a pH of 5 would have 100 times (10×10) the concentration of H+ than a solution with pH of 7, and a pH of 4 would have 1,000 times ($10 \times 10 \times 10$) the concentration of H+ than a solution with pH of 7. Human stomach acid, with a pH of 2, is 100,000 times as acidic as pure water! On the other end of the scale, the pH of household ammonia, which is 13, is 1,000,000 times more basic that pure water.

3.2 SUMMARY

- Elements are distinct substances that cannot be converted into other substances; elements are made of small particles called atoms
- Atoms are made of positively charged protons, neutrally charged neutrons, and negatively charged electrons; the number of protons determines which element an atom is, and the valence electrons are the parts of atoms that form chemical bonds with other atoms
- Covalent bonds are chemical bonds formed by the equal (nonpolar) or unequal (polar) sharing of electrons between atoms; ionic bonds involve the transfer of valence electrons between atoms; hydrogen bonds are weak interactions between molecules resulting from the attraction of a slightly positively charged hydrogen atom in one molecule and a slightly negatively charged atom of another molecule
- Carbohydrates are made of monosaccharides that may be linked by glycosidic linkages to form disaccharides and polysaccharides; the main functions of carbohydrates are energy storage and structural support
- Proteins are made of amino acids linked together by peptide bonds; proteins have a variety of functions, including as components of the immune system, enzymes, transport

molecules, and molecules involved in structure and movement; the side chains of amino acids determine their function and the way a protein is folded

- Lipids are not polymers but rather contain long chains of carbon and hydrogen atoms in varying arrangements; different types of lipids function to store energy, form cellular membranes, and transmit chemical signals as hormones
- Saturated and unsaturated fats are two important types of lipids found in our diet, which differ by whether there are any double bonds between their carbon atoms
- Nucleic acids are made of nucleotides linked by phosphodiester bonds; each nucleotide is composed of a sugar, a phosphate group, and a nucleotide base, and the function of nucleic acids is to store and transmit hereditary information
- Carbohydrates, proteins, and lipids are the main sources of energy in the diets of animals; each type of macromolecule contains a specific amount of energy per gram, which is measured in calories
- The four emergent properties of water that help support life on Earth are cohesion, versatility as a solvent, temperature moderation, and expansion upon freezing

3.3 KNOWLEDGE CHECK

1. Neon is inert because neon atoms will not form chemical bonds with other atoms. Which of the following determines whether an atom will be inert?
   a. the number of valence electrons
   b. the ratio of protons to electrons
   c. the ratio of protons to neutrons
   d. the total number of electrons
   e. the total number of electrons, protons, and neutrons

2. In which type of macromolecule are subunits joined by peptide bonds?
   a. carbohydrates
   b. lipids
   c. nucleic acids
   d. proteins
   e. water

3. All of the following are functions of lipids, EXCEPT:
   a. acting as hormones to direct the function of cells and tissues
   b. catalyzing and increasing the speed of chemical reactions
   c. maintaining cellular membrane structure
   d. providing long-term energy storage in animals
   e. serving as a source of energy in the human diet

4. Which one of the following statements about saturated fats is true?
   a. They are a healthier component of food than unsaturated fats.
   b. They are usually liquid at room temperature.
   c. They contain at least one double bond.
   d. They contain at least one sulfur atom.
   e. They have more hydrogen atoms than an unsaturated fat with the same number of carbons.

5. How much more acidic is a substance with pH 3 than a substance with pH 6?
   a. 2 times
   b. 3 times
   c. 10 times
   d. 1000 times
   e. 1,000,000 times

6. Identify the characteristic(s) of elements used to determine their order in the periodic table of elements.

7. Identify the three components of atoms and their electrical charges.

8. The term **calorie** is defined as the amount of energy needed to raise the temperature of one gram of water by one degree Celsius. Design an experiment that could be used to determine how many calories are contained in a sample of food from an unknown source.

3.4 ACTIVITY

Food, Macromolecules, and Calories

Carbohydrates, lipids, and proteins are often discussed in terms of food and the amount of calories derived from that food. The purposes of this activity are to describe what you know about each type of macromolecule, and to apply your knowledge of calories to calculate the number of calories in different samples of food. The estimated number of calories released by fully breaking down one gram of carbohydrates or proteins is 4, and there are an estimated 9 calories per gram in lipids. Note that the fourth type of macromolecule, nucleic acids, is present at such low concentrations in food that it does not contribute a measurable amount of energy to the diet.

3.5 ACTIVITY

Calories and Food: Further Thought

Calorie labels on food can be helpful for monitoring the amount of energy consumed through eating and subsequently for estimating the amount of energy that must be expended to maintain the same body weight. However, the estimated number of calories per gram of macromolecule listed in Table 3.6 is based on controlled experiments conducted in a laboratory, and laboratories do not perfectly replicate conditions within the human body. It is difficult to predict precisely how many calories an individual will actually absorb from food containing certain macromolecules as a result of the influence of a number of factors involving the physiology of the body, the way the food was prepared, and genetic factors affecting both the organism(s) that the food came from, and the individual consuming the food. The purpose of this activity is to learn more about the factors that affect the amount of calories obtained from eating food. To learn about some of these factors, read the following article, and then answer the questions that accompany this activity.

Dunn, R. (2013). Everything you know about calories is wrong. *Scientific American, 309*(3), 56–9.

3.6 ACTIVITY

Testing for Macromolecules in Food

The presence of different types of macromolecules in food and other substances can be determined by a number of different chemical tests (Table 3.6). This type of testing can be useful for identifying the components of an unknown food sample or determining if a food contains ingredients that are not indicated on the label, such as added sugar. The Benedict's test reacts to the presence of small sugars, including monosaccharides and disaccharides. Benedict's reagent, which is the active chemical used in the Benedict's test, is an aqua-blue color when not exposed to small sugars. However, if exposed to small sugars and heated, Benedict's reagent will change color to indicate a positive test result. The color of a positive Benedict's test result depends on the concentration of sugars present; low sugar concentrations will produce a green color, moderate sugar concentrations will appear orange, and high sugar concentration will appear red. Certain polysaccharides, including starch not by cellulose, can be detected by the iodine test. Iodine (often used in the form of potassium iodide) is a yellow substance that will turn black in the presence of certain polysaccharides. Proteins can be detected with the Biuret test, which uses a blue biuret reagent that changes color to lavender when mixed with proteins and heated. Finally, lipids can be detected with the Sudan IV test. Sudan IV is a red dye that will appear orange when mixed with lipids.

In addition to testing various foods to verify the presence or absence of certain macromolecules, these tests have a number of practical applications. One common example is the use of the iodine test to detect counterfeit United States currency. Authentic United States currency is made of linen, which is produced using cellulose fibers from the flax plant. Although cellulose is a polysaccharide, it is not detected by the iodine test for polysaccharides. Therefore, when a marker containing iodine is used to mark authentic currency, the mark will be yellow, indicating a negative test for polysaccharides. Counterfeit currency is often made from paper, which is easier and cheaper to obtain and use. Most paper is made from wood pulp, which contains starch, the polysaccharide that produces the strongest positive reaction in the iodine test. Because of this positive reaction, an iodine marker will leave a black mark on counterfeit paper-based currency, providing quick and easy fraud detection.

Another real-world application of chemical tests for macromolecules is use of the Biuret test to detect protein in urine. Urine from a healthy person is usually a mixture of water and

TABLE 3.6: Chemical Tests for Macromolecules

Test	Macromolecule tested for	Positive test result	Negative test result
Benedict's	monosaccharides and disaccharides (sugars)	green, orange, or red	blue
Biuret	protein	violet	blue
Iodine	certain polysaccharides (but not fiber/cellulose)	black	yellow
Sudan IV	lipid	orange	red

small molecules that are the waste products from cells and the breakdown of larger molecules. The presence of macromolecules, such as proteins with intact peptide bonds, could indicate a health problem such as kidney disease or sickle cell anemia. Doctors may routinely or selectively screen a patient's urine using the Biuret test to determine if proteins are present and conduct follow-up tests if a positive result is obtained, indicating a potential problem. The purpose of this activity is to apply your knowledge of these tests to identify the types of molecules in a hypothetical sample of food.

ACTIVITY QUESTIONS

3.4

1. Construct a table with three rows and four columns. In the first column, write carbohydrate in the first row, write nucleic acid in the second row, and write protein in the third row. In the second column, list the monomers that comprise each type of macromolecule. In the third column, identify the name of the chemical bonds that connect each type of monomer. In the fourth column, provide an example of each type of macromolecule.
2. Explain why lipids should not be included in the table constructed for Question 1.
3. Compare and contrast saturated fats and unsaturated fats in terms of molecular structure, physical properties, and list the types of foods that are sources of significant amounts of each type of fat. Then, identify which one is a healthier component of the diet and explain why.
4. Indicate how many calories are in one serving of each food shown in Figure 3.13. Show your calculations.
5. Was your result to the previous question consistent with what you expected before completing the calculations? Explain.
6. Describe how to set up an experiment that would measure the amount of calories in a sample of food. Refer to the definition of a calorie for a clue as to what type of data your experiment should measure.

FIGURE 3.13: Food labels

Food #1

Nutrition Facts	
1 servings per container	
Serving size	1 serving
Amount Per Serving	
Calories	
	% Daily Values*
Total Fat 1g	2%
Saturated Fat 0g	0%
Trans Fat 0g	
Sodium 7mg	0%
Total Carbohydrate 36g	12%
Dietary Fiber 11g	44%
Total Sugars 6g	
Includes 0g Added Sugars	0%
Protein 13g	26%

*The % Daily Value (DV) tells you how much a nutrient in a serving of food contributes to a daily diet. 2,000 calories a day is used for general nutrition advice.

Food #2

Nutrition Facts	
1 servings per container	
Serving size	1 serving
Amount Per Serving	
Calories	
	% Daily Values*
Total Fat 11g	17%
Saturated Fat 3g	15%
Trans Fat 0g	
Sodium 148mg	6%
Total Carbohydrate 14g	5%
Dietary Fiber 1g	4%
Total Sugars 1g	
Includes 0g Added Sugars	0%
Protein 2g	4%

*The % Daily Value (DV) tells you how much a nutrient in a serving of food contributes to a daily diet. 2,000 calories a day is used for general nutrition advice.

3.5

1. Two people who eat identical servings of food may each absorb a different amount of calories from that food. Identify five factors that affect how many calories an individual obtains from food that were discussed in the article titled "Everything you know about calories is wrong." Briefly describe how/why each factor affects calorie intake.

3.6

1. If a sample of polysaccharide was analyzed, then reanalyzed after being exposed to an enzyme that broke all of its glycosidic linkages, which of the following test results would change?
   a. The results of the Benedict's test only
   b. The results of the Benedict's test and Biuret test
   c. The results of the Benedict's test and Iodine test
   d. The results of the Biuret test and Iodine test
   e. The results of the Iodine test only

FIGURE 3.14: Food label for 2% milk

Nutrition Facts

1 servings per container	
Serving size	**8 fl. oz.**
Amount Per Serving	
Calories	**130**
	% Daily Values*
Total Fat 5g	**8%**
Saturated Fat 3g	**15%**
Trans Fat 0g	
Cholesterol 15mg	**5%**
Sodium 130mg	**5%**
Total Carbohydrate 12g	**4%**
Dietary Fiber 0g	**0%**
Total Sugars 12g	
Includes 0g Added Sugars	**0%**
Protein 9g	**18%**

*The % Daily Value (DV) tells you how much a nutrient in a serving of food contributes to a daily diet. 2,000 calories a day is used for general nutrition advice.

TABLE 3.7: Expected Results for 2% Milk

Test	Positive or Negative	Appearance
Benedict's test		
Biuret test		
Iodine test		
Sudan IV test		

2. Use the nutrition label for 2% milk shown in Figure 3.14 to complete Table 3.7, indicating what the expected results would look like for the each test if conducted on 2% milk.
3. A mystery food was analyzed in a lab and the following results were obtained:

   Benedict's: blue
   Biuret: violet
   Iodine: red
   Sudan IV: orange
   Using the information provided in Figure 3.15, determine which of the three foods described could have been the mystery food. Describe the reasoning for your response by interpreting the test results and discussing how they could be used to eliminate the other two foods.

FIGURE 3.15: Mystery foods

Food #4: Apple

Nutrition Facts	
1 servings per container	
Serving size	**242g**
Amount Per Serving	
Calories	**130**
	% Daily Values*
Total Fat 0g	**0%**
Saturated Fat 0g	**0%**
Trans Fat 0g	
Sodium 0mg	**0%**
Total Carbohydrate 34g	**11%**
Dietary Fiber 5g	**20%**
Total Sugars 25g	
Includes 0g Added Sugars	**0%**
Protein 1g	**2%**

*The % Daily Value (DV) tells you how much a nutrient in a serving of food contributes to a daily diet. 2,000 calories a day is used for general nutrition advice.

Food #5: Butter

Nutrition Facts	
1 servings per container	
Serving size	**28g**
Amount Per Serving	
Calories	**160**
	% Daily Values*
Total Fat 14g	**22%**
Saturated Fat 3g	**15%**
Trans Fat 0g	
Sodium 0mg	**0%**
Total Carbohydrate 8g	**3%**
Dietary Fiber 1g	**4%**
Total Sugars 0g	
Includes 0g Added Sugars	**0%**
Protein 23g	**46%**

*The % Daily Value (DV) tells you how much a nutrient in a serving of food contributes to a daily diet. 2,000 calories a day is used for general nutrition advice.

Food #6: Chicken breast

Nutrition Facts	
1 servings per container	
Serving size	**112g**
Amount Per Serving	
Calories	**190**
	% Daily Values*
Total Fat 10g	**15%**
Saturated Fat 3g	**15%**
Trans Fat 0g	
Cholesterol 70mg	**23%**
Sodium 70mg	**3%**
Total Carbohydrate 0g	**0%**
Dietary Fiber 0g	**0%**
Total Sugars 0g	
Includes 0g Added Sugars	**0%**
Protein 23g	**46%**

*The % Daily Value (DV) tells you how much a nutrient in a serving of food contributes to a daily diet. 2,000 calories a day is used for general nutrition advice.

4. Which of the four test results would be the same for all three foods described in the previous question?

3.7 REFERENCES

Dunn, R. (2013). Everything you know about calories is wrong. *Scientific American, 309*(3), 56–9.

Pratt, C. W. (2011). A biology laboratory exercise using macromolecule assays to distinguish four types of milk. *Journal of Microbiology and Biology Education*, *12*(1), 44–45

Schakel, S. F., Buzzard, I. M., and Gebhardt, S. E. (1997). Procedures for estimating nutrient values for food composition databases. *Journal of Food Composition and Analysis, 10,* 102–114

Wood, J. D., et al. (2004). Effects of fatty acids on meat quality: A review. *Meat Science*, *66*(1), 21–32.

Yee, W. and Bonney, K. M. (2015). Bonding with the tutor: How to stick together in chemistry. National Center For Case Study Teaching In Science Case Collection. Buffalo, NY: University of Buffalo. http://sciencecases.lib.buffalo.edu/cs/collection/detail.asp?case_id=762&id=762.

FIGURE CREDITS

FIG. 3.1: AhmadSherif, "atom structure," https://commons.wikimedia.org/wiki/File:Atom-struc.svg. Copyright in the Public Domain.

FIG. 3.2A: Copyright © Pumbaa (CC BY-SA 2.0) at https://commons.wikimedia.org/wiki/Electron_shell#/media/File:Electron_shell_006_Carbon.svg.

FIG. 3.2B: Copyright © Pumbaa (CC BY-SA 2.0) at https://commons.wikimedia.org/wiki/File:Electron_shell_020_Calcium.svg.

FIG. 3.3: Copyright © BruceBlaus (CC BY-SA 4.0) at https://commons.wikimedia.org/wiki/File%3ACovalent_Bonds.png.

FIG. 3.4: Copyright © BruceBlaus (CC BY-SA 4.0) at https://commons.wikimedia.org/wiki/File%3AIonic_Bonds.png.

FIG. 3.5: Copyright © OpenStax College (CC BY-SA 3.0) at https://commons.wikimedia.org/wiki/File%3A210_Hydrogen_Bonds_Between_Water_Molecules-01.jpg.

TABLE 3.1: Source: S. L. Seager and M. R. Slabaugh, Chemistry for Today: General, Organic, and Biochemistry, 2014.

FIG. 3.7: NEUROtiker, "glucose structure," https://commons.wikimedia.org/wiki/File:Beta-D-Glucopyranose.svg. Copyright in the Public Domain.

FIG. 3.8: Copyright © Don A. Carlson (CC BY-SA 3.0) at https://commons.wikimedia.org/wiki/File:Sucrose-inkscape.svg.

FIG. 3.9: Lord Skunk, "polysaccharide structure (cellulose)," https://commons.wikimedia.org/wiki/File:HaworthStrukturCellulose.jpg. Copyright in the Public Domain.

FIG. 3.10: Toxicologytoday~commonswiki, "polysaccharide structure (glycogen)," https://commons.wikimedia.org/wiki/File:Glicogeno_ramificato.gif. Copyright in the Public Domain.

FIG. 3.11: Copyright © Johndoct (CC BY-SA 4.0) at https://commons.wikimedia.org/wiki/File:Amino-acid-structure.jpg.

FIG. 3.12: Copyright © OpenStax CNX (CC BY-SA 4.0) at https://cnx.org/contents/s8Hh0oOc@9.21:QhGQhr4x@7/Biological-Molecules#fig-ch02_03_06.

FIG. 3.13: Generated with https://www.onlinelabels.com/label-generator-tools/Nutrition-Label-Generator.aspx

FIG. 3.14: Generated with https://www.onlinelabels.com/label-generator-tools/Nutrition-Label-Generator.aspx

FIG. 3.15: Generated with https://www.onlinelabels.com/label-generator-tools/Nutrition-Label-Generator.aspx

DNA Structure

4.1 INTRODUCTION

Nucleic acids are arguably the most important molecules for sustaining life, as they are responsible for passing on all of the instructions necessary for life from one organism to another. Despite their importance, nucleic acids were not first isolated until the mid-1800s. At that time, discovery of DNA revolutionized the burgeoning fields of genetics and evolutionary biology, and transformed the way scientists understand the natural world. Elucidation of the structure of DNA approximately 100 years later caused another transformational change in biology because of the far-reaching implications of this knowledge in furthering advancements in biotechnology and medicine. This chapter provides a historical background of the most seminal discoveries about DNA, paired with descriptions of the structure and function of nucleic acids.

DNA Structure

Animals, plants, fungi, and other microscopic organisms all rely on **deoxyribonucleic acid (DNA)** for storing the genetic information necessary to produce and sustain life, and for passing that information to their offspring. Therefore, knowing the structure of DNA is critical to understanding many basic aspects of life on Earth, such has how hereditary information is copied and passed from a parent organism to its offspring. Without a clear understanding of DNA structure, it would not be possible to develop many of the life-improving biotechnologies that are widely used today, including many of the tools used in genetic testing for diseases, the production of genetically modified organisms, and the forensic analysis often used in criminal investigations.

DNA is a polymer; a large molecule composed of many repeating subunits, or monomers, that are similar in structure. The monomers from which DNA is built are called **nucleotides**. Each DNA nucleotide is made up of three parts: a sugar called deoxyribose, a nitrogen-containing nucleotide base, and a phosphate group (Figure 4.1). The four types of nucleotide bases found in DNA are adenine (A), cytosine (C), guanine (G), and thymine (T). It is the combination of A, C, G, and T that encodes the genetic information in DNA. Each strand of DNA is held together by a phosphate-sugar

LEARNING OBJECTIVES

- Describe the structure of DNA
- Compare and contrast the structures of DNA and RNA
- Summarize the main contributions of the following scientists to the discovery of the structure of DNA: Erwin Chargaff, Francis Crick, Rosalind Franklin, Linus Pauling, James Watson, Maurice Wilkins
- Justify, argue, and evaluate the ethical implications of the controversy surrounding the role of Rosalind Franklin, James Watson, and Francis Crick in the discovery of DNA structure
- Demonstrate how to extract DNA from cells

"backbone" in which the phosphate group of one nucleotide is connected to the deoxyribose sugar of an adjoining nucleotide by a type of strong chemical bond known as a **phosphodiester bond**. DNA molecules are usually composed of two strands of DNA wound together in a spiral shape. In this formation, known as a double helix, the sugar-phosphate backbone of each DNA strand is located on the outside, and the nucleotide bases of each strand are located on the inside. The chemical properties of each component of DNA provided a clue that helped Watson and Crick solve the structure of DNA. They realized, unlike Pauling, that the phosphate groups have a strong negative charge that would repel any similarly charged portions of other DNA strands, similar to how the negative poles of two magnets repel each other when placed in proximity. Because of this, the phosphate groups must be on the outside of the DNA molecule, pointed away from each other, leaving the nucleotide bases to line the interior of the molecule.

The human genome is approximately three billion DNA nucleotides long, and it is the sequence of these nucleotides that encodes the hereditary information necessary for life. If stretched out, the DNA in each human cell would be a thin strand approximately 2 meters (6 feet) long and less than 2½ nanometers (9.8×10^{-8} inches) wide. It is important to note that up to 98% of our DNA does not directly affect the physical or chemical characteristics of our bodies because it is "noncoding," which means it is not part of the genes that encode instructions for synthesizing biological molecules. The entire set of genetic material, including genes and noncoding sequences, of an organism is known as the organism's **genome**. Within a double helix, the two strands of DNA are connected by a series of weak chemical bonds known as hydrogen bonds. As shown in Figure 4.2, adenine and guanine are both two-ringed structures that are similar in shape and size. Thymine and cytosine are smaller, single-ringed structures, and each is roughly the same size and shape as the other. To maintain the same distance between the two strands of DNA throughout the double helix, adenine always forms a base pair with thymine, and cytosine base pairs with guanine. Because of this, the two strands of a DNA double helix are said to be **complementary**. If the nucleotide

FIGURE 4.1: Nucleotide structure—Nucleic acids are made of repeating subunits called nucleotides. Each nucleotide is composed of a phosphate group, a five-carbon sugar, and a nucleotide base.

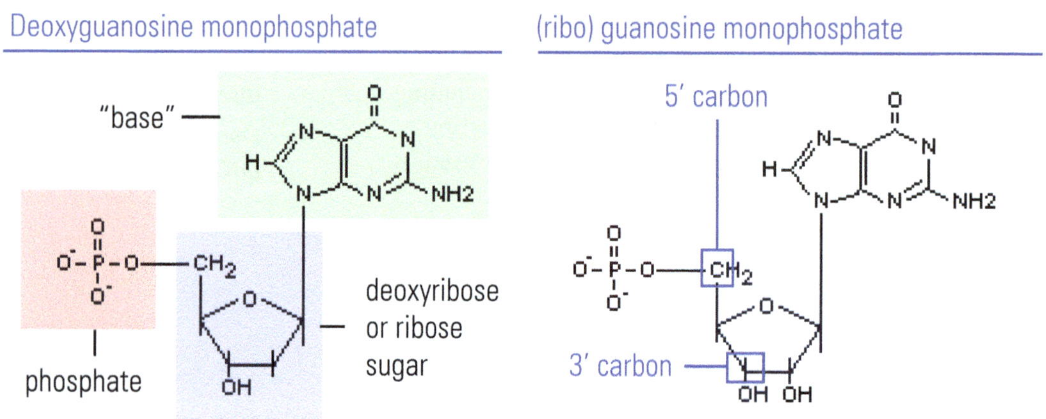

sequence of one strand is known, the sequence of the other strand can be inferred based on the rules of base pairing.

In addition to always being the same distance apart, the two strands of a DNA double helix run in opposite directions and are thus described as being **antiparallel.** The direction that a DNA strand runs in can be observed by looking at which direction the oxygen atom in the deoxyribose sugar faces. The carbon atoms in the deoxyribose sugar are numbered 1', 2', 3', 4', and 5' in a clockwise direction starting at the oxygen atom. The strand that has the 3' carbon on top and the 5' carbon on the bottom is referred to as the 3' to 5' strand, and the strand that has the 5' carbon on top and the 3' carbon on the bottom is said to be 5' to 3' (Figure 4.2). Because the enzymes of cells that copy and "read" DNA can only do so in the 5' to 3' direction, this distinction is important during DNA replication. This directionality is also important for the functioning of enzymes during the process of transcription, which produces a messenger RNA copy of a DNA sequence for use in the production of protein.

RNA Structure

In addition to DNA, organisms rely on another type of nucleic acid known as **ribonucleic acid (RNA)** to transmit genetic information (Table 4.1). Similar to DNA, RNA is a polymer of nucleotides connected by phosphodiester bonds, which is read in the 5' to 3' direction. Both DNA and RNA contain the nucleotides adenine, cytosine, and guanine, but instead of thymine, RNA contains a slightly different molecule called uracil. The sugar base of RNA, called ribose, differs from the deoxyribose sugar found in DNA by the absence of one oxygen atom connected at the 2' carbon position. Finally, unlike DNA, RNA is usually present as a single-stranded molecule. This final difference contributes to important functional variations between DNA and RNA, such as the fact that RNA is less stable than DNA and is usually not used for long-term storage of genetic material by eukaryotic cells. Interestingly, it is thought that RNA was the first type of genetic material to evolve on Earth and was the only type of genetic material present in the first living organisms that existed billions of years ago. The development of DNA occurred later through a series of modifications made to RNA. For example, the deoxyribose sugar present in DNA is identical to the ribose sugar present in RNA except that it is missing an oxygen atom; hence the name "de-oxy ribose."

TABLE 4.1: Comparison of DNA and RNA

Component	DNA	RNA
Bases	adenine, cytosine, guanine, thymine	adenine, cytosine, guanine, uracil
Sugar	deoxyribose	ribose
Structure	double-stranded helix	single-stranded
Function	stores and transmits hereditary genetic information	involved in translating genetic information into proteins

FIGURE 4.2: **DNA structure**—DNA is a double-helix, composed of two antiparallel strands of complementary base pairs.

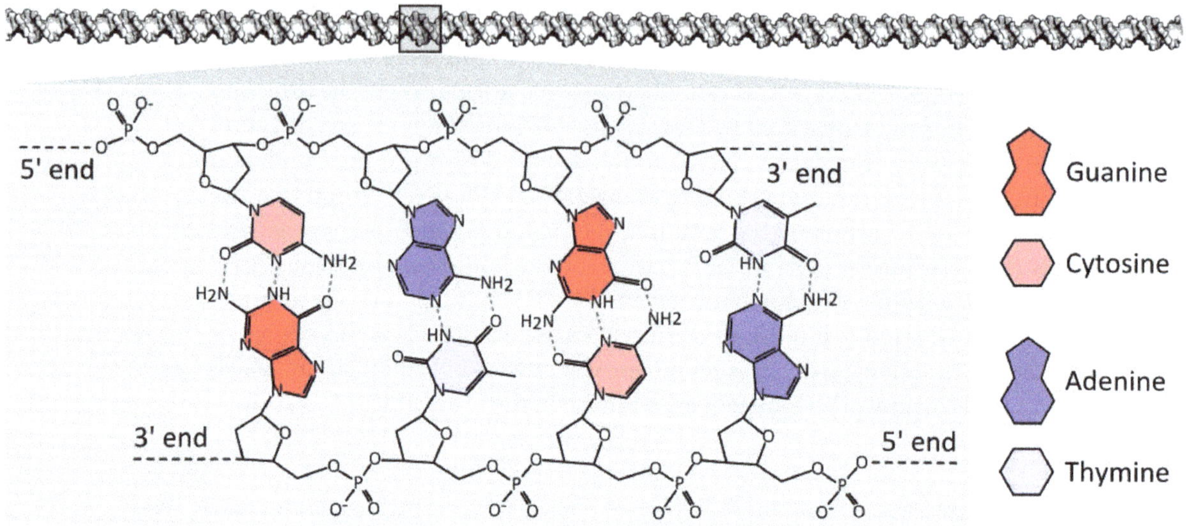

History of DNA Discoveries

Despite its importance, the double-helix structure of DNA was not accurately identified until the mid-1950s, many decades after Charles Darwin put forth his theory of evolution and Gregor Mendel described in detail the basic principles of genetic inheritance. The experiments that identified the DNA double helix were preceded by a series of experiments by other scientists that demonstrated DNA was the molecule that stores hereditary information and provided early clues to what its structure might be. Several of these are identified in Table 4.2.

By 1952, it was known that DNA, found in the nucleus of cells, was the hereditary material organisms used to transfer traits to one another. It was also known that DNA contained nucleotides linked by phosphate bonds, and that the amount of the nucleotide base cytosine is always equal to the amount of guanine, whereas the amount of adenine is similarly equal to the amount of thymine. The relative frequencies of the four types of nucleotide bases, known as Chargaff's rule, suggested that adenine always pairs with thymine and cytosine pairs with guanine. This provided a crucial hint about the way DNA is structured that other scientists later used to construct accurate models of DNA structure. Despite this basic understanding, many people at the time incorrectly believed that DNA was made of three intertwined strands with the nucleotides on the exterior and the phosphate groups pointed inward. Then, in the April 25, 1953, issue of the prominent scientific publication *Nature*, James Watson and Francis Crick (Figures 4.3A and 4.3B) published what would become one of the most famous scientific papers of all time. In it, the authors rightly proclaimed that DNA was actually a double helix with the information-encoding nucleotides paired up along the middle, much like the teeth of a zipper.

At the same time that Watson and Crick were attempting to build a model of the structure of DNA in the Cavendish laboratory at Cambridge University, two other scientists, Rosalind Franklin and Maurice Wilkins, were also analyzing the structure of DNA at a rival lab at nearby King's College in London (Figures 4.3C and 4.5A). Much of the work conducted by Watson, Crick, Franklin, and Wilkins relied on data generated by a technique called x-ray diffraction.

TABLE 4.2: Important Discoveries Leading to the Discovery of DNA Structure

Year	Scientists	Contribution to DNA Research
1869	Friedrich Miescher	First to isolate and characterize DNA; termed the substance "nuclein."
1928	Fred Griffith	Showed experimentally that bacteria are capable of transmitting traits from one another through a process termed transformation.
1929	Phoebus Levene	Discovered that DNA nucleotides are comprised on deoxyribose sugar, nucleotide bases, and phosphate groups; identified adenine, cytosine, guanine, and thymine, and proposed "tetranucleotide hypothesis," that described DNA structure incorrectly and suggested it was too simple to be genetic material.
1944	Oswald Avery, Colin MacLeod, and Maclyn McCarty	Provided experimental evidence that DNA is the "transforming principle" responsible for the passage of traits from cell to cell.
1950	Erwin Chargaff	Proposed that in DNA there is always the same amount of adenine as thymine, and the same amount of cytosine as guanine (Chargaff's Rule).
1952	Alfred Hershey and Martha Chase	Demonstrated that DNA, not protein, is the hereditary material through a convincing experiment using DNA that infect bacteria.
1953	Linus Pauling and Robert Corey	Published an incorrect triple helix model of DNA with the phosphates on the inside, but did make several observations about the structure and bonds of nucleotides that were helpful in solving the structure of DNA.

During **x-ray diffraction**, high-energy electromagnetic beams called x-rays are focused with a screen and directed at the substance to be analyzed (Figure 4.4). Most often, the substance analyzed is a stable, crystalized version of a molecule; when this is the case, the process is referred to as **x-ray crystallography**. When the x-rays hit the molecule, they are diffracted, or scattered, creating a pattern that can be detected on a photographic plate (Figure 4.4). This pattern can be analyzed using mathematical equations based on known diffraction patterns to determine the structure of the molecule that diffracted the x-rays. Linus Pauling used this method to discover the basic three-dimensional structure of proteins years earlier but was unable to apply his knowledge of x-ray diffraction to correctly determine the structure of DNA.

Rosalind Franklin was an expert in the technique of x-ray diffraction, and it is Franklin's Photo 51, which shows the diffraction pattern of the DNA double helix, that has become the most famous image associated with the discovery of DNA (Figure 4.5B). After successfully completing the labor-intensive process of producing x-ray diffraction images, Franklin

FIGURE 4.3A: **James Watson**—James Watson, Francis Crick, and Maurice Wilkins won The Nobel Prize in Physiology or Medicine in 1962 "for their discoveries concerning the molecular structure of nucleic acids and its significance for information transfer in living material."

FIGURE 4.3B: **Francis Crick**—James Watson, Francis Crick, and Maurice Wilkins won The Nobel Prize in Physiology or Medicine in 1962 "for their discoveries concerning the molecular structure of nucleic acids and its significance for information transfer in living material."

FIGURE 4.3C: **Maurice Wilkins** —James Watson, Francis Crick, and Maurice Wilkins won The Nobel Prize in Physiology or Medicine in 1962 "for their discoveries concerning the molecular structure of nucleic acids and its significance for information transfer in living material."

discovered that the DNA double helix could take on two different forms depending on the amount of water contained in the molecule at the time of analysis. DNA with low water content, which was often used for analysis and in attempts to produce x-ray diffraction images, became known as "A-DNA," whereas DNA with higher water content, such as what is usually found inside cells, became known as "B-DNA." Differences in water content affect the three-dimensional shape of DNA, leading A-DNA and B-DNA to produce different images when analyzed using x-ray diffraction (Figure 4.5B). Because x-ray diffraction images cannot be interpreted literally, like photographs, but instead must be analyzed using complicated mathematics, this made characterization of the structure of DNA quite challenging. However, Watson and Crick were able to accurately interpret the patterns depicted in Photo 51 and publish a paper describing their findings. One important observation made by Watson and Crick was that the "X" pattern observed in the x-ray diffraction image of "B-DNA" suggested that DNA is helical. Further analysis of the pattern of smeared lines in the image and the diamond-shaped white spaces flanking the "X" allowed Watson and Crick to determine that DNA was made of repeating subunits of a certain size, providing clues about how the helix was twisted.

FIGURE 4.4: X-ray diffraction schematic

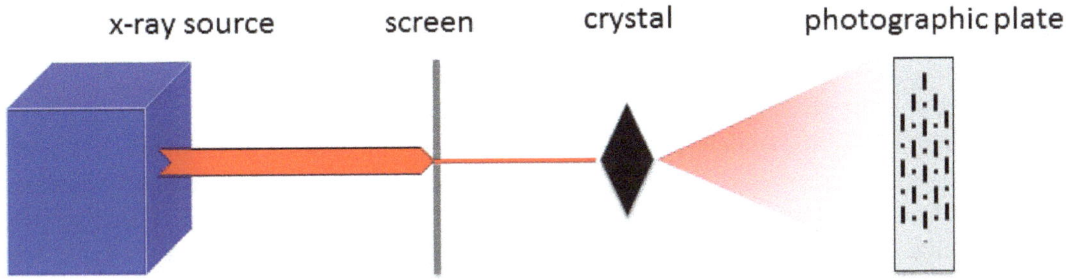

FIGURE 4.5A: **Rosalind Franklin**—Rosalind Franklin used x-ray crystallography to study DNA. Watson and Crick interpreted the image of B-DNA to determine the double-helix model of DNA structure.

FIGURE 4.5B: X-ray image of DNA

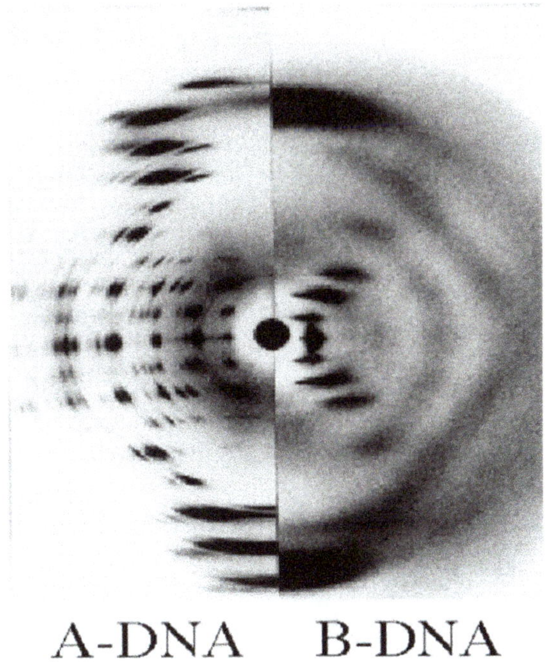

Although Franklin and Wilkins also each published their own paper in the same April 25, 1953, issue of the journal *Nature* as Watson and Crick, the key findings of Rosalind Franklin were overshadowed by the publicity devoted to Watson and Crick's lead article. The interactions between these scientists, and the ways they chose to acknowledge each other's work, were a source of considerable controversy. After Rosalind Franklin died from ovarian cancer in 1958 at the age of 37, some speculated that exposure to carcinogenic x-rays during her laboratory work contributed to her death. It also been suggested that Franklin may have inherited a mutation in her DNA, known as *BRCA*, which is known to significantly increase the risk of developing breast and ovarian cancer. Years later, when only Watson, Crick, and Wilkins were awarded the 1962 Nobel Prize for the discovery of DNA structure, debate over the treatment of Rosalind Franklin by the scientific community broadened and continues to smolder to this day.

DNA and the History of Science

Although scientific work is often discussed as a collection of discrete, finalized facts, scientific knowledge most often results from a series of trials, errors, revisions, and refinements by a number of collaborating and competing scientists. The history of DNA-related discoveries highlights this aspect of the scientific process. The discoveries identified in Table 4.2 also demonstrate the global nature of science. For example, the work of Friedrich Miescher was conducted in Germany. Erwin Chargaff also worked in Germany, as well as Paris, before fleeing the Nazis in 1935 to continue his career at Columbia University in New York City. Phoebus Levene, Oswald Avery, Colin McCleod, and Maclyn McCarty all completed their famous work on DNA in New York City as part of what is now known as The Rockefeller University. Alfred Hershey and Martha Chase also have ties to New York City, though they conducted their most famous experiment in Washington, D.C., while Linus Pauling and Robert Corey were working at the California Institute of Technology. Perhaps, most famously, James Watson, Francis Crick, Rosalind Franklin, and Maurice Wilkins completed their work on the structure of DNA in London, England. Interestingly, Crick and Wilkins both spent part of their later careers at universities in California, whereas Watson worked for many years on Long Island, near New York City.

Finally, California was also the site from which Matthew Meselson and Frank Stahl later experimentally demonstrated the method by which a DNA double helix is replicated in living cells. The model identified by Meselson and Stahl, known as the semiconservative model of DNA replication, is what Watson and Crick hinted to in one of the final sentences of their 1953 paper when they wrote, "It has not escaped our notice that the specific pairing we have postulated immediately suggests a possible copying mechanism for the genetic material." Meselson and Stahl's famous paper proclaiming this finding was published on May 14, 1958, just short of one month after Rosalind Franklin's death. The fact that Meselon and Stahl never received a Nobel Prize for their work on DNA has also left a legacy of controversy.

4.2 SUMMARY

- Before the structure of DNA was known, there was evidence that the DNA found in the nuclei of cells was the hereditary material used to transfer traits from one organism to another
- DNA is composed of repeating subunits called nucleotides; each nucleotide contains deoxyribose sugar, a phosphate group, and a nucleotide base
- Erwin Chargaff provided data that helped James and Watson and Francis Crick realize that the two strands of DNA are complementary, meaning that adenine always pairs with thymine and cytosine always pairs with guanine

- Linus Pauling was a leading DNA researcher in the early 1950s, but the model he proposed incorrectly described DNA as a triple helix with the sugar-phosphate backbones of each DNA strand on the inside of the helix
- Rosalind Franklin, who worked with Maurice Wilkins, used x-ray crystallography to produce an image called Photo 51, which could be interpreted to reveal the structure of DNA
- James Watson and Francis Crick were the first to build a correct model of the DNA double helix, showing two complementary strands with the sugar-phosphate backbones on the outside and the nucleotides on the inside
- A long-lasting controversy has surrounded the way Watson and Crick obtained Franklin's Photo 51, which was important for solving their model of DNA, and the way they interacted with her and acknowledged her for contribution to their discovery

4.3 KNOWLEDGE CHECK

1. Understanding the structure of DNA was important for all of the following, EXCEPT:
   a. development of Darwin's theory of evolution
   b. genetic testing for human diseases
   c. production of genetically modified organisms
   d. production of tools used in forensic investigations of crime scenes
   e. understanding how DNA is copied (replicated)

2. Who incorrectly proposed that DNA was a triple helix with the phosphate-sugar backbones on the inside?
   a. Erwin Chargaff
   b. Francis Crick
   c. Rosalind Franklin
   d. Linus Pauling
   e. James Watson

3. The work of Erwin Chargaff was helpful to Watson and Crick in building their model of DNA structure because:
   a. it suggested that DNA contains phosphate bonds.
   b. it suggested how many nucleotides there are in each turn of the double helix.
   c. it suggested that DNA is a triple-helix rather than a double helix.
   d. it suggested that the nucleotides are found on the outside of the DNA molecule rather than the inside.
   e. it suggested which nucleotides pair together.

4. Which of the following pairs could be complementary sequences in a normal DNA double helix?
   a. 3'-ATGCAA-5' paired with 3'-TACGTT-5'
   b. 3'-ATGCGA-5' paired with 5'-ATGCGT-3'
   c. 5'-TACCAG-3' paired with 5'-ATGGTC-3'
   d. 5'-TACGGA-3' paired with 3'-AUGCCU-5'
   e. 5'-TACTGA-3' paired with 3'-ATGACT-5'

5. Which of the following describes one difference between DNA and RNA?
   a. RNA contains adenine instead of thymine.
   b. RNA contains deoxyribose instead of ribose.
   c. RNA contains hydrogen bonds instead of phosphodiester bonds.
   d. RNA contains uracil instead of deoxyribose.
   e. RNA contains uracil instead of thymine.

6. Sketch a nucleotide and label the three main parts.

7. Describe the process of x-ray crystallography.

8. Describe the contribution of Phoebus Levene to the discovery of DNA structure.

4.4 ACTIVITY

Primary Literature Analysis

Scientific knowledge is often discussed as a series of independent facts without acknowledging the process that led to the discovery of that knowledge. However, a rich history of collaboration and competition among many scientists underlies many of the most well-known scientific facts. The structure of DNA provides an excellent view into the hidden history of science.

To better understand the discovery of DNA structure and its history, it is helpful to read first-hand one of the most famous scientific papers ever written in which Watson and Crick announced to the world their discovery of the DNA double helix. Before reading the paper, it might be helpful to preview the questions listed for section 4.4 at the end of this chapter. As you read the paper, try to focus on the main points, limitations, and implications of the work, trying not to get stymied by specific experimental details that might be confusing and are beyond the focus of this activity. The purpose of this activity is to evaluate a scientific article for accuracy, reliability, and limitations.

Watson, J. D. and Crick, F. H. (1953). Molecular structure of nucleic acids; a structure for deoxyribose nucleic acid. *Nature, 171*(4356), 737–8. http://www.nature.com/nature/journal/v171/n4356/pdf/171737a0.pdf.

4.5 ACTIVITY

Debate (Adapted from Bonney, K. M. 2015)

After you have analyzed Watson and Crick's 1953 *Nature* paper, prepare to debate the ethical issues raised by their interactions with Rosalind Franklin. Watching one or both of the following videos, which describe the scientific and ethical implications of the work of Watson, Crick, Franklin, and Wilkins, can be useful for this activity:

- *The Double Helix.* (2014). Howard Hughes Medical Institute BioInteractive
- *DNA – Secret of Photo 51.* (2012). *NOVA*, Public Broadcasting Service

After watching these video(s) you will be separated into one of two groups or instructed to prepare a response for one or both sides of the argument individually. You will then write a short essay describing your argument in response to the following prompt(s). The objective of this activity is to appraise, criticize, and justify claims about the discovery of the structure of DNA and the ethical implications of the events surrounding this discovery.

If you are in the James Watson and Francis Crick group, synthesize a one-half page argument from their perspective justifying the way they treated Rosalind Franklin during and after the discovery of DNA structure. Provide reasoning for why they deserved to win the Nobel Prize without sharing the award with Franklin, and why the acknowledgment they gave Franklin in their 1953 *Nature* paper and after their acceptance of the Nobel Prize was sufficient. Be prepared to debate your position with the Franklin group.

If you are in the Rosalind Franklin group, synthesize a one-half page argument from your perspective illustrating why you were unfairly treated by Watson and Crick during and after the discovery of DNA structure. Provide reasoning for why you deserved to share the Nobel Prize, and why the acknowledgment Watson and Crick gave you in their 1953 *Nature* paper and after their acceptance of the Nobel Prize was insufficient. Be prepared to debate your position with the Watson and Crick group.

4.6 ACTIVITY

DNA Extraction

Most cells from living organisms contain a complete set of that organism's DNA. In the cells of eukaryotic organisms, such as animals, plants, and fungi, most of the DNA is located inside a structure called the nucleus (Figure 4.6). Within the nucleus, DNA and some associated proteins called histones are organized into structures called **chromosomes. Histones** are much like spools upon which long strands of thread-like DNA are wound. In human cells, each strand of DNA is approximately 6 feet (2 meters) long. The cell and nucleus are both surrounded by lipid membranes that must be ruptured to release the DNA from the cell. Unlike the animal cell shown in Figure 4.6, the cells of plants and fungi are also protected by a tough cell wall adjacent to the exterior side of the cell membrane. Whereas DNA is a polar molecular, due mainly to the many negatively charged phosphate groups contained in sugar-phosphate backbone, the lipid membranes of the cell and nucleus are nonpolar. This difference in polarity between DNA and other cellular components can be exploited to extract DNA from cells. Although a single

FIGURE 4.6: **Basic cell structure**—DNA is packaged into structures called chromosomes within the nucleus of plant and animal cells.

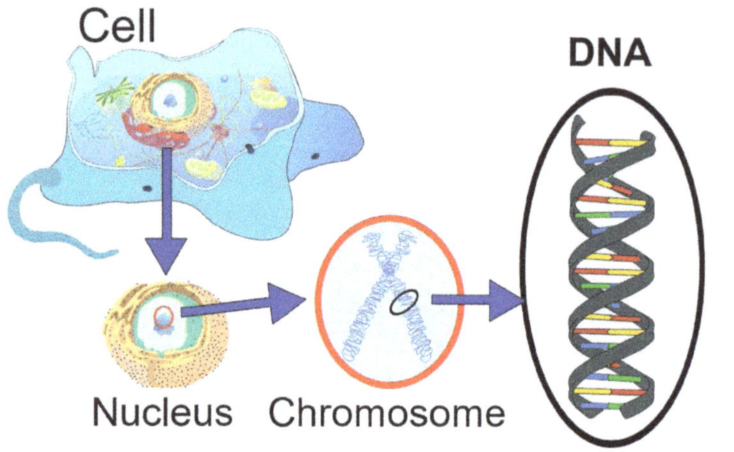

strand of DNA could measure 6 feet long if stretched out, the DNA double helix is so narrow that the DNA of a single cell can only be visualized using powerful microscopes. However, if DNA from a large number of cells is extracted and combined, it can be seen with the naked eye.

In this activity, wheat germ (the embryo, or DNA-containing portion of the seed of the wheat plant) will be used a source of DNA because it is safe and easy to work with. Although the materials used in this experiment are household chemicals, they are similar to those used in professional laboratories. The procedure described here is similar to the procedure scientists would use to obtain DNA for genetic testing or a forensic science investigation, except that the relatively large amount of DNA available from the wheat germ allows us to significantly simplify our procedure.

Materials

- Wheat germ (1 teaspoon)
- Detergent (1 teaspoon)
- Water (25 mL)
- Isopropyl alcohol (10 mL)
- Test tube (1)
- Wooden stir stick (1)
- Teaspoon (1)

Procedure

1. Before beginning, read over Steps 2–8 and write a hypothesis for when during the procedure DNA will first become visible and why it will become visible at that time.
2. Add 1 teaspoon of wheat germ to the test tube.
3. Add 25 mL of warm water to the test tube (fill to line 1).
4. Using the wooden stir stick, gently stir the wheat germ for 3 minutes.
5. Add 1 teaspoon of detergent.
6. Using the wooden stir stick, gently stir the wheat germ for 5 minutes. Be careful not to create bubbles or foam.
7. Tilt the test tube at a slight angle. Slowly add 10 mL of isopropyl alcohol by pouring it down the side of the test tube. The alcohol should form a layer on top of the water. Be careful not to mix the alcohol with the water.

8. Place the test tube on the desktop or in a holding rack and allow settling for 10-15 minutes without disturbing.
9. Record the appearance of the contents of the tube and of the DNA when it appears.

4.7 ACTIVITY

DNA Extraction Virtual Lab

The purpose of this activity is to provide a virtual lab alternative to the hands-on DNA extraction described in Activity 4.6. However, it may be useful to complete both activities to compare DNA extraction from plant cells (in Activity 4.6) with DNA extraction from human cells shown in the virtual lab.

To start this activity, open the virtual lab by copy/pasting the following link into your web browser. Note: You may need to authorize usage of Flash Player on your computer to complete the virtual lab. Access here: http://learn.genetics.utah.edu/content/labs/extraction/

Once open, follow the instructions on the screen to complete the virtual lab. Answer the accompanying activity questions as you complete the virtual lab.

ACTIVITY QUESTIONS

4.4

1. What are the main claims about DNA structure made in the paper?
2. What evidence is used to support the claims made in the paper?
3. What evidence is missing or unclear in the paper? In other words, what else might you want to know before accepting the claims that are made?
4. What work from other scientists was probably helpful for Watson and Crick?
5. How did Watson and Crick acknowledge the work of others that might have been helpful to them? Was this level of acknowledgment appropriate, or was it too much or too little?
6. In this paper, Watson and Crick state: "It has not escaped our notice that the specific pairing we have postulated immediately suggests a possible copy in mechanism for the genetic material." Describe what Watson and Crick mean by "the specific pairing." What "possible copying mechanism" is suggested by this finding?

4.5

1. Describe the specific contributions of Francis Crick and James Watson, and of Rosalind Franklin, to the discovery of DNA structure.
2. Summarize the arguments presented in the debate between the Watson and Crick group and the Rosalind Franklin group. Indicate which side of the argument is best supported, and briefly discuss your reasoning.

4.6

1. State whether your initial hypothesis and reasoning were correct. If not, propose an alternative explanation for why DNA appears at a certain stage of the procedure.
2. Identify the purpose of adding (i) water, (ii) detergent, and (iii) alcohol during this activity.
3. Describe the visual characteristics of the DNA that were made visible in this activity.
4. In some DNA extraction procedures, meat tenderizer or contact lens solution is added to the DNA extraction protocol. Both meat tenderizer and contact lens solution contain enzymes that break down proteins. Why would this be helpful during DNA extraction?

4.7

1. List three reasons why scientists isolate DNA, as identified in the virtual lab.
2. Identify the type of cell used for DNA extraction in the virtual lab.
3. Identify two components of the "lysis buffer" and describe the function of each.
4. Discuss what would happen if the concentrated salt solution was not added to the tube of DNA.
5. After the tube was centrifuged, where was the DNA?
6. At what step did DNA become visible to the naked eye? Why?
7. Describe the last step of the DNA extraction procedure that was shown in the virtual lab.

4.8 REFERENCES

Avery, O. T., MacLeod, C. M., & McCarty, M. (1944). Studies on the chemical nature of the substance inducing transformation of pneumococcal types: induction of transformation by a deoxyribonucleic acid fraction isolated from pneumococcus type III. *The Journal of Experimental Medicine, 79*(2), 137–158.

Bonney, K. M. (2015). *Putting the pieces together: The discovery of DNA structure and replication. National Center For Case Study Teaching In Science Case Collection.* Buffalo, NY: University of Buffalo. http://sciencecases.lib.buffalo.edu/cs/files/dna_structure_flip.pdf.

Dahm, R. (2005). Friedrich Miescher and the discovery of DNA. *Developmental Biology, 278,* 2:274–288.

Levene, P. A. (1919). The structure of yeast nucleic acid: IV. Ammonia hydrolysis. *Journal of Biological Chemistry, 40,* 415–424.

Maddox, B. (2002). *Rosalind Franklin: The dark lady of DNA.* New York: HarperCollins.

Pauling, L. and Corey, R. B. (1953). A proposed structure for the nucleic acids. *Proceedings of the National Academy of Sciences of the United States, 39*(2), 84–97.

Travers, A. and Muskhelishvili, G. (2015). DNA structure and function. *FEBS Journal, 282,* 2279–2295.

Watson, J. D. and Crick, F. H. (1953). Molecular structure of nucleic acids; a structure for deoxyribose nucleic acid. *Nature, 171*(4356), 737–8.

FIGURE CREDITS

FIG. 4.1: Adapted from "Nucleotide Structure," https://commons.wikimedia.org/wiki/File:Nucleotides.gif. Copyright in the Public Domain.

FIG. 4.2: Copyright © Thomas Shafee (CC BY-SA 4.0) at https://commons.wikimedia.org/wiki/File:DNA_chemical_structure_2.svg.

FIG. 4.3A: National Cancer Institute (NCI), "James Watson," https://commons.wikimedia.org/wiki/File:James_Dewey_Watson.jpg. Copyright in the Public Domain.

FIG. 4.3B: Copyright © Materialscientist (CC BY 4.0) at https://commons.wikimedia.org/wiki/File:Francis_Crick.jpg.

FIG. 4.3C: Wikimedia commons, "Maurice Wilkins," https://commons.wikimedia.org/wiki/File:Maurice_Wilkins_nobel.jpg. Copyright in the Public Domain.

FIG. 4.5A: JWSchmidt, "Rosalind Franklin," http://collaboration.wikia.com/wiki/File:Rosalind_Franklin.jpg. Copyright in the Public Domain.

FIG. 4.5B: Copyright © I.C. Baianu et al. (CC BY-SA 3.0) at https://commons.wikimedia.org/wiki/File:ABDNAxrgpj.jpg.

FIG. 4.6: Copyright © Tiger66 (CC BY-SA 3.0) at https://commons.wikimedia.org/wiki/File:Eukaryote_DNA-en.svg.

DNA Replication

5.1 INTRODUCTION

Living cells divide to make more cells for a variety of reasons. In multicellular organisms, cells divide through the process of mitosis to grow and to repair old and damaged cells. Eukaryotic organisms that undergo sexual reproduction also engage in another process of cell division called meiosis to produce reproductive cells called gametes. Bacteria and other single-celled organisms must also divide as part of their own reproductive process. Before the process of cell division can begin, the cell must produce a copy of its DNA to give to the newly produced cell. This chapter explains how DNA is copied, with a focus on the process that occurs in eukaryotes, such as humans.

DNA Replication

In eukaryotes, the process of **DNA replication** occurs during the part of the cell cycle known as the synthesis phase, or S phase. The way that DNA is replicated is intricately related to its structure. DNA consists of a double helix made of two complementary, antiparallel strands of adenine, cytosine, guanine, and thymine nucleotides. Strong phosphodiester bonds that connect each subunit along the outside of the double helix hold each strand together. The two strands are connected to each other by weaker hydrogen bonds that form between the nucleotide bases located on the inside of the molecule. This much is known about the structure of DNA from the paper titled "Molecular structure of nucleic acids; a structure for deoxyribose nucleic acid" published by James Watson and Francis Crick in 1953. Once the structure of DNA was known, Watson and Crick, and others with sufficient knowledge of the topic, began to speculate about how living organisms copy, or replicate, their DNA. In the conclusion of their aforementioned paper, Watson and Crick state: "It has not escaped our notice that the specific pairing we have postulated immediately suggests a possible copying mechanism for the genetic material." In a follow-up paper published the next year, the two further explained that:

LEARNING OBJECTIVES

- List and describe the three hypothetical models of DNA replication investigated by Meselson and Stahl
- Describe the DNA replication experiment conducted by Meselson and Stahl and discuss the data they obtained
- Identify the model of DNA replication supported by the data from the Meselson-Stahl experiment
- Describe the steps of the process of DNA replication
- Identify the function of each of the following: DNA ligase, DNA polymerase, helicase, single-strand binding proteins, and topoisomerase

> *We imagine that prior to duplication the hydrogen bonds are broken, and the two chains unwind and separate. Each chain then acts as a template for the formation onto itself of a new companion chain, so that eventually we shall have two pairs of chains, where we only had one before. Moreover, the sequence of the pairs of bases will have been duplicated exactly.*
>
> *(Crick, F. H. C., and Watson, J. D. (1954). The complementary structure of deoxyribonucleic acid.* Proceedings of the Royal Society of London A, 223, *80.)*

In other words, Watson and Crick suggested that the DNA double helix splits apart, and each strand is used as a template for making a second strand, which results in the formation of two identical DNA double helices from the one original molecule. This is, in fact, the correct explanation for how DNA is replicated. Even though their hypothesis was correct, the papers published by Watson in Crick did not present convincing experimental data to support their claims. In the case of DNA replication, the key experimental data were not collected until several years later by two other scientists named Matthew Meselson and Frank Stahl, and published in 1958. The experiment conducted by Meselson and Stahl to demonstrate how DNA was replicated was technologically innovative and brilliantly simple at the same time.

Meselson and Stahl Experiment

When Meselson and Stahl planned their experiment to determine the mechanism by which DNA is replicated, three hypothetical models had been proposed to explain how living organisms copy a single DNA molecule to make two. The three models that Meselson and Stahl investigated were the conservative model, the dispersive-model, and the semiconservative model (Figure 5.1). The conservative model of DNA replication proposes that both original (parental) strands of a DNA double helix are used as templates for making new strands of DNA, and that somehow both original template DNA strands remain in one double helix after replication, whereas a second double helix is formed from the two newly synthesized strands of DNA. This method of replication is so-named because it conserves, keeping the same, original DNA molecule. The dispersive model states that the original molecule of DNA undergoes double-strand breaks to form double-stranded pieces that are

FIGURE 5.1: Hypothetical models of DNA replication—The three hypothetical models proposed to explain how DNA is replicated are the conservative, dispersive, and semiconservative models.

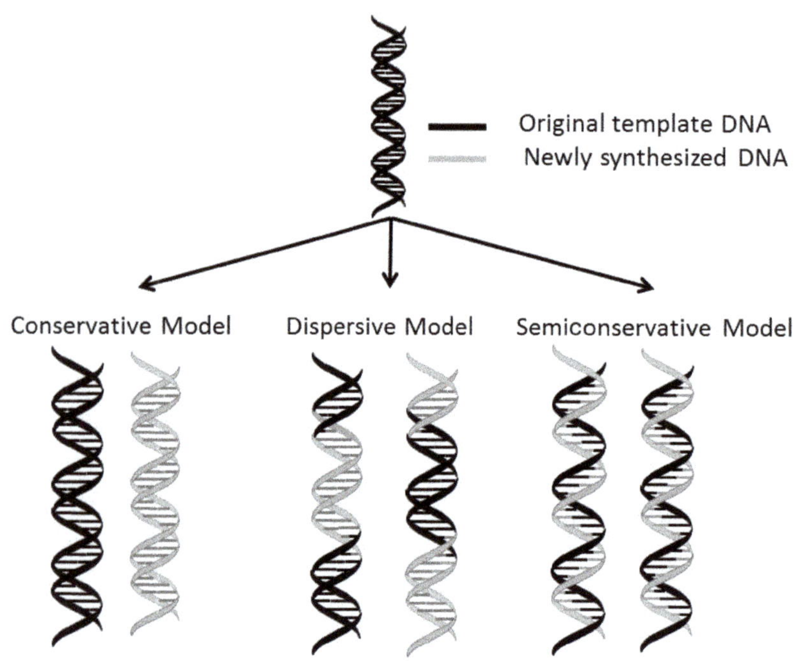

then distributed, or dispersed, between the two DNA double helices present after one round of DNA replication. The segments of original DNA somehow act as templates to direct the synthesis of the missing portions of each newly forming DNA molecule. Conversely, according to the semiconservative model, the complementary strands of the original DNA molecule are separated from each other, then each of the original strands serves as a template for synthesizing a copy of the complementary strand needed to complete each new double helix. The semiconservative model indicates that each of the two complete DNA molecules formed will contain one complete strand of original DNA and one complete strand of new DNA. Because each original strand of DNA remains fully intact, this method of replication is said to be semiconservative.

Although they likely could have predicted which model of DNA replication was correct based on information gathered from Watson and Crick, Meselson and Stahl wanted to demonstrate which model was correct by conducting an experiment and collecting data. Meselson and Stahl operated under the assumption that all living organisms replicate their DNA in a similar manner. This meant that if they could demonstrate how DNA is replicated in one species, they would know how DNA is replicated in every species. To study DNA replication, they chose to work with bacteria. One reason for choosing bacteria is that they divide rapidly, and thus must replicate their DNA frequently. Bacteria are also small, simple, well characterized, and can be grown and studied easily and inexpensively, making them an ideal model organism for the study of DNA replication. Meselson and Stahl chose a strain of *Escherichia coli* bacteria known to replicate its DNA once every 20 minutes. Therefore, for every one DNA molecule present at the start of the experiment, time 0, there would be two identical DNA molecules present after 20 minutes, and four identical DNA molecules present after 40 minutes. Because the goal of the experiment was to determine how one molecule of DNA is used to form two identical DNA molecules, Meselson and Stahl needed a way to distinguish the nucleotides present in the original parental DNA from those that were newly added during DNA synthesis. One commonly used method of tracking microscopic molecules like DNA is to integrate radioactive atoms, known as isotopes, into the molecules, and then expose the molecules to a reactive film. This results in dark shadows appearing on the film wherever the isotopes are located. For their experiment, Meselson and Stahl chose to label the nitrogen atoms in DNA with two different nitrogen isotopes known as ^{14}N and ^{15}N. These isotopes are identical, except that ^{15}N contains one more neutron than ^{14}N, so DNA containing ^{15}N is denser and heavier than DNA containing ^{14}N. Therefore, if DNA containing these isotopes is placed in a cylindrical tube and centrifuged, DNA containing ^{15}N will sink lower in the tube than DNA containing ^{14}N.

A method called density gradient centrifugation was used to separate DNA containing the heavier ^{15}N from DNA containing the lighter ^{14}N. To do this, DNA was placed in a cylindrical tube filled with a material called cesium chloride (CsCl). The tube was spun at high speed by a machine called a centrifuge, which caused the heavier DNA to move toward the bottom of the tube while the lighter DNA stayed closer to the top of the tube. Even though DNA is too small to be seen with the naked eye, the nitrogen isotopes produced a signal that could be detected to determine the position of the DNA in the tube. To correctly interpret the results of their experiment, Meselson and Stahl needed to determine precisely where in the tube DNA containing ^{14}N and DNA containing ^{15}N would be found. To do this, they first grew one flask of bacteria for many generations in a nutrient broth containing only ^{14}N, then grew a second flask of bacteria for many generations in a nutrient broth containing only ^{15}N (Figure 5.2A).

This resulted in the DNA of those bacteria containing only ^{14}N or ^{15}N, respectively. Next, they isolated and centrifuged DNA from each flask of bacteria and recorded where in the tube the signals from the nitrogen isotopes were detected (Figure 5.2B). These flasks served as controls, and the data collected from these controls allowed the researchers to interpret experimental data from bacteria with unknown DNA composition. If the DNA from an experimental group of bacteria yielded a signal at exactly the same place in the tube as the ^{14}N control, it could be presumed to contain only ^{14}N. Similarly, if the DNA from an experimental group of bacteria produced a radioactive signal at the same place in the tube as the ^{15}N control, it must contain only ^{15}N. If DNA from bacteria was found to produce a signal exactly halfway between the ^{14}N control signal and the ^{15}N control signal, it could be reasoned that the DNA was composed of half ^{14}N and half ^{15}N

Next, Meselson and Stahl conducted an experiment to determine how DNA is replicated. To do this, bacteria were first grown for a number of generations in nutrient broth containing only ^{15}N so that all of the nitrogen atoms contained in their DNA would be ^{15}N isotopes. Next, the bacteria were washed and transferred to a nutrient broth containing only ^{14}N. Because nitrogen is required for synthesizing DNA, any new DNA made in the nutrient broth containing only ^{14}N would also contain ^{14}N. After 20 minutes, when the bacteria had undergone one round of DNA replication, their DNA was isolated and analyzed (Figure 5.2C). The bacterial DNA was analyzed again after 40 minutes when two rounds of DNA replication had been completed (Figure 5.2C). This experimental data was then compared with the control data and the hypothetical models to determine which of the three models of DNA replication was consistent with the data.

As expected, after 0 minutes, 100% of the bacterial DNA was located exactly where the ^{15}N control signal was located. After 20 minutes, 100% of the DNA was located exactly halfway between the ^{15}N control signal and the ^{14}N control signal. After 40 minutes, 50% of the DNA

FIGURE 5.2A: Meselson and Stahl experiment— Bacteria grown in either ^{15}N or ^{14}N incorporated these nitrogen isotopes into their DNA.

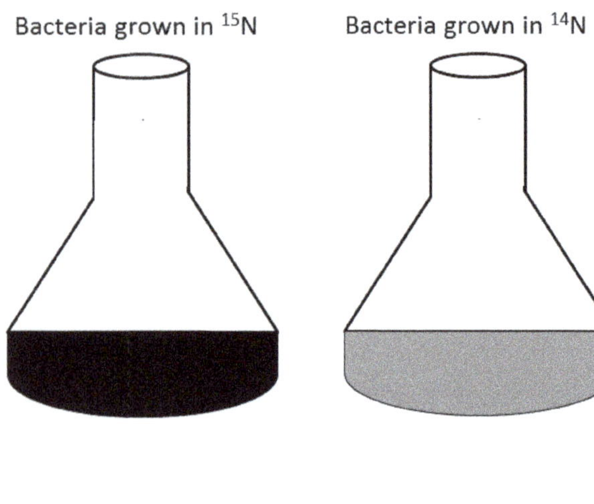

FIGURE 5.2B: Meselson and Stahl experiment—DNA containing ^{15}N was more dense and detected lower in a CsCl gradient than ^{14}N-containing DNA after centrifugation.

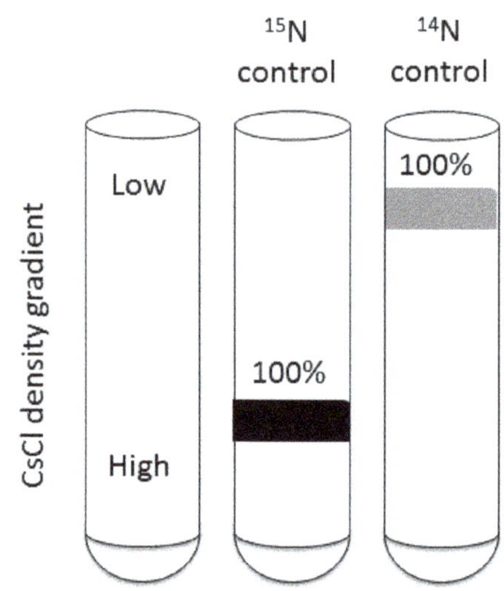

was halfway between the ^{15}N control signal and the ^{14}N control signal, and 50% was located at the position of the ^{14}N control signal. Based on this data, Meselson and Stahl identified the semiconservative model of DNA replication as the correct model. The data were consistent with what would be expected if, during each round of replication, the two strands of DNA separate, and one strand of old DNA was paired with one strand of new DNA. In contrast, if conservative replication had been occurring in the bacteria, some amount of DNA containing only ^{15}N would always be present. After 20 minutes, 50% of the DNA would be located at the ^{15}N position, and 50% would be located at the ^{14}N position; after 40 minutes, 25% of the DNA would be located at the ^{15}N position, and 75% would be located at the ^{14}N position. The proportion of ^{15}N-containing DNA would be halved with every cycle of replication, as only ^{14}N would be available for synthesizing new molecules of DNA. If the dispersive model has occurred, a different pattern would have been observed in which every molecule of DNA was located somewhere between the ^{14}N position and the ^{15}N position because of the dispersal of old and newly synthesized DNA into each strand during replication.

FIGURE 5.2C: Meselson and Stahl experiment—DNA was collected after 0, 20, and 40 minutes, and separated using a CsCl gradient. The percentage of total DNA detected at various points in the gradient is indicated.

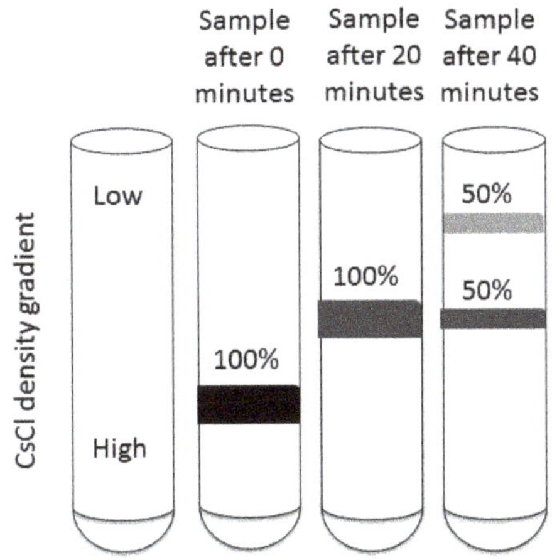

DNA Replication: A Closer Look

Although Meselson and Stahl demonstrated how DNA replication occurs conceptually, many of the specific details for how cells accomplish the replication of genetic material remained to be elucidated. Years of investigation by numerous scientists since Meselson and Stahl have produced a well-defined understanding of how DNA replication occurs. DNA replication in eukaryotes is a complex process driven by several enzymes and accessory molecules. This process is divided into three main phases: initiation, elongation, and termination. Because DNA sequences are very long and it is in the cell's interest to complete DNA replication as quickly as possible, replication usually takes place simultaneously at multiple sites along the DNA sequence. The sites at which initiation occur are called origins of replication. These sites contain certain DNA sequences recognized by the proteins that initiate replication. Upon binding to the DNA, these proteins separate the two strands of the double helix forming a structure called a replication bubble. At each edge of the replication bubble, where the two strands separate from each other, is a Y-shaped structure known as the **replication fork** (Figure 5.3). At the replication fork, an enzyme called **helicase** unwinds the double helix, separating the two strands of DNA so that each can serve as a template for the synthesis of new, complementary strands. **Single-strand binding proteins** hold the two strands apart until they have been replicated. Another enzyme called **topoisomerase** facilitates the unwinding of the double helix by moving along the DNA ahead of the replication fork and reducing the twisting and physical strain on the molecule that would otherwise be caused by the actions of helicase.

FIGURE 5.3: DNA replication fork—Enzymes and other proteins separate double-strand DNA and synthesize new DNA at the replication fork.

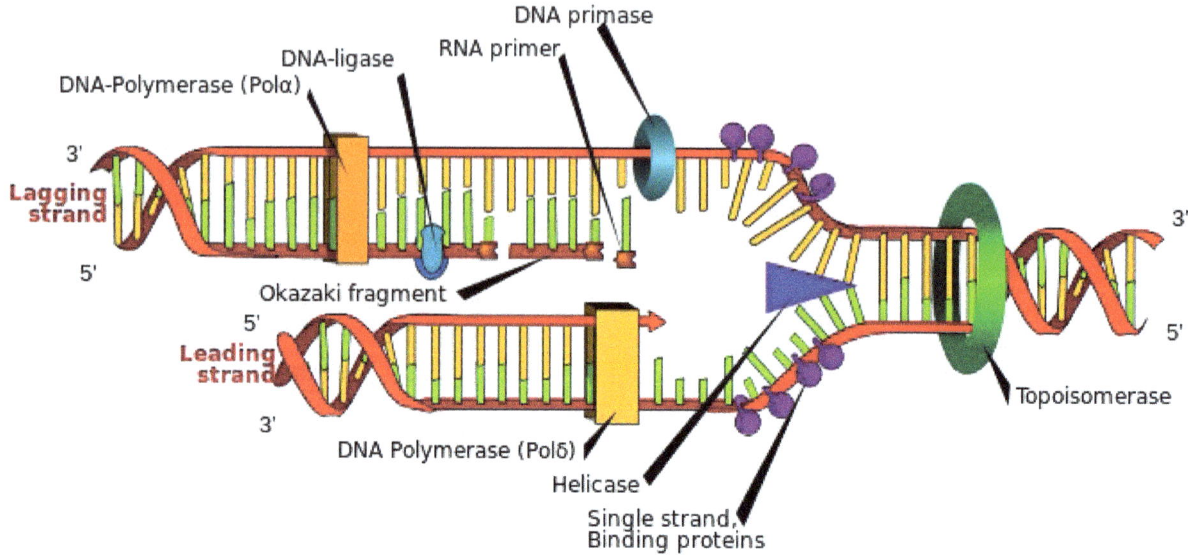

Initiation of DNA replication also involves the attachment of an RNA primer to the DNA template. This is required because the enzymes that synthesize DNA are only capable of adding nucleotides to an existing chain of nucleotides. These enzymes cannot actually initiate the synthesis of a nucleic acid. However, once an enzyme called primase adds a short stretch of complementary RNA nucleotides to the DNA template at the origin of replication, DNA nucleotides can be attached to this **primer**. DNA nucleotides are added to the pre-existing chain by an enzyme called **DNA polymerase**. After initiation, DNA polymerase will move down the length of the template DNA in the 5' to 3' direction, adding complementary nucleotides to the newly synthesized strand of DNA during the process of elongation. Because the strands of the original DNA are antiparallel, the process of elongation cannot unfold in exactly the same way along each template strand. One of the template strands is orientated with its 3' end at the replication fork, so DNA polymerase can easily move along this strand, continuously synthesizing a new, complementary strand in the 5' to 3' direction. The DNA produced in this way is known as the **leading strand**. The other template strand is invariably oriented in the opposite direction, with its 5' end toward the replication fork. Because DNA polymerase can only add nucleotides in the 5' to 3' direction, it can only add small stretches of nucleotides at a time, working backward from the replication fork toward the origin of replication. It must then wait for the helicase to move further down the double helix and expose another section of the template strand before it can bind to it and continue synthesizing the new strand. Each time synthesis of a new section begins, primase must first add an RNA primer. The DNA strand that is elongated in this discontinuous manner is known as the **lagging strand**. The lagging strand is synthesized in small pieces known as **Okazaki fragments**, named after the scientist who discovered them. Once elongation is complete, the RNA primers are removed. The gaps in the newly synthesized DNA left by removal of the primers are then filled in with DNA nucleotides by an enzyme called **DNA ligase**.

DNA Proofreading and Repair

The human genome is more than three billion base pairs long, and the sequence of the average human gene contains more than 8,000 base pairs. To prepare for cell division, DNA polymerase works incredibly quickly, adding up to 100 nucleotides per second! Although DNA replication is a high fidelity process, and DNA polymerase has a low rate of error, mistakes do inevitably occur during the replication process. To prevent the propagation of these errors and any deleterious effects they may cause, cells employ a multistep process to proofread and repair mistakes. DNA polymerase immediately checks each newly added base against the template strand before proceeding to add more bases. Most mistakes are corrected immediately; DNA polymerase removes the incorrect nucleotide base and replaces it with the correct one before moving on. A second line of protection, known as the mismatch repair system, checks double-stranded DNA for incorrectly paired bases, then removes and replaces any mismatches. **Mismatch repair** is useful for correcting errors made during replication as well as DNA damage caused by mutagens such as x-rays and harmful chemicals. As part of a third mechanism, known as **nucleotide excision repair**, enzymes called nucleases remove and repair DNA damaged by ultraviolet radiation from sunlight or tanning beds. Ultraviolet radiation is known to cause a type of mutation called dimerization, which involves the formation of a covalent chemical bond between two adjacent nucleotides that causes them to improperly stick together in a way that interferes with the function of DNA polymerase. Inherited genetic mutations that render the enzymes involved in mismatch repair or nucleotide excision repair nonfunctional have been linked to certain cancers and other diseases that result from the inability to correct potentially harmful mutations.

Telomerase Replication

Even if no mistakes are made during DNA replication, the inherent limitations of DNA polymerase present another problem for living organisms. DNA polymerase can only add nucleotides to an existing chain of nucleotides, and can only do so in the 5' to 3' direction. In the leading strand, replication proceeds continuously until the very end of the template without an issue. However, near the end of the lagging strand there is not enough room to place a primer for DNA polymerase to hook on to, so a small number of nucleotides at the end of the sequence cannot be replicated. As a cell repeatedly divides over time, the lagging strand gets progressively shorter as more and more tiny pieces are left off the end of the sequence. Eventually this could lead to the loss of an important gene sequence, rendering a crucial protein non-functional, and creating a potentially serious and even fatal deficiency. To prevent this from happening, the ends of eukaryotic chromosomes are lined with repetitive sequences called **telomeres** that are not part of a gene and exist solely as a buffer against the loss of important genetic information from the end of the lagging strand. In humans, telomeres consist of a six base sequence, TTAGGG, which is repeated between 100 and 3000 times depending on the individual. The length of telomeres is inherited, with some families known to have longer telomeres than others. There is mounting evidence that telomere length is linked to aging and longevity. The implication being that the longer the telomeres at birth, the longer the cells of that organism can continue dividing to replace damage and worn out cells without risking the incursion of potentially deleterious genetic mutations. As one ages, telomeres grow shorter and shorter, increasing the chance that cells become damaged and abnormal.

FIGURE 5.4: Replication of telomeres—Telomerase adds repeating telomere sequences to the ends of linear chromosomes.

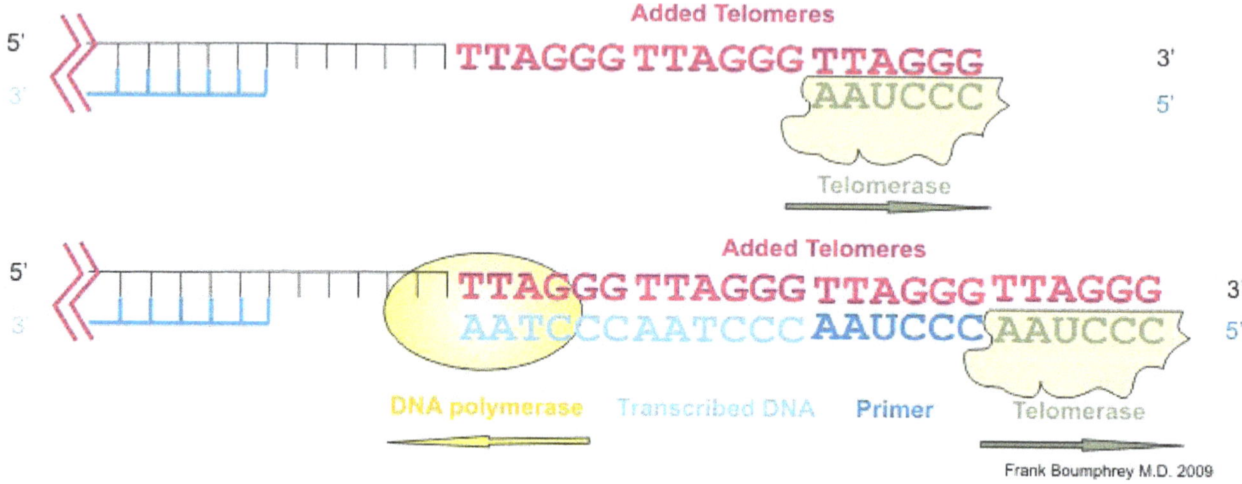

Most cells lack the capability to replace lost telomere length lost resulting from repeated rounds of cell division, but in some cells it is crucial that this be avoided. The germ cells that produce gametes must have a way to replace lost telomere sequence after each round of DNA replication. Otherwise the gametes they produce would eventually be missing important gene sequences. To this, germ cells use an enzyme called telomerase to catalyze the lengthening of telomeres. After DNA replication, **telomerase** elongates the end of the lagging strand of the template DNA to make room for a primer and allow DNA polymerase to add complementary bases until the end of the template sequence (Figure 5.4).

Because all nucleated cells in a eukaryotic organism contain an identical set of DNA, all cells possess the genes to make telomerase. However, in most somatic (non-germ line) cells of adult organisms, this gene is not active. Although the inability for most cells to restore their telomeres may ultimately lead to aging, the inactivation of telomerase is also an important protective mechanism. Organisms develop and function properly within the constraint of telomere length. One type of cell known to diverge from this fate is the cancer cell. When analyzed, the cells of many cancerous tumors are found to have reactivated their telomerase genes after the point when they should no longer be capable of directing the production of telomerase. This results in the ability of cancerous cells to divide essentially limitlessly, until the organism is treated for the cancer or dies. Cancer cells are even said to be immortal because of this ability.

Prokaryotic DNA Replication

DNA replication is highly conserved, or very similar, among different species of eukaryotes and prokaryotes. However, the implications of telomerase are not an issue for prokaryotes, because their circular chromosomes do not have ends like linear eukaryotic chromosomes do. DNA replication in prokaryotes takes place from a single origin of replication but can occur up to 10 times as fast as in eukaryotes, at a pace of 1,000 nucleotides per second. At this rate, even the largest prokaryotic genomes, which contain up to 10 million base pairs, can be replicated in less than 3 hours.

5.2 SUMMARY

- DNA replication results in two DNA double helices identical to the original DNA
- Meselson and Stahl provided experimental evidence that DNA replication is semiconservative rather than conservative or dispersive
- During replication, the DNA double helix is unwound by helicase with the help of topoisomerase and kept apart by single-strand binding proteins; at the origin of replication, primase adds short RNA primers so that DNA polymerase can add DNA nucleotides complementary to the template strand
- DNA replication occurs at the replication fork in the 5' to 3' direction; the leading strand is replicated continuously, while the lagging strand is produced in small pieces called Okazaki fragments that must later be connected by DNA ligase
- Telomeres protect the ends of eukaryotic chromosomes from degradation during repeated rounds of DNA replication; telomerase is an enzyme that lengthens telomeres and is often improperly active in cancer cells
- Newly replicated DNA is carefully proofread by DNA polymerase; errors that remain are fixed through mismatch repair and nucleotide excision repair

5.3 KNOWLEDGE CHECK

1. Identify the bonds broken by helicase during DNA replication.
   a. covalent bonds
   b. hydrogen bonds
   c. ionic bonds
   d. peptide bonds
   e. phosphodiester bonds

2. What is the complementary DNA sequence of 3'-AATGGC-5'?
   a. 3'-AATGGC-5'
   b. 3'-TTACCG-5'
   c. 3'-UUACCG-5'
   d. 5'-AATGGC-3'
   e. 5'-TTACCG-3'

3. Which of the following fills in the gaps of the newly synthesized lagging strand?
   a. DNA ligase
   b. DNA polymerase
   c. primase
   d. single-strand binding proteins
   e. telomerase

4. Which model of DNA replication was supported by the Meselson-Stahl experiment?
   a. conservative model
   b. dispersive model
   c. liberal model
   d. Okazaki model
   e. semiconservative model

5. Which of the following would correct DNA damage caused by ultraviolet radiation?
   a. helicase
   b. mismatch repair
   c. nucleotide excision repair
   d. replication fork
   e. telomerase

6. Explain why Okazaki fragments are produced during synthesis of the lagging strand.

7. Describe the experiment conducted by Meselson and Stahl to identify the method of DNA replication.

8. Propose an explanation for what would happen to an organism with a mutation that resulted in abnormally short telomeres. Compare this to what would happen to an organism with a mutation that resulted in abnormally long telomeres.

5.4 ACTIVITY

Explaining DNA Replication

The purpose of this activity is to demonstrate understanding of DNA replication by writing a letter to a friend explaining key components of the process. This letter should be clear, accurate, and thorough, yet written in an informal tone using words that a friend who has not taken a class like this would understand. Try using analogies, metaphors, and similes to compare technical terms with objects and concepts familiar to nonscientists. For example, it might be helpful to compare the replication fork to a zipper and describe helicase as being like the zipper pulls that unzips the DNA so that the nucleotide "teeth" can be accessed by the other proteins involved in DNA replication. In your letter, be sure to clearly explain the meaning and/or specific role of each of the following components in the replication of DNA: DNA ligase, DNA polymerase, helicase, lagging strand, leading strand, Okazaki fragments, primer, replication fork, single-strand binding proteins, and topoisomerase.

5.5 ACTIVITY

Putting the Pieces Together

The purpose of this activity is to visualize the three hypothetical models of DNA replication that were investigated by Matthew Meselson and Frank Stahl. Several historical events that provided key foundational knowledge about DNA structure and how DNA molecules are replicated, including Meselson and Stahl's famous experiment, are discussed in the case study titled "Putting the pieces together: The discovery of DNA structure and replication" published in the National Center For Case Study Teaching In Science (NCCSTS) Case Collection.

> "Putting the pieces together: The discovery of DNA structure and replication"
> http://sciencecases.lib.buffalo.edu/cs/collection/detail.asp?case_id=786&id=786

For this activity, start by reading the narrative of Part III of the case study titled "Building a model of DNA replication." If instructed do so, construct the three models of DNA replication using gumdrops, marshmallows, and toothpicks as described in the case study. Alternatively, you may be directed to view a video demonstrating the building of these models. This video is titled *Building a model of DNA replication* and can be viewed at https://youtu.be/SAXHwkKe6Bc.

Once you have completed the aforementioned activity as directed by your instructor, complete the data tables found in "Part III: Building a model of DNA replication," and answer Questions 1–5 listed in that part of the case study.

5.6 ACTIVITY

Modeling DNA Replication

Use the 3D Molecular Design model kit to make a physical model that demonstrates what occurs during DNA replication. For more background information, consult the following resource: 3D molecular designs. (2016). *DNA replication activity guide*. 3dmoleculardesigns.com.

Note: The nucleotide pieces with rounded bottoms represent DNA nucleotides, and the pieces with square bottoms are RNA nucleotides. The arrows on all of the nucleotides point toward the 3' direction. Do not use the gray nucleotides for Part 1 of this activity.

1. Using the kit, construct a model of the following DNA sequence:

   5'-CCGCATGTGTGAGATACATTGGCCAAGACACTGTTAGCTC-3'

2. Construct a model of the complementary sequence to the sequence you just made, producing a double-stranded DNA molecule.

3. Place the double-stranded DNA you just made next to the right side of the "DNA Replication" mat.
4. Slide the DNA through the "helicase" as instructed by the markings on the mat.
5. Once 10 nucleotides from each DNA strand have emerged from the left side of the helicase, set the two "DNA polymerase" molecules onto the left edge of the mat.
6. Start sliding each strand of DNA through one of the DNA polymerase molecules as indicated. Note which strand is the leading strand and which strand is the lagging strand.
7. Add complementary DNA nucleotides to each strand as each nucleotide passes through the DNA polymerase. Note that DNA polymerase can only synthesize DNA in the 5' to 3' direction.
8. Once 10 nucleotides have been added to each of the two newly synthesized strands of DNA, take a photo or make a sketch showing the entire "DNA Replication" mat as well as both DNA polymerase molecules and both newly synthesized DNA molecules.
9. Continue adding DNA nucleotides until the end of the template strand is reached.
10. Submit the photo/sketch made in Step 8 along with the answers to the Activity questions.

5.7 ACTIVITY

Modeling the Meselson and Stahl Experiment

Use the 3D Molecular Design model kit to demonstrate what occurs during DNA replication.

Note: The nucleotide pieces with rounded bottoms represent DNA nucleotides, and the pieces with square bottoms are RNA nucleotides. The arrows on all of the nucleotides point toward the 3' direction. The colored nucleotides represent the original DNA, and the gray nucleotides represent the newly synthesized DNA.

1. The purpose of this lab is to create physical representations of the three hypothetical models of DNA replication investigated by Matthew Meselson and Frank Stahl. Review Figures 5.1 and 5.2 before proceeding.
2. Using multicolored DNA nucleotides, form a model of a double-stranded DNA molecule that is six nucleotide base pairs in length. The multicolored nucleotides represent the original DNA labeled with ^{15}N. When the DNA is replicated, no additional multicolored nucleotides may be used. Use only gray nucleotides to represent the ^{14}N-labeled nucleotides incorporated into newly synthesized DNA during DNA replication.
3. Using the appropriate nucleotides, demonstrate one round of the conservative model of DNA replication.
4. Take a photo or make a sketch showing both strands of DNA.
5. Using the appropriate nucleotides, demonstrate a second round of the conservative model of DNA replication.
6. Take a photo or make a sketch showing all four strands of DNA.
7. Repeat Steps 3–6 for the dispersive and semiconservative models of DNA replication.
8. Submit the photo/sketch made in Steps 4, 6, and 7 along with the answers to the Activity questions.

ACTIVITY QUESTIONS

5.4

1. Following the prompt provided in the activity description, write a one- to two-page letter to a friend clearly explaining the meaning and/or specific role of each of the following components in the replication of DNA: DNA ligase, DNA polymerase, helicase, lagging strand, leading strand, Okazaki fragments, primer, replication fork, single-strand binding proteins, and topoisomerase.

5.5

1. Reproduce and complete the data tables in Part III of the case study, and answer Questions 1–5 from that part of the case study.

5.6

1. The two strands of a DNA double helix are antiparallel and complementary. Describe how this could be observed in the DNA model.
2. Describe what happened to the DNA double helix model as it slid through the helicase.
3. Compare and contrast replication of the lagging strand of the DNA model with replication of the leading strand of the model.

FIGURE 5.5: Data expected from Meselson and Stahl experiment

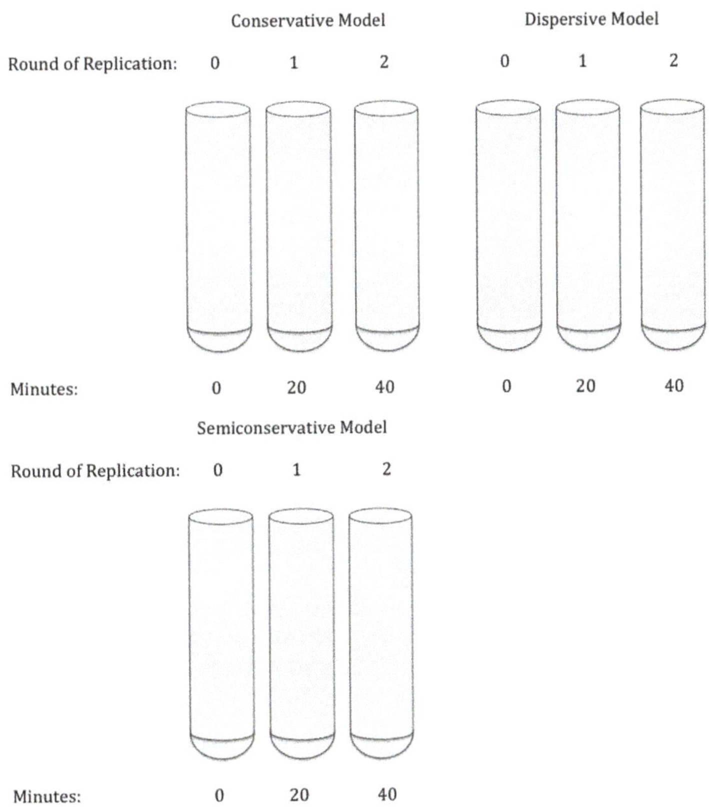

5.7

1. Using the data you collected in Steps 3–7, label where in the centrifuge tubes illustrated in Figure 5.5 ^{15}N and ^{14}N would be found, then draw lines to

indicate where DNA would be expected, after each round of replication, according to each model. You may wish to trace or redraw Figure 5.5 on your answer sheet.

2. Note that the bacteria used in the Meselson-Stahl experiment completed one round of replication every 20 minutes. Compare the data you recorded for Question 1 to the data shown in Figure 5.2. After one round of replication, which model(s) of DNA replication were consistent with the data obtained from the Meselson-Stahl experiment? Which model(s) could be eliminated as potentially accurate? Why?
3. After two rounds of replication, which model(s) of DNA replication were consistent with the data obtained from the Meselson-Stahl experiment? Which model(s) could be eliminated as potentially accurate? Why?
4. When James Watson and Francis Crick published their double-helix model of DNA structure, they stated: "It has not escaped our notice that the specific pairing we have postulated immediately suggests a possible copying mechanism for the genetic material." Describe how the structure of DNA itself suggests how it is replicated.
5. Predict the data that would be obtained from a third round of replication, according to the model you identified in Question 3. Describe how it would be similar and/or different from the data obtained after the second round of replication.

5.8 REFERENCES

Bonney, K. M. (2015). Putting the pieces together: The discovery of DNA structure and replication. National Center For Case Study Teaching In Science Case Collection. Buffalo, NY: University of Buffalo. http://sciencecases.lib.buffalo.edu/cs/files/dna_structure_flip.pdf.

Burgers, P. M. J. (2009). Polymerase dynamics at the eukaryotic DNA replication fork. *Journal of Biological Chemistry, 284*(7), 4041–4045

Huberman, J. A., & Riggs, A. D. (1968). On the mechanism of DNA replication in mammalian chromosomes. *Journal of Molecular Biology, 32*(2), 327IN15335–334IN22341.

Levy, Michael Z., et al. (1992). Telomere end-replication problem and cell aging. *Journal of Molecular Biology, 225*(4), 951–960.

Meselson, M., & Stahl, F. W. (1958). The replication of DNA in *Escherichia coli*. *Proceedings of the National Academy of Sciences of the United States of America, 44*(7), 671–682.

FIGURE CREDITS

FIG. 5.3: Mariana Ruiz, "DNA Replication Fork," https://commons.wikimedia.org/wiki/File:DNA_replication_en.svg. Copyright in the Public Domain.

FIG. 5.4: Copyright © Boumphreyfr (CC BY-SA 3.0) at https://commons.wikimedia.org/wiki/File:Telomerase.png.

Transcription and Translation

LEARNING OBJECTIVES

- Summarize the main steps of transcription
- Describe the purpose of RNA processing
- Apply knowledge of transcription to synthesize an RNA sequence from a DNA sequence
- Identify the enzyme that adds nucleotides during RNA transcription and the direction in which RNA is transcribed
- Summarize the main steps of translation
- Identify the purpose of the following terms during translation: amino acid, anticodon, codon, mRNA, ribosome, peptide bond, polypeptide, tRNA
- Apply knowledge of translation to determine the amino acid sequence encoded by a given mRNA sequence

6.1 INTRODUCTION

Living organisms store and transmit hereditary genetic information using nucleic acids. Most of this information is contained in individual units of DNA called genes. Genes often contain the instructions for building a protein or protein subunits. First, DNA is converted into mRNA through a process called transcription, then the mRNA sequence directs synthesis of a polypeptide chain during the process of translation. The various structures and functions of proteins produced by translation are integral in producing the observable and non-observable traits of the living organism. The preceding description, though accurate, is highly simplified. Many additional factors that affect gene expression, mRNA formation, and protein folding affect the way genes contribute to traits. This chapter discusses some of these factors and provides a description of the processes of transcription and translation.

Living organisms use DNA to transmit the instructions necessary for life from parents to offspring. For most cell types, every time a new cell is made, it receives a complete set of DNA, containing all of the instructions that the cells of that organism possess. Not all DNA encodes instructions that determine the characteristics of cells and organisms. The distinct sequences of nucleotides that do determine hereditary characteristics are called **genes**. To transform the basic information contained in genes into the observable structures and traits of a cell or organism requires a complex multistep process that may be affected by many internal and external factors. The process through which information contained in a gene is used to produce a functional product is called **gene expression**. For most genes, this process starts by using the instructions contained in DNA to make an RNA copy, which is then used to produce a protein. The function of proteins ultimately contributes to observable characteristics. To understand the process of making a protein from DNA, it may help to start with a metaphor.

For example, consider the process of making proteins from DNA as similar to making food from a recipe (Figure 6.1). When you want to make a new type of food, you might start by opening a cookbook or accessing a cooking website that contains recipes for many different types of

FIGURE 6.1: **DNA is the cookbook of life**—The process of translation is similar to reading the recipe and combining all of the right ingredients in the correct order.

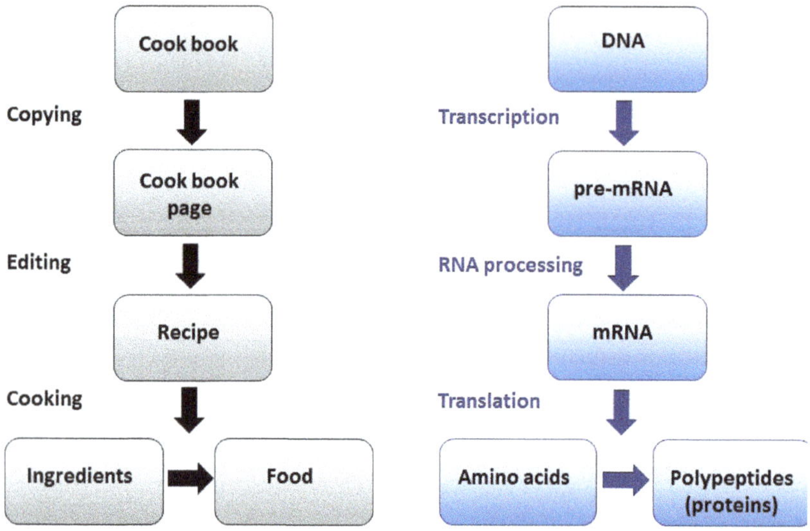

food. Because you don't want to risk damaging the recipe book or computer in the kitchen, you would copy or print the recipe you wanted onto a piece of paper. The page in the book that the recipe is on might contain an advertisement or other information that you do not actually need to make the food, so you would omit that from the recipe that you copy down. You may also add some extra notes that you find helpful, such as directions where to find all of the ingredients that you will need. Then you would take just the information you need into the kitchen and start adding the ingredients together.

Overview of Gene Expression

The transfer of information from DNA to RNA to protein is called the central dogma of molecular biology. Francis Crick proposed the central dogma in the years after he and James Watson published their article on the double helix structure of DNA. Crucial details of this process were gradually elucidated over the next two decades through the painstaking work of many other scientists, including Robert W. Holley, Har Gobind Khorana, and Marshal W. Nirenberg, who were awarded the 1968 Nobel Prize in Physiology or Medicine for their work with DNA and protein synthesis. We now know that the process of making a protein in a cell starts with enzymes unwinding the DNA double helix in the nucleus of the cell to make the instructions needed accessible to synthesize the protein. The DNA sequence is then copied into **messenger RNA (mRNA),** which serves as a portable "recipe" for making the protein. This process is called transcription. Transcription is followed by RNA processing, during which the cell removes unnecessary portions from the sequence of nonfunctional "pre-mRNA" and adds an additional sequence to produce a functional mRNA. Finally, the mRNA is transported from the nucleus to the **ribosome**, which is a specialized cellular structure that "reads" the mRNA. During the process of translation, which occurs in the ribosome, amino acids are joined together in the order specified by the mRNA to form a polypeptide. Once completed, the polypeptide is folded into a functional protein. To reference the earlier metaphor, amino acids are like ingredients, ribosomes are like a kitchen, and polypeptides and proteins are like the food that is made. The processes of transcription and translation will be described in more detail in the following pages.

Transcription

During transcription, an RNA sequence is synthesized from one strand of the DNA double helix called the template strand. The DNA sequence of each gene is much longer than the portion that will be transcribed and is surrounded by the DNA sequence of other genes as well as noncoding DNA. To facilitate the process of transcription, genes contain a specific sequence called a **promoter** that signals where the enzymes and proteins involved in transcription should bind. In the cooking analogy described earlier, the promoter could be thought of as a bookmark that signals where the recipe is located and speeds up the process of locating it. The strength of the promoter attracting these enzymes determines how often a particular gene will be transcribed. Genes with strong promoters will be transcribed more often than genes with weak promoters, leading to different levels of protein production. Proteins required by the cell in large quantities are often encoded in genes with strong promoters. In eukaryotes, many genes have a very similar promoter sequence called a TATA box. It is called that because it begins with the nucleotide base sequence TATA. At the start of transcription, several proteins called transcription factors bind to the TATA box. **Transcription factors** regulate transcription of genes by carrying out several important functions such as mediating the binding of important enzymes to the DNA double helix.

One of the first enzymes that bind to DNA under the direction of transcription factors is RNA polymerase (Figure 6.2). **RNA polymerase** is the enzyme that synthesizes RNA during

FIGURE 6.2: The role of RNA polymerase during transcription—After transcription, enzymes remove introns from the newly transcribed pre-mRNA. Sequences are also added to the ends of the mRNA to help it exit the nucleus and move to a ribosome.

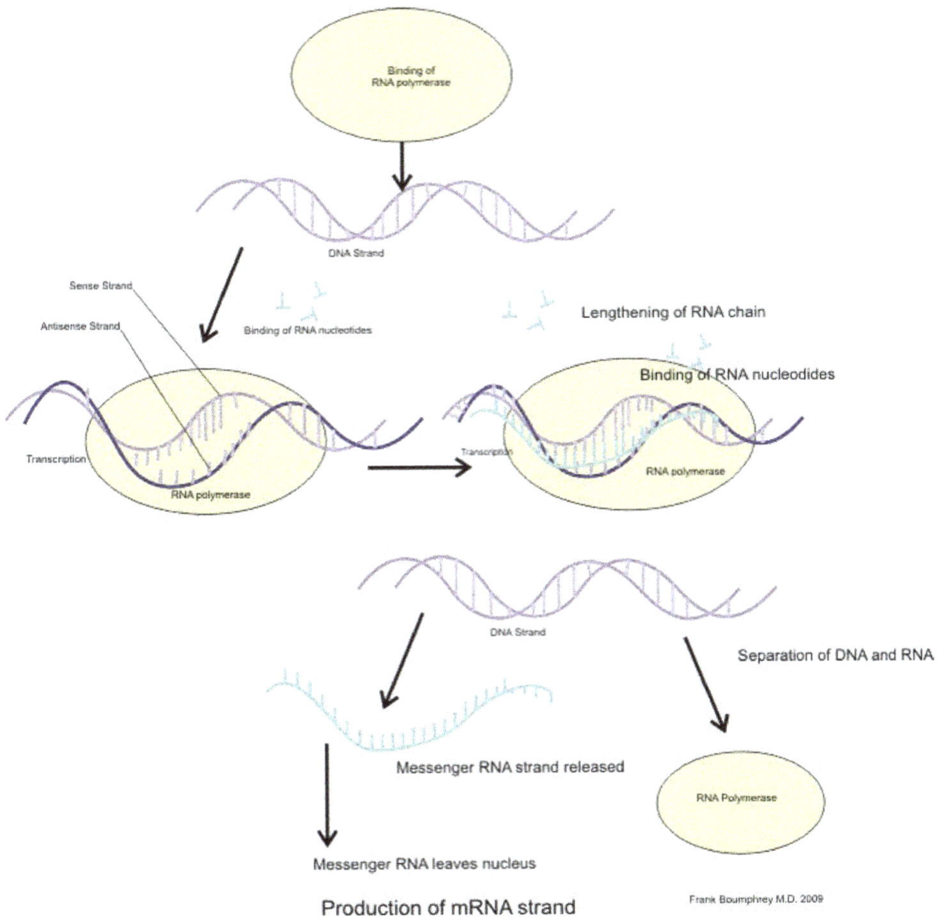

transcription by linking together a chain of RNA nucleotides that corresponds to the template DNA sequence. Because the nucleotide base pairs that encode the genetic information in DNA are located on the inside of the DNA double helix, an enzyme called **DNA helicase** must first unwind the double helix. Collectively, all of the enzymes and transcription factors involved in starting transcription are called the transcription initiation complex.

The process of transcription consists of three main phases. The first phase, which is called initiation, involves binding of RNA polymerase and associated proteins to the promoter region of the DNA template. The elongation phase occurs directly after initiation, during which the RNA polymerase enzyme moves down the DNA template and adds complementary bases of RNA to synthesize a strand of mRNA. RNA is always transcribed in the 5' to 3' direction, using the 3' to 5' strand of DNA as the template. RNA and DNA are structurally similar molecules, with several key differences (Table 6.1). One difference is that RNA contains ribose instead of deoxyribose. The two other key differences that are that RNA is single-stranded instead of double-stranded, and contains A, G, C, but instead of T, RNA contains U (uracil). Therefore, if the nucleotide sequence TAC appeared in the DNA template, the RNA polymerase would add the nucleotide AUG to the nascent RNA strand. An example of an RNA sequence that would be transcribed from a given DNA sequence is illustrated in Figure 6.3.

TABLE 6.1: Differences Between DNA and RNA

	DNA	RNA
Structure	Double helix	Single stranded
Nucleotide bases	Guanine, Cytosine, Adenine, Thymine	Guanine, Cytosine, Adenine, Uracil
Sugar	Deoxyribose	Ribose
Function	Hereditary material	Copy genetic material, site of protein synthesis

FIGURE 6.3: Production of mRNA from DNA template during transcription—First, a 5' to 3' mRNA sequence is transcribed from a 3' to 5' template strand of DNA.

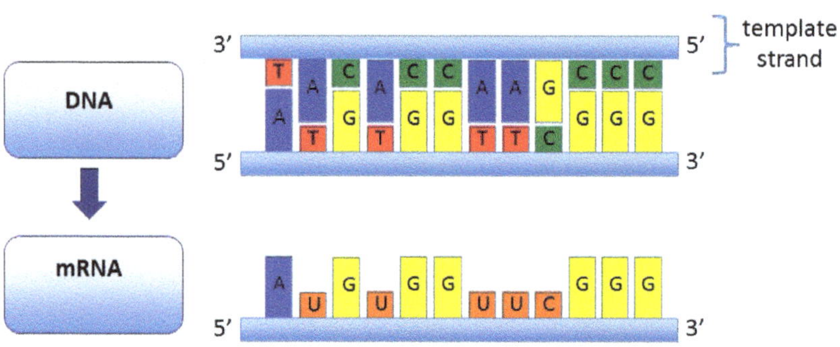

Once transcription is completed, the RNA polymerase leaves the site of transcription, and the DNA and RNA dissociate from each other. This final phase of transcription is called termination. After termination, the DNA double helix closes and takes on its original conformation, and the RNA polymerase is free to initiate transcription at a different promoter. If the newly produced RNA transcript is mRNA, it will now leave the nucleus and enter the cytoplasm, where it will direct the synthesis of proteins by ribosomes. In addition to mRNA, the process of transcription may also produce two other types of RNA. One type, **ribosomal RNA (rRNA)**, is the material that forms the structure of ribosomes. The other type, **transfer RNA (tRNA)**, plays a crucial role in translation by recognizing which amino acids are specified by the mRNA sequence and positions them in the ribosome to be added to the growing polypeptide.

RNA Processing

FIGURE 6.4: Transcription and RNA processing—After transcription, the mRNA is modified to remove introns and add flanking sequences.

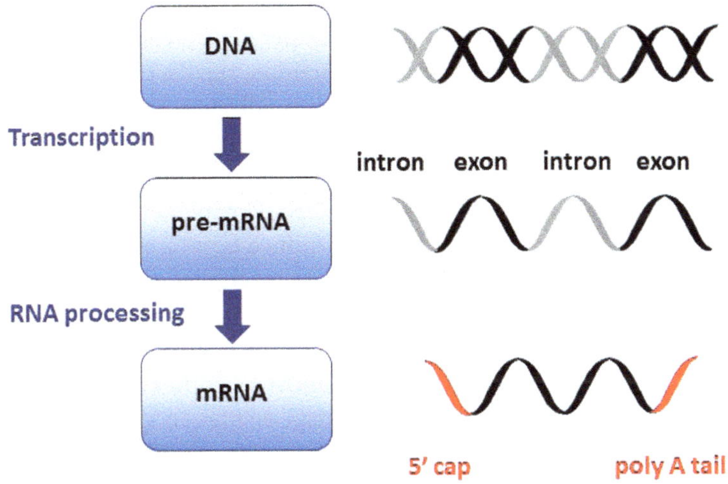

In eukaryotic cells (cells that have a nucleus), there is one more step before translation, which is called **RNA processing.** RNA processing involves removing parts of the DNA that are not needed for the mRNA because they do not provide instructions for making the protein (Figure 6.4). The cell actually transcribes the entire DNA sequence associated with the gene into mRNA first, producing what is called "pre-mRNA." Next, the parts of the pre-mRNA that do not specify instructions for making the protein are removed, leaving behind only the parts that describe how to make the protein. The parts of the gene sequence kept in the mRNA and expressed as part of a protein are called **exons.** The intervening parts of the pre-mRNA that do not encode instructions for making part of the protein that are removed during RNA processing are called **introns.**

Some introns are known to contain sequences that regulate gene expression, but the specific function of most introns is not yet known. Even if individual introns have no function, the presence of introns may have roots in evolution. Perhaps most importantly, the presence of introns allows more than one polypeptide to be encoded by a single gene. This allows an organism to produce many more protein products than would be possible if each gene encoded only one protein. The process through which multiple polypeptides are produced from a gene sequence is known as alternative RNA splicing. The term "splicing" refers to the process of connecting, or splicing, exons together after the introns are removed.

The final step of RNA processing is to add short sequences of RNA nucleotides to the ends of the mRNA transcript. On the 5' end, a modified form of guanine nucleotides called the **5' cap** is added. On the 3' end, between 50 and 250 adenine nucleotides are added to form the **poly A tail**. These sequences protect the RNA from degradation and help direct it to the proper location in the cell.

Prokaryotes (organisms with cells that do not have a nucleus) do not undergo RNA processing. Prokaryotes differ from eukaryotes in several other ways during the process of producing proteins using the genetic information encoded in DNA. One important difference is that, in prokaryotes, translation of an mRNA can begin before transcription of that mRNA is completed, but in eukaryotes, transcription and translation are separated by both space and time. This is because transcription occurs only inside the nucleus, but the mRNA must be transported into the cytoplasm for translation to begin. Another important difference between prokaryotes and eukaryotes is that gene expression in prokaryotes is regulated primarily at the transcription level through interactions of other proteins with and near the promoter. In eukaryotes, regulation of gene expression is much more complex and occurs through a number of different processes.

Translation

After transcription, the information encoded by the newly synthesized mRNA sequence is used to produce protein precursors called polypeptides through the process of **translation**. Translation occurs in specialized cellular structures known as ribosomes. To form a polypeptide, ribosomes facilitate the linking of amino acids in the order specified by the mRNA template. Cells are continually engaging in the transcription and translation of numerous gene products. A single mammalian cell may contain up to 10 million ribosomes, which are capable of working simultaneously to fulfill the cell's protein production needs. Different portions of a single mRNA molecule may be translated by several individual ribosomes at the same time. The ability to produce large amounts of protein quickly is essential to the functioning of most cells because of the wide variety of essential functions that proteins perform. Proteins also account for a greater proportion of the mass of living organisms than any other type of molecule except water. To maintain necessary protein levels, cells devote more energy to translation than to almost any other metabolic activity.

RNA is like a language written in an alphabet that only contains four letters: A, C, G, and U. Each of these letters represents one of the four nucleotides of RNA: adenine, cytosine, guanine, or uracil. This is similar to the language of DNA, except that DNA contains thymine instead of uracil. In contrast to transcription, translation involves converting the message contained in genes into a much different language: a language consisting of amino acids.

In the language of RNA, each word contains three nucleotide letters and is called a **codon**. Each codon instructs the ribosome to add a particular amino acid to the chain of amino acids (polypeptide) that will be folded into a protein. In other words, amino acids are the "definitions" of the words in the RNA language. For the correct series of amino acids to be incorporated into the polypeptide being synthesized, the ribosome and other cellular machinery involved in translation must appropriately interpret which sets of three nucleotides in the genetic code should be grouped into codons. The way that the nucleotide sequence is divided into codons is known as the reading frame. The beginning point of the reading frame is determined by the placement of a special codon, known as the start codon. The start codon is always AUG, which also specifies the addition of the amino acid methionine (Met). An mRNA sequence may begin with nucleotides other than AUG, but translation will not begin until the start codon reaches the appropriate position in the ribosome.

Each codon always codes for the same amino acid. In other words, the "definition" of AUG is always methionine, and never glycine, or any other amino acid. Because of this, it is said that the genetic code is unambiguous. Because each codon is three letters long and there four possible letters to choose from, there are 43, or 64, possible codon combinations. However, there are only 20 amino acids. This is because some amino acids are encoded by multiple codons. Put another way, some codons are like synonyms; different words have the same meaning. Because of this, it is said that the genetic code is redundant.

Most amino acids are encoded by multiple codons. Only tryptophan (Trp) and methionine (Met) are encoded by a single codon. This type of variability is possible because there is flexibility in how anticodons on tRNA pair with codons in mRNA. In most cases, the first two bases of the anticodon must be complementary to the first two bases on the codon for binding of the tRNA to occur. However, base paring in the third position is more versatile. This versatility, called wobble, allows the 5' end of the anticodon to form hydrogen bonds with more than one different type of nucleotide base in the 3' end of the codon.

Memorizing which amino acid is encoded by each codon can provide many hours of entertainment. However, there is a more practical alternative to memorizing which amino acid is encoded by each of the 64 codons, and it involves the use of a codon chart. Codon charts can be formatted in a number of different ways. The codon chart shown in Figure 6.5 is organized with the first letter of the codon in the center. Each potential second letter of the codon is indicated in the first concentric circle outside the middle, and the third potential letter of the codon is located in the outer concentric circle. The three-letter abbreviations of the corresponding amino acids are indicated around the outside of the chart. For example, all codons that start with the letter U are in the upper-right quadrant of the chart, as indicated by the letter U in the upper right of the inner-most part of the circle. The codons that start with UU are on the left-most portion of the upper right quadrant, as indicated by the letter U in the second concentric circle. The third and fourth rings of the chart indicate that the codons UUU and UUC both code for the amino acid phenylalanine (abbreviated as Phe or F), and the codons UUA and UUG code the amino acid leucine (Leu or L). The full name, three-letter abbreviation, and one-letter abbreviation of each amino acid are listed in Table 6.2.

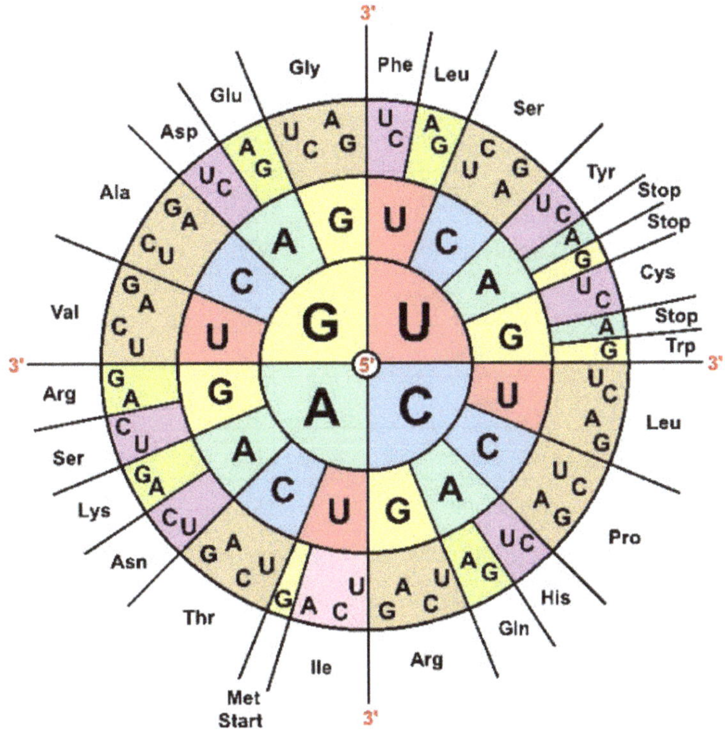

FIGURE 6.5: **Codon chart**—The amino acid encoded by each codon can be determined using this chart.

Note that three codons do not code for an amino acid and are labeled "stop" codons. Unlike the start codon, which signals the addition of a methionine into the polypeptide, stop codons do not code for an amino acid. When the cellular translation machinery reaches a stop codon, a water molecule will be added to the polypeptide chain instead of an amino acid, causing the chain to break and effectively stopping translation at that point. The redundant and unambiguous nature of the genetic code and the significance of stop codons have important implications for genetic mutations, which will be discussed in a later chapter. Thinking ahead, imagine how the effect of a mutation that changed the codon "UAU" to "UAC" would be different from a mutation that changed "UAU" to "UCU" or to "UAG," even though all three changes involve replacing just one letter with a different letter.

Ribosomes are composed of two subunits. The larger subunit is often depicted in illustrations as being on top of the small subunit. Both subunits are made of a type of RNA called ribosomal RNA (rRNA). Unlike mRNA, the purpose of rRNA is not to specify the sequence of amino acids in a polypeptide. Instead, rRNA forms a structure that holds and reads the

TABLE 6.2: Amino Acid Abbreviations

Amino acid	3-letter Abbreviation	1-letter Abbreviation
Alanine	Ala	A
Arginine	Arg	R
Asparagine	Asn	N
Aspartic Acid	Asp	D
Cysteine	Cys	C
Glutamic Acid	Glu	E
Glutamine	Gln	Q
Glycine	Gly	G
Histidine	His	H
Isoleucine	Ile	I
Leucine	Leu	L
Lysine	Lys	K
Methionine	Met	M
Phenylalanine	Phe	F
Proline	Pro	P
Serine	Ser	S
Threonine	Thr	T
Tryptophan	Typ	Y
Tyrosine	Tyr	W
Valine	Val	V

mRNA template and provides a platform for linking together the corresponding amino acids. During translation, the small subunit of the ribosome holds the mRNA template, whereas the large subunit binds to a series of transfer RNAs (tRNAs) (Figure 6.6A). Each tRNA is a small RNA-based structure that contains a region known as an anticodon on one end and is bound to an amino acid on the other end (Figure 6.6B). **Anticodons** are sets of three nucleotides on a tRNA molecule that bind to the complementary nucleotides of a codon in mRNA during translation.

Like DNA, mRNA is read in the 5' to 3' direction. Translation of an mRNA template begins at the part of the large ribosomal subunit known as the aminoacyl-tRNA binding site, or "**A site.**" After this, the mRNA moves to the peptidyl-tRNA binding site, or "**P site,**" then to the exit site, or "**E site.**" For the purpose of remembering the role of each site, it may be helpful to

FIGURE 6.6A: Translation of mRNA into a chain of amino acids—During translation, mRNA moves through the ribosome in the 5′ to 3′ direction.

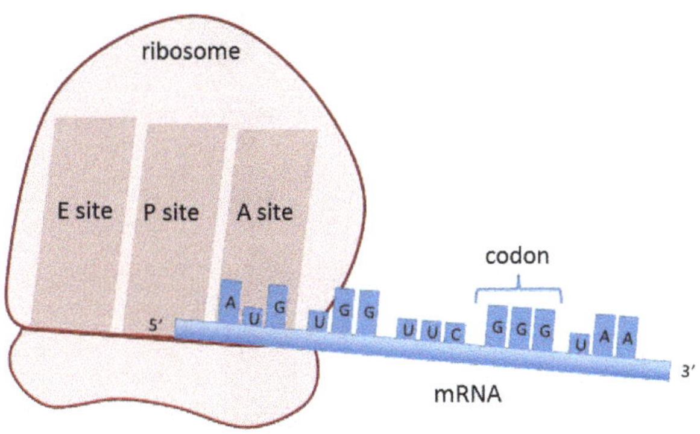

FIGURE 6.6B: Translation of mRNA into a chain of amino acids—When a start codon reaches the A site, the ribosome locates the tRNA with the complementary anticodon.

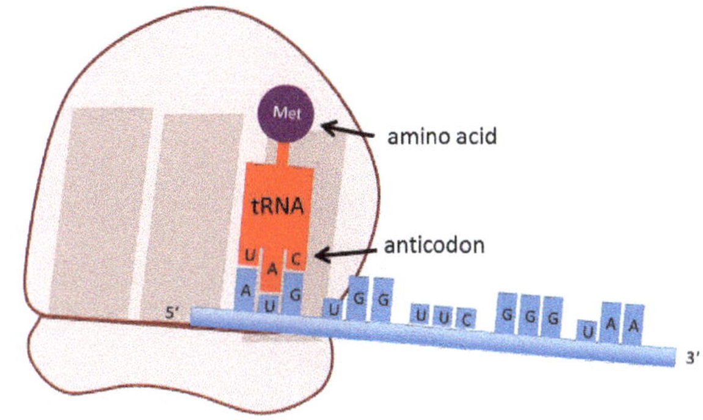

FIGURE 6.6C: Translation of mRNA into a chain of amino acids—After the first codon moves from the A site to the P site, the second codon moves into the A site where the corresponding tRNA is attached.

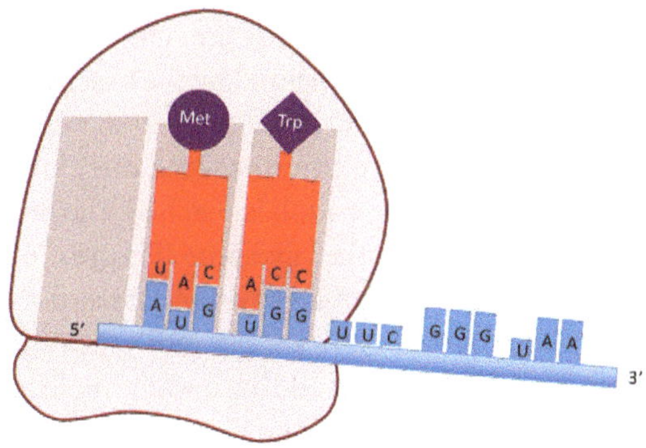

FIGURE 6.6D: Translation of mRNA into a chain of amino acids—Once in the E site, the first tRNA exits the ribosome. The second tRNA, now in the P site, attaches its amino acid to the amino acid left behind by the first tRNA, forming a peptide bond.

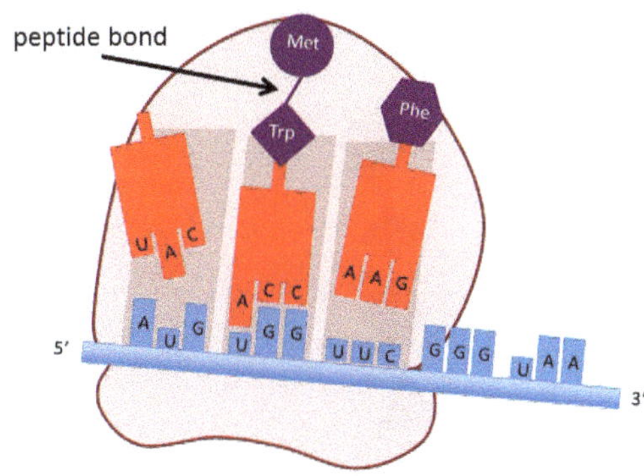

think of the A site as the "attachment site," the P site as the "polypeptide site," and the E site as the "exit site." Translation can be divided into three phases: initiation, elongation, and termination. Initiation begins when a start codon reaches the A site, and the ribosome matches it with the tRNA carrying the complementary anticodon (Figure 6.6B). The tRNA then attaches to the codon that it recognizes and transfers the corresponding amino acid to the newly forming polypeptide. After the first codon moves from the A site to the P site, the second codon will move into the A site, and the corresponding tRNA will attach (Figure 6.6C). The first codon, having moved to the P site, will add its amino acid to the newly forming polypeptide chain (Figure 6.6D). Amino acids in the polypeptide chain are connected to each other by a type of covalent chemical bond known as a peptide bond, hence the names "polypeptide" and "peptidyl

FIGURE 6.6E: Translation of mRNA into a chain of amino acids—As the mRNA moves through the ribosome, the polypeptide chain continues to elongate.

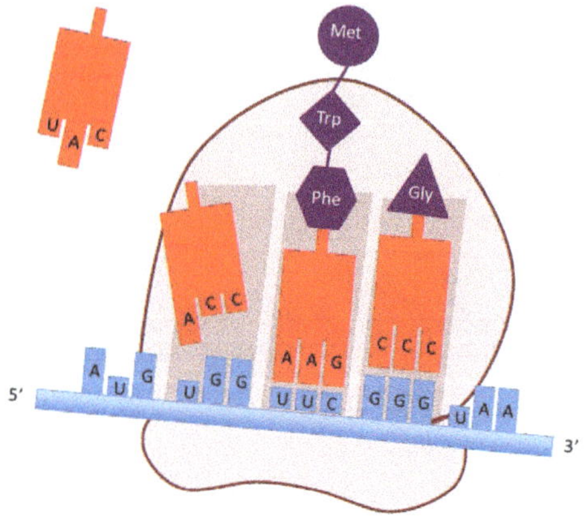

FIGURE 6.6F: Translation of mRNA into a chain of amino acids—When a "stop" codon is reached, the process of translation terminates.

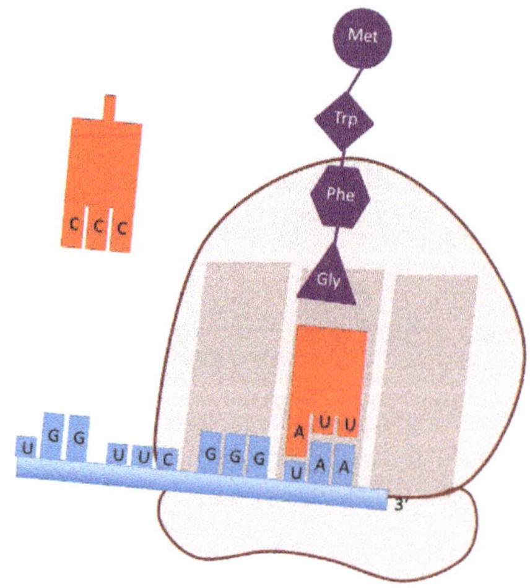

FIGURE 6.6G: Translation of mRNA into a chain of amino acids—After termination, the polypeptide is released into the cytoplasm. The ribosome and tRNA molecules may now be used for the translation of another polypeptide.

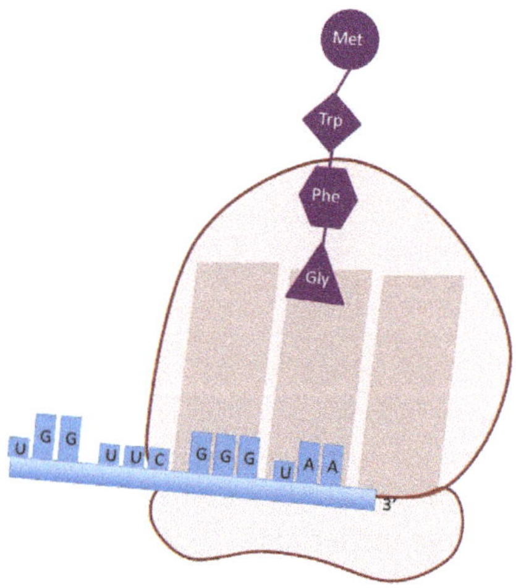

tRNA binding site." Once in the E site, the first tRNA will exit the ribosome, having left its amino acid behind (Figure 6.6E). The second tRNA, now in the P site, will attach its amino acid to the amino acid left behind by the first tRNA. Amino acids will continue to be added to the polypeptide as the mRNA moves through the ribosome during the elongation phase. Once a stop codon is reached, the polypeptide will be released from the ribosome (Figure 6.6F). Stop codons are similarly recognized by their corresponding tRNAs, but those tRNAs carry a water molecule rather than an amino acid. The effect of adding a water molecule to the polypeptide is breaking of the final peptide bond and release of the polypeptide into the cytoplasm (Figure 6.6G). This constitutes the termination phase of translation. However, the tRNA molecules are often recycled after they leave the exit site. Another molecule of the same amino acid is added to the uncharged tRNA so that the same tRNA molecule can be used again and again.

Even after translation is complete, the newly formed polypeptide must undergo additional processing and folding to transform it into a functional protein. These changes, known as posttranslational modifications, may involve modification of certain amino acids through the addition of other molecules, removal of part of the polypeptide, cleavage of the polypeptide into two or more smaller peptides, or the adjoining of multiple polypeptide subunits to form one larger protein. The protein hemoglobin, for example, consists of four polypeptide subunits that must be joined together after translation before gaining functionality.

6.2 SUMMARY

- When making a protein, DNA is first transcribed into mRNA, then edited through RNA processing to remove unnecessary sequences called introns, and add protective sequences to the remaining exons
- Only one strand of the DNA double-helix, the 3' to 5' template strand, is transcribed into RNA so that the newly made RNA can be read in the 5' to 3' direction
- For every A in the template DNA, a U will placed into the transcribed RNA; for every C, a G will be added; for every G, a C will be added, and for every T, an A will be added
- The enzyme that adds nucleotides to the newly formed RNA molecule during transcription is called RNA polymerase
- Translation is the process of making a polypeptide from an mRNA transcript, which occurs in cellular structures called ribosomes
- During translation, the codons of mRNA are recognized by the complementary anticodons of tRNA molecules
- On the opposite end of the tRNA from the anticodon is the corresponding amino acid, which is brought into the ribosome at the A, and attached to the growing polypeptide chain by a peptide bond in the P site
- After leaving attaching their amino acid to the polypeptide, tRNAs leave the ribosome from the E site, and reenter the cytoplasm where they can become recharged with another amino acid and used again for translation

6.3 KNOWLEDGE CHECK

1. What is the name of the process of producing mRNA from a DNA template?
   a. DNA replication
   b. mutagenesis
   c. RNA replication
   d. transcription
   e. translation

2. Which of the following occurs during RNA processing?
   a. exons are removed
   b. introns are removed
   c. template DNA is copied into mRNA
   d. thymines are replaced with uracils
   e. uracils are replaced with thymines

3. If a coding sequence is 21 codons long, including the start codon and the stop codon, how many amino acids would be included in a peptide translated from this coding sequence?
   a. 7
   b. 19
   c. 20
   d. 21
   e. 63

4. Which of the following steps in the making of a protein from a DNA sequence requires anticodons?
   a. mRNA leaves the nucleus and attaches to a ribosome
   b. pre-mRNA transcript undergoes RNA processing
   c. RNA is a transcribed from a DNA template
   d. polypeptide is released from ribosome
   e. succession of tRNAs add their amino acids to the polypeptide chain

5. Identify the amino acid sequence encoded by the following DNA template sequence:

   3'-TACGAGCACACAATTTAC-5'
   a. methionine-leucine-valine-cysteine
   b. valine-asparagine-cysteine-valine-leucine-valine
   c. methionine-leucine-valine-cysteine-isoleucine-methionine
   d. tyrosine-glutamic acid-histidine-isoleucine-tyrosine
   e. methionine-leucine-valine-cysteine-methionine

6. Identify two changes made to RNA during RNA processing.

7. List the main steps that occur during transcription and translation.

8. Predict a gene that is likely to have a strong promoter.

6.4 ACTIVITY

Using the Codon Chart

The RNA codon chart is a useful tool for determining which series of amino acids is specified by the genetic code. Most mRNA sequences that code for polypeptides are more than 1,000 nucleotides long, and in general begin with a start codon and end with a stop codon. However, to gain familiarity with the codon chart, it may be useful to practice with a short sequence. To begin, complete the activity below using the codon chart shown in Figure 6.5 and Table 6.1.

1. First, practice using the codon chart by answering Questions 1–3.
2. Next, answer Question 4 by transcribing the following gene sequence into an mRNA sequence.

   TAA GAG CAA TGC AGT TTT TAC CTC GTA GAG TTT TTT CGT CTA TTT TAA CAG TAA GAG GTA CTC CCC AGT AGT ATG AAA ATG CCC TGC AAA AGT CTC GAG GTC TTG GAG GTC GGA CTA AAA AGT AGT TTT GAG TAC
3. Now, answer Question 5 by translating the mRNA sequence you figured out in the previous step into the corresponding amino acid sequence. Use the one-letter amino acid abbreviations found in Table 6.1.
4. Note: Steps 2 and 3 represent only a small portion of the sequence of a gene. To find out what protein your amino acid sequence is a portion of, you can use the online tool called BLAST. BLAST stands for Basic Local Alignment Search Tool and is a free, online, searchable database containing millions of nucleotide base sequences of genes and amino acid sequences of proteins.

To search BLAST follow these steps:

- Go to https://blast.ncbi.nlm.nih.gov/Blast.cgi.
- Select *protein blast*. In the box labeled "Enter query sequence" enter the one-letter abbreviations of your amino acid sequence. Hit the button labeled BLAST.
- Wait for your results. Be patient it may take a few minutes. You will need to scroll down the page where you will see a list of results to find out what protein your amino acid sequence is a part of.

6.5 ACTIVITY

Modeling Transcription

The purpose of this activity is to demonstrate the process of transcription. Use the 3D Molecular Design model kit to provide a physical representation of the process of transcription through completion of the following hands-on activity. For more background information, consult the following resource: 3D molecular designs. (2016). *DNA replication activity guide.* 3dmoleculardesigns.com.

Note: The nucleotide pieces with rounded bottoms represent DNA nucleotides and the pieces with square bottoms are RNA nucleotides. The arrows on all of the nucleotides point toward the 3' direction.

1. Using the kit, construct a model the following DNA sequence:

   5'-CCGCATGTGTGAGATACATTGGCCAAGACACTGTTAGCTC-3'

2. Construct a model of the complementary sequence to the sequence you just made, producing a double-stranded DNA molecule.

3. Place the double-stranded DNA you just made next to the right side of the "Transcription of DNA into RNA" mat.
4. Slide the double-stranded molecule you just made through the "RNA Polymerase" as instructed by the markings on the mat.
5. Identify which strand is the template strand. Note that RNA polymerase can only assemble mRNA in the 5' to 3' direction.
6. As the template strand emerges from the left side of the RNA polymerase, add complementary RNA nucleotides to form the corresponding mRNA strand.
7. Once the mRNA strand is 10 nucleotides long take a photo or make a sketch showing the entire "Transcription of DNA into RNA" mat, including both strands of DNA, the mRNA, and the RNA polymerase.
8. Continue adding RNA nucleotides until the end of the template strand is reached.
9. Separate the mRNA from the template DNA, and reconnect the two complementary strands of double stranded DNA.
10. Take a photo or make a sketch showing the entire complete strand of mRNA.

6.6 ACTIVITY

Modeling Translation

The purpose of this activity is to demonstrate the process of translation. Use the 3D Molecular Design model kit to provide a physical representation of the process of translation through completion of the following hands-on activity. For more background information, consult the following resource: 3D molecular designs. (2016). *DNA replication activity guide*. 3dmoleculardesigns.com.

Note: The nucleotide pieces with rounded bottoms represent DNA nucleotides, and the pieces with square bottoms are RNA nucleotides. The arrows on all of the nucleotides point toward the 3' direction.

1. Place the mRNA you made in Part 2 next to the right side of the "Translation of RNA into Proteins" mat.
2. Locate all of the tRNA molecules and amino acids in the kit. Attach the amino acids to the tops of the correct tRNAs.
3. Begin sliding the mRNA through the "Ribosome" as instructed by the markings on the mat.
4. Once a start codon reaches the "A" site of the ribosome, find the tRNA with the matching anticodon. Connect the anticodon to the codon.
5. Slide the mRNA through the ribosome until the codon after the start codon is in the A site. Attach the anticodon of the corresponding tRNA to that codon.
6. Slide the mRNA through the ribosome until the next codon reaches the A site. Attach the corresponding tRNA to this codon.

7. The first codon should now be in the "E" site and the second codon should be in the "P" site. Remove the amino acid from the first tRNA and connect it to the second amino acid still attached to the second tRNA. This simulates the formation of a peptide bond between the first two amino acids.
8. Take a photo or make a sketch showing the entire "Translation of RNA into Proteins" mat showing with the first three tRNAs and amino acids.
9. Slide the mRNA through the ribosome until the next codon reaches the A site. Attach the corresponding tRNA. Remove the tRNA attached to the first codon as that codon leaves the "E" site.
10. Continue sliding the mRNA through the ribosome, repeating Steps 7 and 9 with each codon. Because more than one codon can code for the same amino acid, one or more tRNAs may have to be reused. Before reusing a tRNA it must be recharged (an amino acid must be added).
11. When a stop codon is reached, add the corresponding tRNA. Note there is no amino acid attached to this tRNA. Release the mRNA and polypeptide from the ribosome.
12. Take a photo or make a sketch showing the entire polypeptide.

ACTIVITY QUESTIONS

6.4

1. List all of the codons that code for the amino acid serine (Ser).
2. List all of the stop codons.
3. Complete Table 6.3 with the corresponding information.

TABLE 6.3: Understanding Codons

DNA codon	mRNA codon	tRNA anticodon	amino acid
ATC			
	UAC		
		GUA	
			Trp

4. Write the mRNA sequence obtained by transcribing the following DNA sequence:

TAA GAG CAA TGC AGT TTT TAC CTC GTA GAG TTT TTT CGT CTA TTT
TAA CAG TAA GAG GTA CTC CCC AGT AGT ATG AAA ATG CCC TGC AAA
AGT CTC GAG GTC TTG GAG GTC GGA CTA AAA AGT AGT TTT GAG TAC

5. Write the amino acid sequence obtained by translating the mRNA sequence from the previous question.
6. Write the name of the protein (from BLAST results).

6.5

Submit the photos/sketches made in Steps 7 and 10.

1. How can you identify which strand of DNA is the template strand? Why can mRNA only be transcribed from one strand of a DNA double helix?
2. What occurs during the three phases of transcription: initiation, elongation, and termination?

6.6

Submit the photos/sketches made in Steps 8 and 12.

1. Describe what happens in each of the three sites (A, P, E) of the ribosome.
2. Identify the tRNA that was used first in this activity by stating the anticodon and amino acid it was attached to.
3. Which tRNA had to be used more than once during this activity? Why?
4. Why did the tRNA for histidine not need to be reused even though the amino acid chain contained two histidine residues?

6.7 REFERENCES

Clancy, S. & Brown, W. (2008). Translation: DNA to mRNA to protein. *Nature Education, 1*(1), 101.

Knowlton, A. A. (1995). Current concepts in transcription, translation, and the regulation of gene expression. *CHEST Journal, 107*(1), 241–248.

Singer, M. F. (1968). 1968 Nobel Laureate in Medicine or Physiology. *Science, 162*(3852), 433–436.

FIGURE CREDITS

FIG. 6.2: Copyright © Boumphreyfr (CC BY-SA 3.0) at https://commons.wikimedia.org/wiki/File:Mrna_trans2.png.

FIG. 6.5: Copyright © Onie~commonswiki (CC BY-SA 3.0) at https://commons.wikimedia.org/wiki/File:Codons_aminoacids_table.png.

Genetic Mutations and Protein Structure

LEARNING OBJECTIVES

- List common causes of genetic mutations
- Identify examples of different types of genetic mutations, including point mutations, substitution mutations, missense mutations, nonsense mutations, silent mutations, insertions, deletions, and frameshift mutations
- Define the terms hydrophilic and hydrophobic
- Describe the characteristics and chemical bonds associated with the four levels of protein structure
- List factors that may affect protein structure
- Identify an example of how a change in protein structure may affect protein function

7.1 INTRODUCTION

In addition to the numerous ways that cells normally regulate gene expression, genes may be affected by alterations to the genetic code that are not part of normal cellular regulatory mechanisms. These alterations are called mutations. Genetic mutations may be caused by a number of different factors, including mistakes made during DNA replication and as the result of physical or chemical mutagens. The effect of genetic mutations varies greatly, from no noticeable effect on the function or appearance of an organism to catastrophic changes that result in death. This chapter discusses some of causes and effects of genetic mutations, especially as they relate to changes in protein structure.

Causes and Consequences of Genetic Mutations

The term genetic **mutation** refers to any alteration in the nucleotide sequence of an organism's DNA or in the RNA sequence of a virus. Mutations that occur during the lifetime of an individual organism are known as acquired mutations. Many acquired mutations occur randomly as a result of mistakes made during DNA replication. Cells continually employ a variety of methods for detecting and correcting genetic mutations, and most of the time a mutation occurs, it will be successfully fixed or eliminated from the body. However, given the fact that the genetic code of most organisms contains billions of base pairs, and multicellular organisms like humans contain at least one copy of this entire code in trillions of cells within their body, a number of mutations inevitably manage to persist. Acquired mutations can also be caused by environmental factors called mutagens. A **mutagen** is any physical or chemical agent that causes a genetic mutation. Examples of mutagens include ultraviolet radiation from the sun, x-rays, gamma rays, certain chemicals such as asbestos, and even some viruses. In contrast, hereditary mutations are inherited from parents and are passed on through sperm and egg cells (gametes). Because all cells in the body originate from one fertilized egg, any mutations present in that fertilized egg are copied into the DNA of newly arising cells.

The most well-known genetic mutations are those that negatively impact health and decrease the affected individual's fitness or ability to survive and reproduce. Thousands of human diseases are partially or completely the result of genetic mutations. There are also mutations, such as the lactose persistence mutation that allows adult humans to digest the sugar lactose found in dairy products, which actually increase fitness. However, the majority of genetic mutations have no discernible effect on an organism's health or development. There are several reasons for this. First, cells have a number of different mechanisms in place to check for and repair mutations before they disrupt important functions of the cell. Cells containing potentially harmful mutations that cannot be repaired are often eliminated from the body through a process of programmed cell death called **apoptosis**. Second, the majority of an organism's DNA sequence is not necessary for the function of genes or is involved in a redundant or inconsequential function. Mutations that occur in these regions of DNA cause no noticeable effect on observable traits of the organism. Third, many mutations do not result in noticeable alteration of the structure or function of the molecule they encode. Further, genetic mutations may only affect a small number of cells within an organism, depending on where and when the mutation originated. For example, if a skin cell or another cell in the body develops a mutation during the lifespan of the organism, only cells descendent from the cell with the mutation will also have the mutation. Finally, depending on the effect of the mutation, other genetically normal cells with the same or a similar function may be able to compensate for the lack of function from cells containing a mutation. To understand the effect of genetic mutations, it is important to examine mutations in terms of their effect on protein structure, because most genes encode proteins (or portions of proteins). In turn, it is these proteins that direct the vast majority of life-sustaining chemical reactions in the cell, including the synthesis of carbohydrates, lipids, nucleic acids, and other types of molecules.

Types of Mutations

Mutations range in size from those that affect only a single nucleotide base to those that affect large portions of chromosomes containing multiple genes. In terms of the effect on the nucleotide sequence, point mutations are the smallest. **Point mutations** are changes of a single nucleotide pair within a gene. There are number of ways to classify point mutations based on the cause and effect of the mutation. One type of point mutation is a nucleotide-pair substitution. A **nucleotide-pair substitution**, also known as a base-pair substitution or simply a substitution mutation, occurs when a single nucleotide is replaced by a different nucleotide, along with its corresponding partner in the complementary strand of DNA. This type of mutation can be further categorized as missense, nonsense, or silent. A **missense mutation** results in one amino acid being changed to a different amino acid (Figure 7.1). This term refers to the fact that the nucleotide sequence that results from a missense mutation still makes "sense" to the cellular machinery involved in translation; it just makes the wrong type of sense. This may or may not have a noticeable effect on the protein made from that gene, depending on the role of that particular amino acid in maintaining the normal structure and function of the protein. Sickle cell disease, which is described later in this chapter, is one example of a serious and potentially fatal genetic disorder caused by a missense mutation.

If a point mutation changes a codon that codes for an amino acid into a stop codon, it is referred to as a **nonsense mutation** (Figure 7.1). When a nonsense mutation occurs, not only is the amino acid that corresponds to that codon not added to the polypeptide chain, the entire process of translation is terminated prematurely and no more codons are added after the stop

FIGURE 7.1: Examples of point mutations

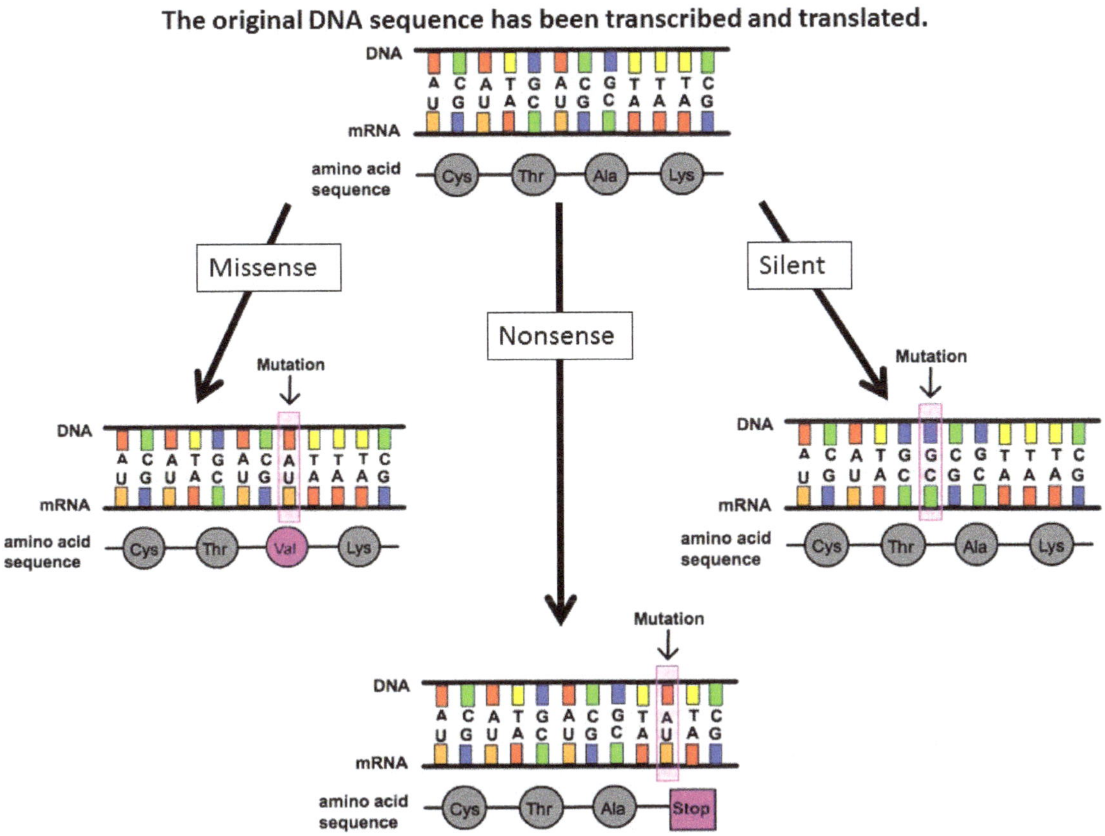

codon. Because this may result in the loss of many important amino acids, nonsense mutations usually result in production of a nonfunctional protein. Alternatively, some point mutations have no observable effect on the amino acid sequence or the resulting protein function or phenotype. These mutations, called **silent mutations,** occur when a codon is changed for another codon that encodes the same amino acid (Figure 7.1). Oftentimes if the third nucleotide base of a codon is different, it will result in a silent mutation. Silent mutations are possible because of a property of the genetic code known as redundancy. This means that, for many amino acids, there are multiple codons that code for it. Redundancy of the genetic code likely evolved as a mechanism to protect against harmful mutations. The fewer times an organism is adversely affected by a genetic mutation, the more likely it is to survive and reproduce.

Two other types of common mutations are insertions and deletions. **Insertions** involve the addition of one or more nucleotide base pairs that do not belong in the sequence, whereas **deletions** involve the removal of one or more nucleotide base pairs that are supposed to be contained in the sequence (Figure 7.2). Like nonsense mutations, insertions and deletions usually have a serious effect on the production of a functional protein because they not only affect the codon in which the mutation occurred but the codons that come afterward as well. This is because insertions and deletions usually disrupt the way the nucleotide code is grouped into codons. Codons are groups of three nucleotide bases that signal which amino acid should

FIGURE 7.2: Example of insertion and deletion mutations

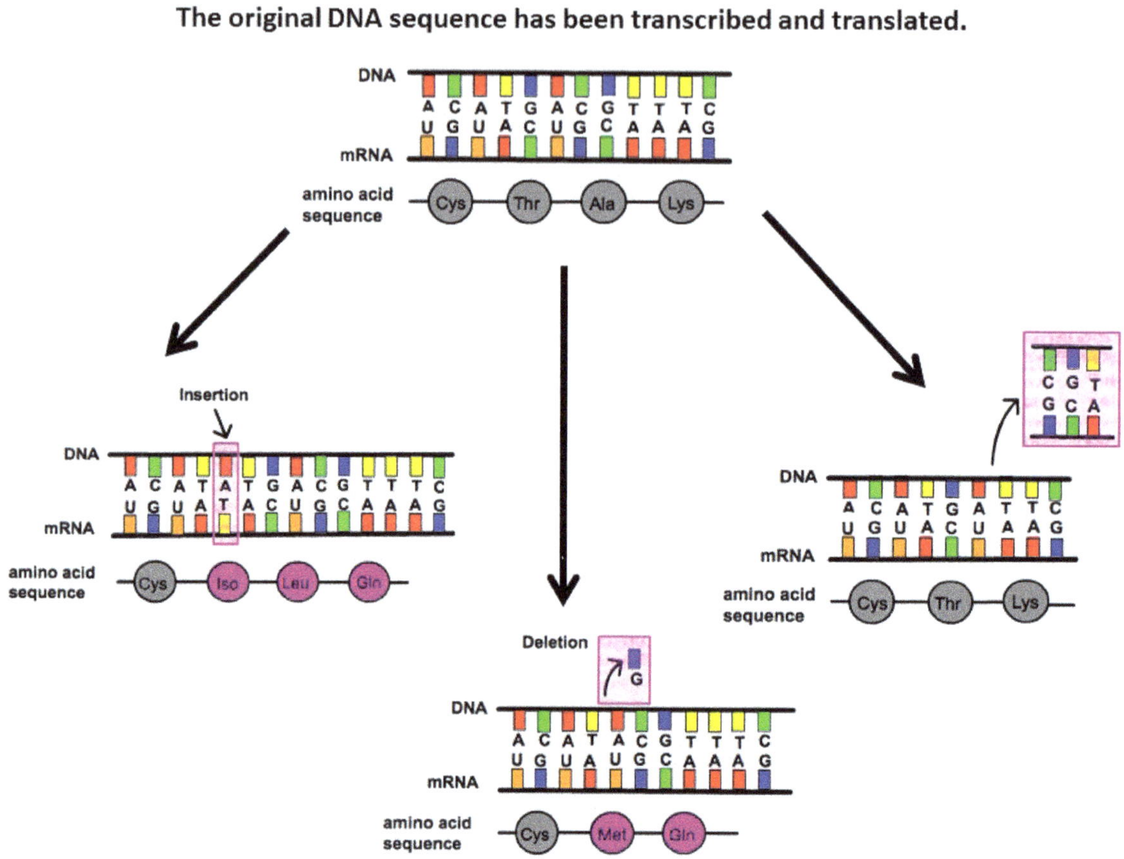

be added to the growing polypeptide during the translation of mRNA into protein. When the grouping of codons is disrupted, it is known as a **frameshift** mutation, because it changes the reading frame in which the genetic code is translated. Unless it occurs near the very end of a coding sequence, a frameshift mutation usually results in a nonfunctional protein. One exception to this rule is when a multiple of three nucleotides is added or deleted in between codons preserving the reading frame. In these situations, the only change is an addition or removal of one or a small number of codons, which only affects one or a few amino acids.

Levels of Protein Structure

Although the majority of genetic mutations do not exert any noticeable change to the organism, some mutations do result in harmful or beneficial changes to the function of the affected organism. This effect begins when changes in the nucleotide sequence of a gene lead to changes in the amino acid sequence of the polypeptide produced from that gene. If the change to the polypeptide affects the structure of the protein, it will likely affect the function of the protein. The three-dimensional shape of a protein is a critical determinant of its function because most proteins work by recognizing or attaching to the specific three-dimensional shape of other molecules. The shape of all proteins is based on three superimposed levels of structure, known as the

FIGURE 7.3: **Four levels of protein structure**—Primary, secondary, tertiary, and quaternary protein structure represent increasingly complex three-dimensional formations stabilized by various chemical bonds.

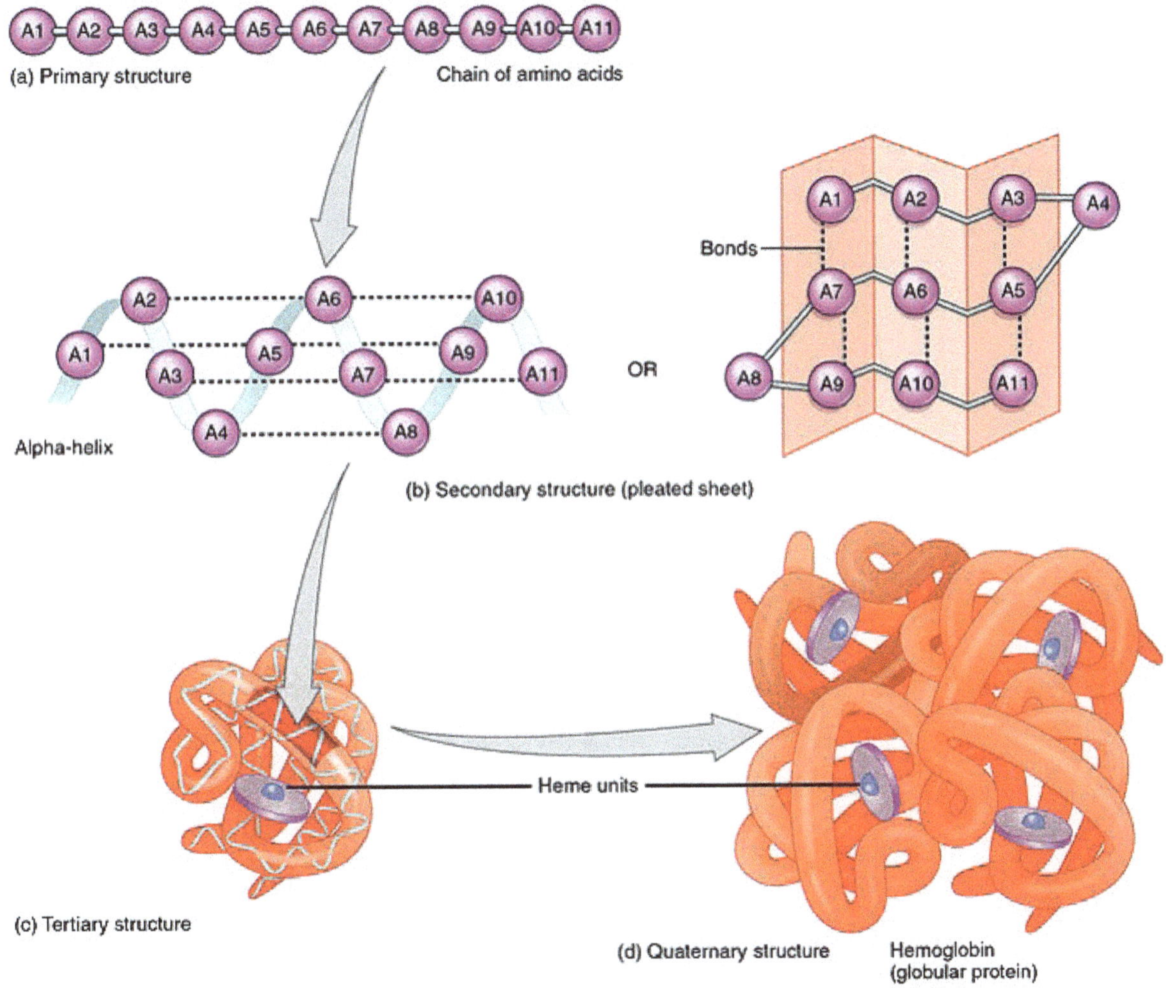

primary structure, secondary structure, and tertiary structure (Figure 7.3). Some proteins can also be described as having a fourth, or quaternary, structure (Figure 7.3). Each level of protein structure has particular physical characteristics stabilized by specific types of chemical bonds.

The **primary structure** of proteins consists of a linear sequence of amino acids, each joined together by strong type of covalent bond known as a peptide bond (Figure 7.4). The sequence of amino acids is directly determined by the sequence of nucleotides present in the gene. Therefore, it is the primary structure of a protein most directly affected by missense and nonsense mutations. Peptide bonds form between the carbon in a carboxyl group of one amino acid and the nitrogen of an amino group of another amino acid (Figure 7.4). Each amino acid also contains an R group, which is a unique configuration of atoms that confers specific chemical properties to the amino acid (Figure 7.5). Amino acids are classified into three main groups based on the chemical properties conferred by their R groups: nonpolar, polar, and electrically charged (Figure 7.5).

FIGURE 7.4: **Primary structure of a protein is held together by peptide bonds**—The primary structure of a protein is a chain of amino acids held together by peptide bonds.

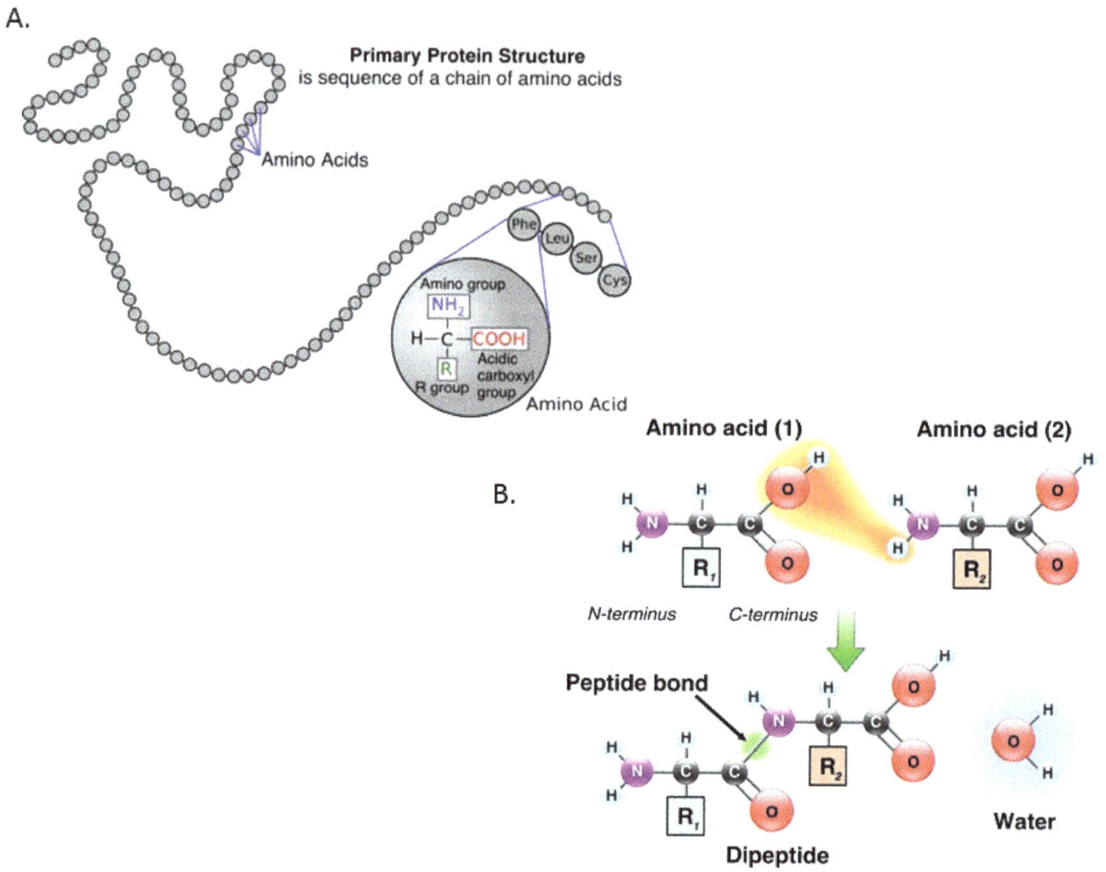

Amino acids with R groups in which there is no equal sharing or transfer of electrons between atoms are nonpolar. Nonpolar molecules are **hydrophobic,** meaning they tend to repel polar molecules like water and will not mix well with water. The R groups of polar amino acids contain some atoms that unequally share electrons within the chemical bonds that hold them together. This unequal sharing of electrons causes the part of the R group that attracts negatively charged electrons more strongly to take on a slightly negative charge, whereas the part of the R group that attracts the electron less strongly takes on a partial positive charge. The negatively charged portions of polar molecules are attracted to the positively charged portions of other polar molecules, and vice versa. Because of the unequal sharing of electrons between oxygen atoms and hydrogen atoms, water molecules have both a partially negative side (near the oxygen atom) and a partially positive side (near the hydrogen atoms).

As a result of this attraction, polar amino acids will mix with, or dissolve in, polar water molecules and are therefore referred to as **hydrophilic**. Amino acids with R groups in which atoms have transferred electrons between each other are electrically charged. Atoms that

FIGURE 7.5: **Amino acid structures**—The chemical properties of amino acids are derived from their unique R group side chains.

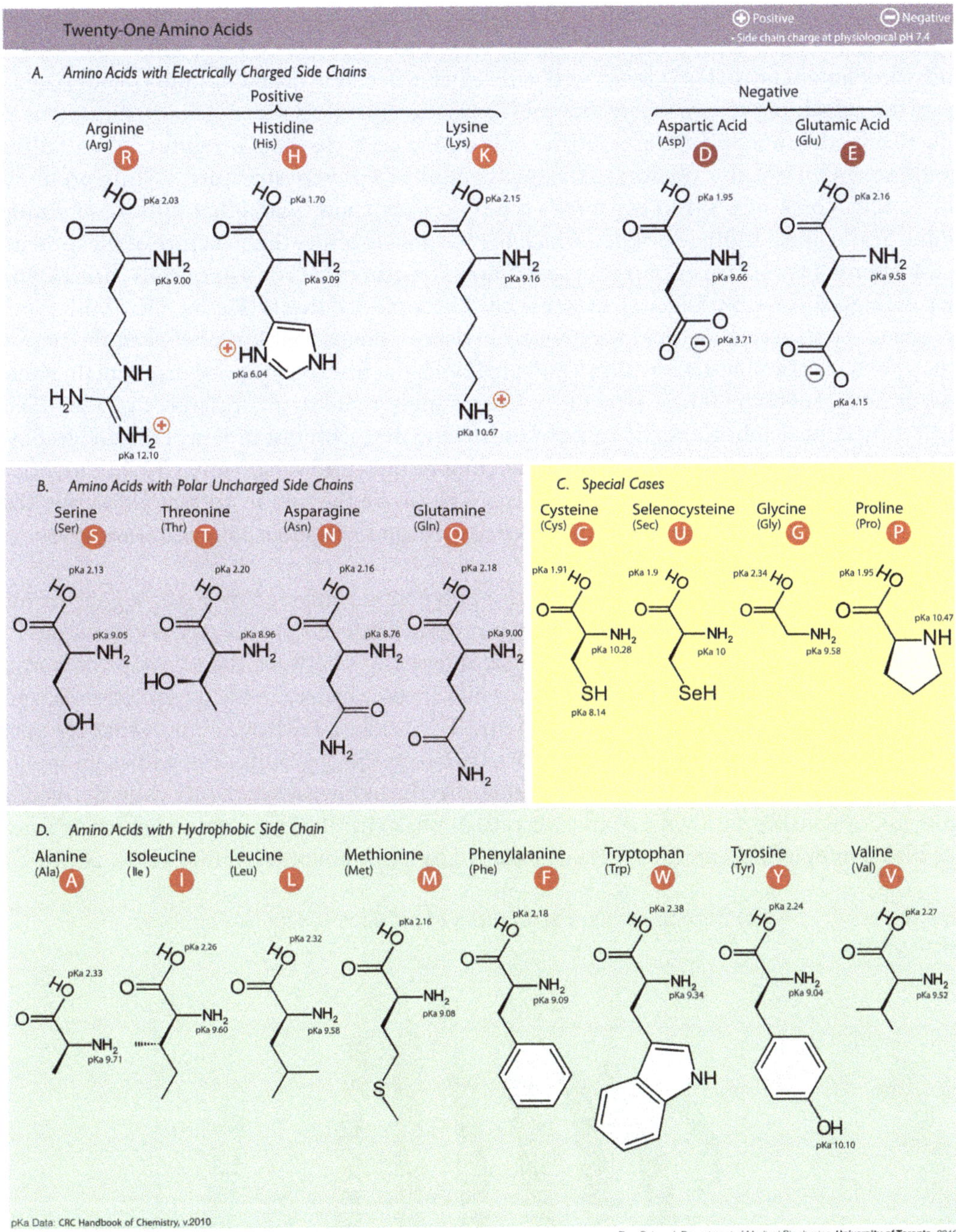

Dan Colocari, Department of Medical Biophysics, **University of Toronto**, 2010

have lost an electron to another atom are positively charged because of the loss of a negatively charged particle from a molecule that had an overall neutral charge. Conversely, atoms that have gained an electron from another atom are positively charged. In addition to being electrically charged, all positively and negatively charged amino acids are hydrophilic, because they are attracted to the oppositely charged portion of water molecules.

Because water is nearly ubiquitous in both the interior and exterior of cells within living organisms, nonpolar amino acids are usually clustered on the inside of proteins because they are hydrophobic and do not interact well with water, whereas hydrophilic amino acids tend to be on the outside of proteins where they are likely to interact with water. It is also important to note that methionine and cysteine are the only amino acids that contain sulfur atoms. Sulfur atoms are important contributors to the stabilization of a protein structure. Within proteins, amino acids containing sulfur can form a type of covalent bond with other sulfur-containing amino acids called disulfide bridges, which help to hold together the structure of the protein.

Most proteins contain examples of **secondary structure** that consists of coils called alpha helices (Figure 7.6A) and folded structures called beta-pleated sheets (Figure 7.6B). Both types of secondary structure are held together by hydrogen bonds that form between the non-R group portions of amino acids within a polypeptide chain. Each polypeptide may contain none, one, or several of each of these structures. Interestingly, these structures were first identified in 1951 by Linus Pauling and his collaborators using the technique of x-ray crystallography. Around the same time, Pauling's team was competing with Watson and Crick to identify the structure of DNA. Although Watson and Crick were the first to correctly describe the DNA double helix, their seminal work relied heavily on knowledge obtained from studying Pauling's work.

The three-dimensional shape of an entire polypeptide encoded by a single gene forms what is known as the **tertiary structure of a protein** (Figure 7.3). The tertiary structure of proteins is formed and stabilized by chemical interactions between the R groups of amino acids within a polypeptide. These interactions may include hydrogen bonds, ionic bonds, and various types of covalent bonds, including disulfide bridges. Hydrogen bonds that are part of tertiary protein structure most often form between the R groups of nonadjacent, polar amino acids. Ionic bonds can form between positively and negatively electrically charged R groups. Covalent bonds, as well as another type of weak chemical interaction known as a van der Waals interaction, can form between the R groups of nonpolar amino acids. However,

FIGURE 7.6A: Secondary structures of proteins—Alpha helices

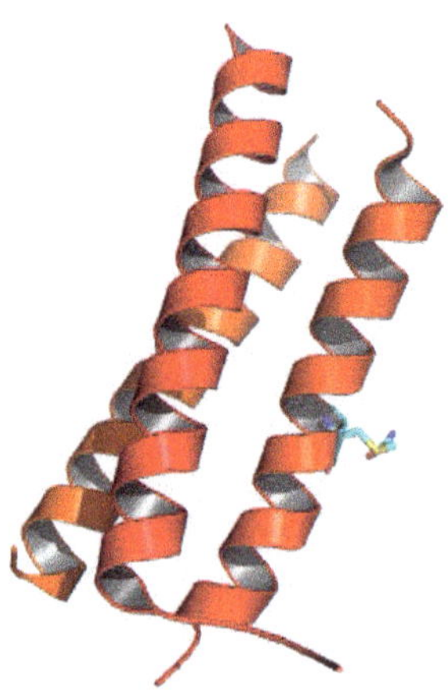

FIG 7.6B: Secondary structures of proteins—Beta-pleated sheets

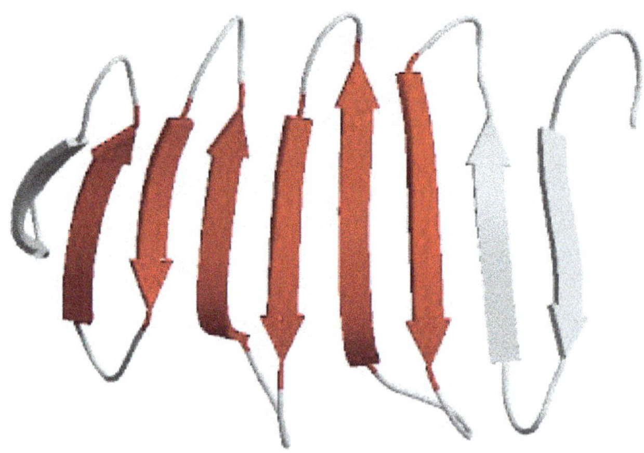

disulfide bridges can only form between the R groups of cysteine and methionine, because those are the only amino acids that contain sulfur.

For proteins that consist of a single polypeptide chain, tertiary structure represents the highest level of structural organization. However, some proteins are formed by multiple polypeptide chains chemically bonded together to function as a single unit (Figure 7.3). In these instances, each polypeptide is referred to as a subunit. The hydrogen, ionic, and covalent bonds that stabilize the interactions between subunits form the **quaternary structure** of a protein.

Changes in Structure Affect Protein Function

One example of how even a small change in protein structure can exert a significant effect on human health is the case of sickle cell disease. Sickle cell disease is an inherited disorder characterized by malformed red blood cells. Normal red blood cells are similar to disk-shaped water balloons, or beanbag chairs, filled with many individual hemoglobin proteins that do not associate with each other. Sickle cell disease results from a single base pair substitution in the DNA sequence that encodes part of the hemoglobin protein known as the beta subunit. This leads to a change of one codon in the mRNA that is transcribed from that DNA, and the subsequent substitution of one amino acid for a different amino acid. This is an example of a missense mutation. Although only 1 out of more than 100 amino acids is affected, this change is significant because it involves the replacement of a hydrophilic glutamate on the exterior of the hemoglobin beta subunit by a hydrophobic valine. Because proteins are surrounded by water both in and outside of cells, polypeptides are typically folded into tertiary and quaternary structures that conceal hydrophobic amino acids in the interior of the protein, leaving only hydrophilic amino acids exposed to water on the protein's exterior.

In individuals with sickle cell disease, hemoglobin molecules form long chains to cover the hydrophobic exterior of a beta subunit on one hemoglobin molecule with the hydrophobic region of another hemoglobin molecule. These long chains deform the shape of red blood cells, much like putting long poles inside a balloon or beanbag chair would deform the shape of those objects. These cells take on a crescent, or sickle-shape, and may clog and damage tiny blood vessels (Figure 7.7). This disrupts the function of hemoglobin, which is to transport oxygen molecules through the bloodstream from the lungs to cells throughout the body, and may lead to potentially painful and serious tissue damage. Symptoms of full-blown sickle cell disease include pain, weakness, damage to the brain and other organs, and paralysis. Further, pregnant women are at increased risk of having a miscarriage if they have sickle cell disease. Only individuals who have inherited two copies of the sickle cell mutant allele experience the severe symptoms of full-blown sickle cell disease. Heterozygotes who have one normal copy of the hemoglobin gene and one mutated copy usually experience mild or no symptoms and are generally healthy. These individuals are said to have the sickle cell trait but not sickle cell disease.

Because sickle cell disease may lower the ability of individuals to survive and reproduce, one might expect the disease-causing allele to be increasingly rare or to even to have been eliminated from the population entirely through the selective pressures of evolution. However, this mutation is relatively common in certain populations. For example, approximately 10% of African-Americans carry the sickle cell trait, and one out of 400 have sickle cell disease. A potential explanation for the unexpected frequency of this mutation is that it may confer some evolutionary advantage that counteracts its negative effects. The most likely benefit conferred by the sickle cell trait is that being heterozygous for the mutation leads to increased resistance

FIGURE 7.7: Sickle cell anemia—Sickle cell anemia is caused by a missense mutation that affects the hemoglobin protein.

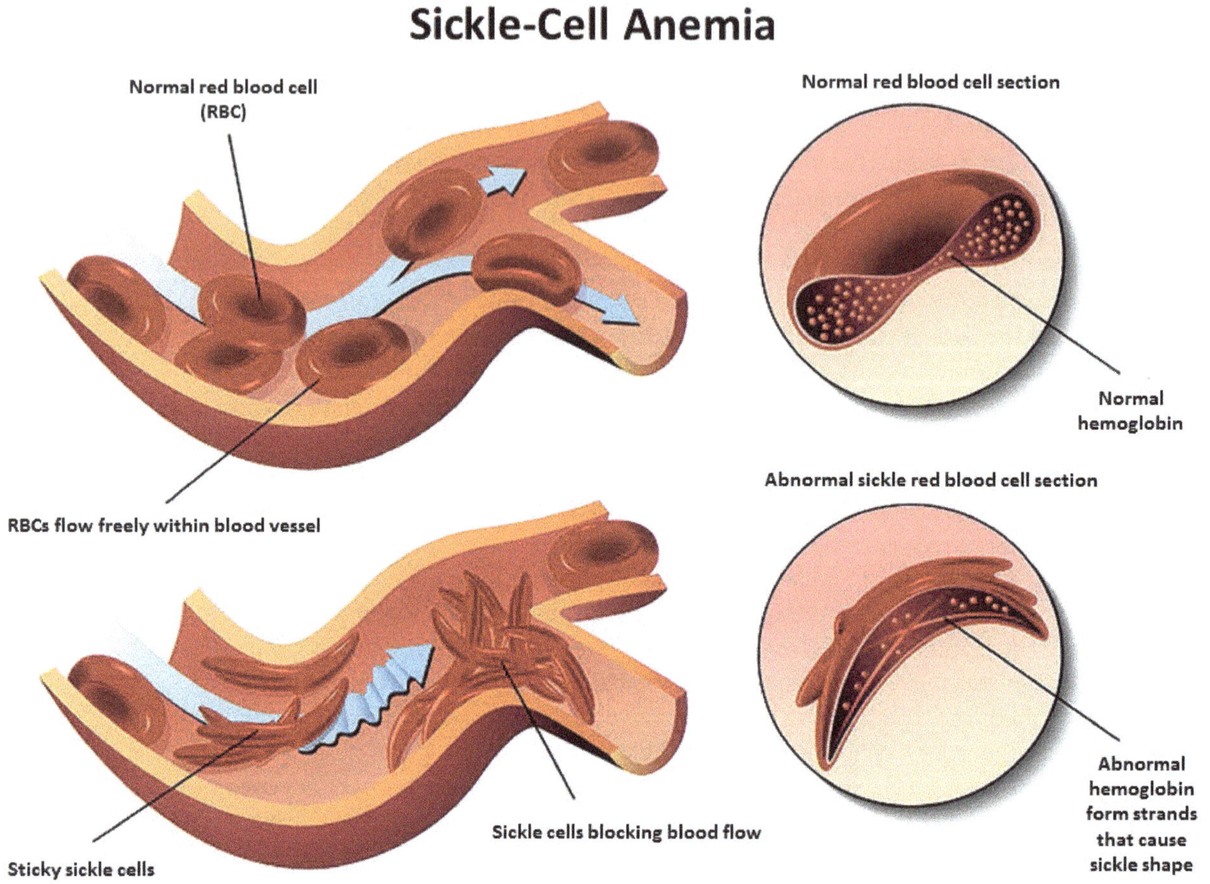

to malaria, especially in children. The connection between sickle cell and malaria lays in the fact that malaria-causing parasites carry out part of their life cycle in red blood cells but have difficulty doing so if those cells are sickle-shaped. This relationship explains why the sickle cell trait is most common in populations whose ancestors lived in Africa, where malaria has long since been common.

Other Factors That Affect Proteins

Even if there are no mutations in the genetic code, protein structure and function can still be affected by many factors within the intracellular and extracellular environment. Proteins are usually folded into the proper conformation immediately after synthesis within cells. Protein folding can occur spontaneously or may require the action of other proteins. Cells have evolved to produce proteins that function optimally within the normal intracellular or extracellular environment. Changes to the pH, salt concentration, and temperature of the local environment can destroy the weak chemical interactions that stabilize tertiary and quaternary protein structures, resulting in alteration or inhibition of protein function. Depending on the function of the protein, alterations in protein function can have significant effects on the health of the organism that relies on the protein. When proteins lose their natural quaternary, tertiary, or secondary structure because of a change in external conditions, a process known as **denaturation** occurs. Denaturation can be observed when cooking meat or egg whites. The structures of proteins found in these foods formed within a narrow range of external conditions, including temperature. When heated beyond their normal temperature range, the chemical bonds that hold the proteins together break, causing the protein to lose its natural three-dimensional structure, or denature. Denaturation often causes hydrophobic amino acids, which are normally concealed on the interior of a protein, to become exposed to the water-containing external environment. The hydrophobic amino acids will clump together to shield themselves from water molecules. This results in the change in color and texture of meat and eggs that occurs during cooking.

7.2 SUMMARY

- Mutations may occur spontaneously during DNA replication or can be caused by exposure to mutagens such as radiation and certain chemicals
- Substitution mutations can be silent (do not change amino acid), missense (change to a different amino acid), or nonsense (change to a stop codon)
- Insertion or deletion of nucleotide bases often results in a frameshift that changes all codons after the mutation
- Protein function is dependent upon structure
- The primary level of protein structure is composed of the amino acid sequence; secondary, tertiary, and quaternary levels of protein structure include increasingly complex three-dimensional formations stabilized by various types of chemical bonds

7.3 KNOWLEDGE CHECK

1. Which of the following statements about genetic mutations is true?
   a. genetic mutations always result in a decrease in fitness
   b. genetic mutations always result in an increase in fitness
   c. genetic mutations never affect fitness
   d. genetic mutations may decrease or have no effect on fitness, but never increase fitness
   e. genetic mutations may decrease, increase, or have no effect on fitness

2. A mutation that changes a codon that codes for one amino acid into a codon that codes for a different amino acid is called a:
   a. duplication
   b. frameshift mutation
   c. missense mutation
   d. nonsense mutation
   e. silent mutation

3. Identify the type of mutation illustrated here.

   Original gene: 3'-ACAGACTTTCAC-5'
   Gene with mutation: 3'-ACAGAGTTTCAC-5'
   a. missense mutation
   b. silent mutation
   c. nonsense mutation
   d. frameshift mutation
   e. translocation

4. Which amino acid side chain has chemical properties most similar to the properties of the side chain of serine?
   a. arginine
   b. leucine
   c. lysine
   d. threonine
   e. valine

5. Which of the following characterizes primary protein structure?
   a. alpha helices and beta pleated sheets
   b. amino acid sequence
   c. DNA double helix
   d. hydrogen, ionic, and covalent bonds that stabilize subunit
   e. three-dimensional shape of an entire polypeptide

6. List three causes of genetic mutations.

7. Rank the following in order from most advantageous to least advantageous and explain your reasoning: homozygous normal for hemoglobin gene, homozygous for sickle cell trait, heterozygous for sickle trait.

8. Some genetic mutations are silent, meaning they do not result in a change in the amino acids sequence. Explain how this is possible. Then, predict how this property of the genetic code might be advantageous evolutionarily.

7.4 ACTIVITY

Researching a Genetic Disorder

Genetic disorders are diseases that result from mutations that can be passed from parent to child. Different kinds of mutations, such as point mutations, insertions, and deletions, can change the amino acid sequence, leading to a structural change in the protein. If the structure of a protein is changed, then the function is also likely to change. Not all mutations result in disease, but they are often examined together because they demonstrate the connections between genes and protein structure and function. Select a genetic disorder from the following list and prepare a short presentation or paper focusing on the connection between the gene mutation, the structure, and function of the protein involved, and how that affects the function of a person with the disorder. Follow all guidelines provided.

List of genetic disorders:

Tay-Sachs (early onset)
Cystic fibrosis
Huntington Disease
Duchenne Muscular Dystrophy
Marfan Syndrome
Huntington Disease
Edwards Syndrome
Hemophilia A

Be sure to include the following information in your paper or presentation:

- What gene is affected and where is it located? OR what chromosome is affected?
- What kind of mutation is involved?
- What is the effect on the protein's amino acid sequence? OR what happens to the number or structure of the chromosome?
- What is the normal function of the affected protein(s) in the body? How does the gene mutation specifically affect the protein's structure and function?
- What are the symptoms of the disorder and how is it diagnosed? How common is the disorder and is the frequency of occurrence associated with a particular religious or racial group or sex?
- What, if any, treatments are available?
- What are three other interesting facts about this disorder (e.g., who first described it, associations with historical figures, etc.)?

Feel free to include diagrams, tables, or figures in your paper to enhance your discussion of structures and/or data. Be sure to cite your sources used. One reliable Internet source to get you started is: Genetics Home Reference (https://ghr.nlm.nih.gov/).

7.5 ACTIVITY

Modeling the Levels of Protein Structure

A protein's shape ultimately determines its function. Enzymes, like RNA polymerase, are proteins that are able to recognize and attach to specific molecules so they can help break down or build other molecules. If the shape of an enzyme is altered, it will not be able to recognize the molecule it needs to work on and cannot do its job properly or at all. This holds true for all proteins, not just enzymes. If the overall shape of any protein is altered because of a gene mutation, then the function will be altered in some way. The relationship between structure and function is an important theme that shows up in biology time and time again, from the molecular level as we have discussed here, up to the level of organs and organ systems.

The original version of this activity was developed at the Milwaukee School of Engineering Center for BioMolecular Modeling (http://www.3dmoleculardesigns.com/3DMD-Files/4-Toober--Mini-Toober/MiniToober Handout.pdf?). The 4-foot long, flexible foam Modeling Mini Toobers used in this activity are available for purchase here: http://www.3dmoleculardesigns.com/Education-Products/Modeling-Mini-Toobers.htm.

Work in groups of two to complete this activity.

1. Obtain a protein folding kit, which contains a mini toober and 11 colored pushpins (four yellow, three white, two green, one red, and one blue).
2. Pick one of the short gene sequences here, and enter it into the data table. Be sure to indicate the letter of the gene sequence you chose in the data table.
   a. TACGGACGTCTGCCCAGAACATTTTAAACAGTGACT
   b. TACAGGCCACTTGGAACAGCCACCGTGACGCATACT
   c. TACTTTACGGAAGTATCACCACTAACACGACAAATT
   d. TACACAGGGGGGACGCATCTAAGGTGAGCCACCATC
3. Transcribe and translate the gene sequence into the amino acid sequence of the protein. It will be helpful to put three nucleotide bases in a box for codons and amino acids. Use the genetic code wheel provided to translate the codons to amino acids (Figure 7.8).
4. Use the information provided in Table 7.1 to complete the data table.
5. Now you can build your protein. The toober represents the backbone of your protein and the peptide bonds holding together your amino acid sequence. Stretch out your toober and, starting at the left end, push the colored pushpin that corresponds to the color of your first amino acid into the toober. Don't poke yourself! Continue to add the pushpins that correspond to all your amino acids. Be sure to use the full length of the toober.
6. Just like a real protein, you will need to fold your protein by following the rules listed in Table 7.2. These rules are based on the properties of the amino acid side chains and how they react to each other and the watery environment the protein is located in. Remember your protein is a 3D object, so you can fold it in any direction as long as you follow the rules.
7. When you have finished folding your protein, take a picture so you can remember the shape of your protein. Compare your protein with another group in your class.

FIGURE 7.8: **Genetic code wheel**—Start in the center with the first nucleotide base of the codon, then out to next circle for the second base and the outermost circle for the third base.

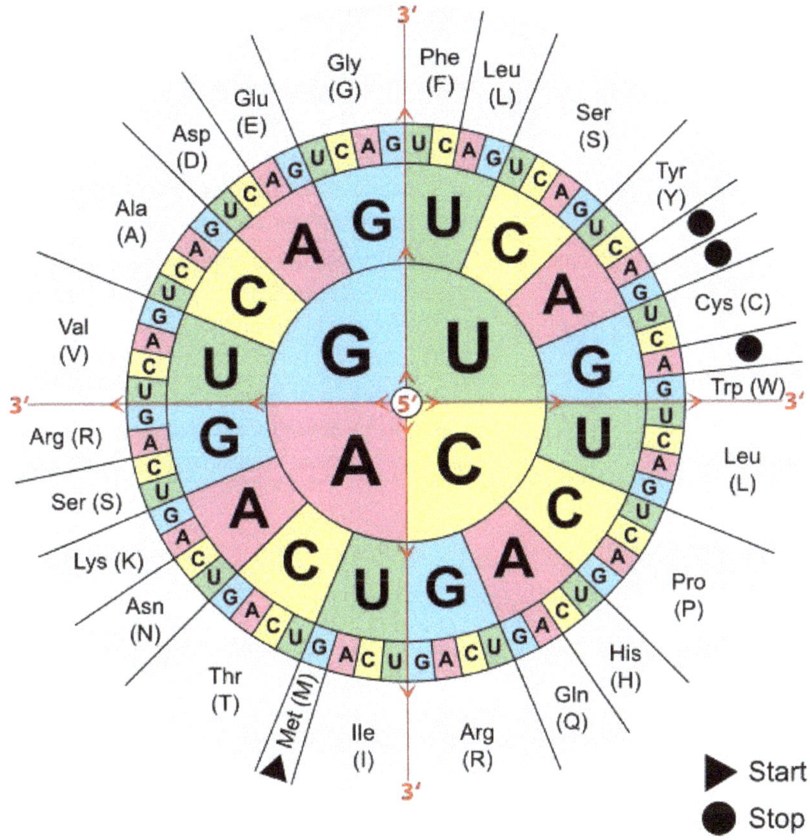

TABLE 7.1: Color-coding Amino Acids by Chemical Property

Amino acid	Property	Color	Amino acid	Property	Color
Alanine (Ala)	hydrophobic	yellow	**Leucine (Leu)**	hydrophobic	yellow
Arginine (Arg)	basic	blue	**Lysine (Lys)**	basic	blue
Asparagine (Asn)	hydrophilic	white	**Methionine (Met)**	hydrophilic	white
Aspartic acid (Asp)	acidic	red	**Phenylalanine (Phe)**	hydrophobic	yellow
Cysteine (Cys)	cysteine	green	**Proline (Pro)**	hydrophobic	yellow
Glutamic acid (Glu)	acidic	white	**Serine (Ser)**	hydrophilic	white
Glutamine (Gln)	hydrophilic	red	**Threonine (Thr)**	hydrophilic	white
Glycine (Gly)	hydrophobic	yellow	**Tryptophan (Trp)**	hydrophobic	yellow
Histidine (His)	hydrophilic	white	**Tyrosine (Tyr)**	hydrophilic	white
Isoleucine (Ile)	hydrophobic	yellow	**Valine (Val)**	hydrophobic	yellow

TABLE 7.2: Protein Folding Rules

Rule	Reason
Yellow will be located on the inside of the protein, away from the surface.	Hydrophobic amino acids "hate" water so they need to be as far away as possible from the watery environment surrounding the protein.
White will be on the surface of the protein.	Hydrophilic amino acids "like" water so they need to be on the surface of the protein, which is close to the watery environment surrounding the protein.
Red and blue are attracted to each other.	Red (acidic, positive charge) and blue (basic, negative charge) are attracted to the opposite charges. They will also be found on the surface of the protein since they are hydrophilic. Their pairs form what is called a salt bridge.
Green will pair up with other greens.	The amino acid cysteine contains Sulfur in its side chain. The sulfur atoms are attracted to other sulfur atoms and form pairs making what is called a disulfide bridge.

ACTIVITY QUESTIONS

7.4

Before beginning your paper or presentation, prepare an outline that answers each of the following questions.

1. What gene is affected and where is it located? OR what chromosome is affected?
2. What kind of mutation is involved?
3. What is the effect on the protein's amino acid sequence? OR what happens to the number or structure of the chromosome?
4. What is the normal function of the affected protein(s) in the body? How does the gene mutation specifically affect the protein's structure and function?
5. What are the symptoms of the disorder and how is it diagnosed? How common is the disorder and is the frequency of occurrence associated with a particular religious or racial group or sex?
6. What, if any, treatments are available?
7. What are three other interesting facts about this disorder (e.g., who first described it, associations with historical figures etc.)?

7.5

1. Complete Table 7.3.
2. What levels of protein structure did you model during the protein folding activity?

TABLE 7.3: Data Table

Gene Sequence												
mRNA Codons												
Amino Acid												
Side Chain Property												
Side Chain Color												

3. For the level of protein structure you did not model, describe a way it could be modeled.
4. What kinds of bonds did you simulate holding together your protein?
5. Describe what would happen to the shape and function of your protein if there was a point mutation that resulted in a stop codon in the middle of your amino acid sequence.

7.6 REFERENCES

Aidoo, M., et al. (2002). Protective effects of the sickle cell gene against malaria morbidity and mortality. *The Lancet, 359*(9314), 1311–1312.

Bethel, C. M., and Lieberman, R. L. (2014). Protein structure and function: An interdisciplinary multimedia-based guided-inquiry education module for the high school science classroom. *Journal of Chemical Education, 91*(1), 52–55.

Chial, H. (2008). Rare genetic disorders: Learning about genetic disease through gene mapping, SNPs, and microarray data. *Nature Education, 1*(1), 192.

Clancy, S. (2008). Genetic mutation. *Nature Education, 1*(1), 187.

Elmore, S. (2007). Apoptosis: A review of programmed cell death. *Toxicologic Pathology, 35*(4), 495–516.

Knowlton, A. A. (1995). Current concepts in transcription, translation, and the regulation of gene expression. *CHEST Journal, 107*(1), 241–248.

FIGURE CREDITS

FIG. 7.3: Copyright © OpenStax College (CC by 3.0) at https://commons.wikimedia.org/wiki/File:225_Peptide_Bond-01.jpg.

FIG. 7.4A: National Human Genome Research Institute, "Primary structure of a protein is held together by peptide bonds," https://commons.wikimedia.org/wiki/File%3AProtein_primary_structure.svg. Copyright in the Public Domain.

FIG. 7.4B: YassineMrabet, "Primary structure of a protein is held together by peptide bonds," https://commons.wikimedia.org/wiki/File%3APeptidformationball.svg. Copyright in the Public Domain.

FIG. 7.5: Copyright © Dan Cojocari (CC BY-SA 3.0) at https://commons.wikimedia.org/wiki/File%3AAmino_acids.png.

FIG. 7.6A: Copyright © Paul RHJ Timmers (CC by 4.0) at https://commons.wikimedia.org/wiki/File:Alphabody1.png.

FIG. 7.6B: Xenonblast, "Secondary structures of proteins," https://commons.wikimedia.org/wiki/File%3ABeta-meander1.png. Copyright in the Public Domain.

FIG. 7.7: Copyright © Diana grib (CC BY-SA 4.0) at https://commons.wikimedia.org/wiki/File:Risk-Factors-for-Sickle-Cell-Anemia_(1)2.jpg.

FIG. 7.8: Mouagip, "Genetic code wheel," https://commons.wikimedia.org/wiki/File%3AAminoacids_table.svg. Copyright in the Public Domain.

Mendelian Genetics

LEARNING OBJECTIVES

- Describe the details and significance of the experiments Gregor Mendel conducted with pea plants
- Identify the genotypic and phenotypic ratios of the F1, F2, and P generation in Mendel's experiments
- Define the following terms: allele, carrier, heterozygous, homozygous, locus, and true-breeding
- Discuss the meaning of Mendel's laws of dominance, independent assortment, and segregation
- Calculate the predicted frequency of a Mendelian trait in a future generation using a Punnett square
- Apply the multiplication rule to determine the probability of inheriting more than one Mendelian trait
- Identify examples of physical traits in humans that follow Mendelian patterns of inheritance
- Identify an example of an autosomal dominant disorder or disease, and an example of an autosomal recessive disorder or disease
- List reasons why some traits do not follow the rules of Mendelian inheritance

8.1 INTRODUCTION

Gregor Johann Mendel was an Augustinian friar, scientist, and teacher who lived in the town of Brno, located in what is now known as the Czech Republic, or Czechia (Figure 8.1). Although not well known during his lifetime, Mendel is now widely recognized as the Father of Genetics. In the 1850s and 1860s, he demonstrated that the inheritance of observable characteristics in pea plants follows predictable patterns. Based on his observations, Mendel proposed that the transmittance of traits from one generation to the next results from the inheritance of discrete factors, now known as genes, which follows certain patterns. Remarkably, Mendel proposed his highly accurate model of inheritance without knowing about chromosomes or DNA, which were not identified until decades later. Much of the knowledge gained from Mendel's work is still useful today and helps explain how a variety of traits are inherited in many different organisms, including humans. This chapter discusses Mendel's experiments, how they shaped our knowledge of genetics, and how we can apply this knowledge to predicting inheritance patterns for certain genetic traits.

FIGURE 8.1: Gregor Mendel—
Gregor Mendel, 1822–1884

Gregor Mendel's Model

During his life, Mendel's work was largely ignored by many scientists who incorrectly believed that inheritance of traits involved a blending of parental characteristics into an intermediate form observed in offspring. This idea is similar to what is observed when dark-colored paint is mixed with white paint to produce a shade of paint somewhere between the two original shades. This explanation often seemed plausible when observing children whose appearance seemed to be a mix of each of their parents. However, this observation was predominantly based on a subset of traits, such as height and skin color, that exhibit what is known as continuous variation. Continuous variation produces a spectrum of small differences among individuals, which may give the appearance of the development of intermediate forms of traits. One key to the success of Mendel's experiments was his focus on traits that exhibit discontinuous variation. The traits that Mendel analyzed exist only as two distinct forms, with no intermediates to obscure inheritance patterns. The underlying cause of discontinuous variation among hereditary traits is the existence of genes. Mendel's observations demonstrated that an individual's set of genes is more like a deck of playing cards than a set of paints. Genes, like hands of playing cards, can be passed from generation to generation in different combinations, but each remains intact and two genes are not blended together to form a new intermediate version.

Mendel developed a five-part model to describe the patterns of inheritance he discovered while analyzing the data from his experiments. First, Mendel reported that variations in inherited traits are caused by inheritance of different versions of the factor, or gene, that controls that trait. The alternative versions of a gene responsible for production of different observable traits are called **alleles**. Next, Mendel stated that each individual inherits two copies of each gene, one from each parent. The third part of Mendel's model is known as the law of dominance. According to the **law of dominance**, one version of a gene completely masks the other version of the gene when both alleles are present. When an individual inherits two different alleles for the same gene, one of them (the **dominant allele**) determines the organism's appearance, and the other (the **recessive allele**) has no noticeable effect on appearance. For each gene, an individual may inherit two copies of the dominant allele, one copy of the dominant allele and one copy of the recessive allele, or two copies of the recessive allele. In the first two scenarios, the individual would express the dominant version of the trait, and only in the latter case would the individual express the recessive version of the trait. The observable physical and biochemical characteristics determined by an individual's genes comprise its **phenotype**. Therefore, the phenotype of heterozygous individuals is indistinguishable from the phenotype of individuals who are homozygous dominant. The terms "dominant" and "recessive" refer only to this pattern of inheritance and do not reflect which version of the gene is most common, most advantageous, or most desirable. It is a common misconception that the dominant allele is the most common within a population. The frequency of each allele is actually the product of a number factors that determine how likely individuals with a certain allele combination are to survive and reproduce and pass on their alleles to future generations.

The fourth part of Mendel's model is known as the **law of segregation**. According to the law of segregation, parents have two copies of each gene but can only pass on one copy to each offspring; therefore, the pair of alleles must separate (segregate) from each other during gamete formation. If an individual has two copies of the same allele, that allele will be passed on to all of their gametes. If an individual has one copy of the dominant allele and one copy of the recessive allele, each allele will be present in half of that individual's gametes. When an individual has two copies of the same allele for a given gene, it is said to be **homozygous**. When an individual has different alleles for a gene, it is **heterozygous**. The set of alleles that an individual possesses for their genes is known as their **genotype**. Genotypes are customarily indicated by a combination of letters representing each allele. Often the first letter of the word that describes the dominant phenotype is used for writing the genotype, but any letter can be used as long as a unique letter is consistently used to represent each gene. One important rule that is always followed is use of the capital version of the letter to represent the dominant allele and use of the lower-case letter for the recessive allele. A short description of the trait observed is used to indicate an individual's phenotype. For example, a pea plant that is heterozygous for the gene that determines pea color has the genotype Yy, which represents one copy of the dominant allele (Y) and one copy of the recessive allele (y). Having yellow peas describes the phenotype of this plant, conferred by the presence of the dominant allele for yellow-colored peas.

The final part of Mendel's model states that, during gamete formation, the pair of alleles for each gene separate independently from the pair of alleles for every other gene. In other words, the inheritance of alleles for one gene is not influenced by and does not affect the inheritance of alleles for other genes. This is similar to how the outcome of one coin toss does not affect and is not affected by the outcome of a previous or following coin toss. This is referred to as the **law of independent assortment**. This law only applies to genes located on different chromosomes or located far apart on the same chromosome. The specific location of a gene on a chromosome is known as its **locus**. It should be noted that genes with loci close together on the same chromosome do not follow the inheritance patterns described by Mendel. In fact, Mendel only described the inheritance of a small number of traits likely because he could not explain the more complex and difficult-to-predict inheritance patterns of other traits.

Mendel's Experiment

Mendel's five-part model of inheritance was developed from his study of the common garden pea plant, *Pisum sativum* (Figure 8.2). One advantage of using this particular species of plant is that a number of its observable traits exist in different varieties, facilitating tracking of inheritance of different traits from generation to generation. These plants also have a short generation time and produce a large number of offspring from each mating, which provides the large sample size required for accurate statistical analysis. Finally, the mating of these plants can be easily controlled to ensure findings are not confounded by the introduction of genetic material of unknown individuals. Importantly, these pea plants are capable of self-fertilization whereby pollen released by the male parts of an individual plant's flowers fertilize the eggs in the carpel, or female part, of its own flowers. Working with plants that self-fertilize ensured that genetic material from unknown sources was not introduced into the experiment. To ensure the plants would self-fertilize, rather than be cross-pollinated by other plants, Mendel meticulously transferred pollen from one flower to another by hand.

Mendel began his experiment by selecting plants that were **true-breeding**, which means they always produced offspring that look identical to the parent plant for certain characteristics when they self-fertilize. These true-breeding parent plants were obtained by following individual plants through many generations of self-fertilization during which only individuals that retained the characteristics of their parents were selected for the following generation. This smart choice was derived from the foresight that some plants with the same phenotype might result from different genotypes, a factor that could potentially obscure inheritance patterns. Specifically, this is because heterozygous individuals and individuals who are homozygous

FIGURE 8.2: **Garden pea (*Pisum sativum*)**—Gregor Mendel conducted his experiments with the garden pea.

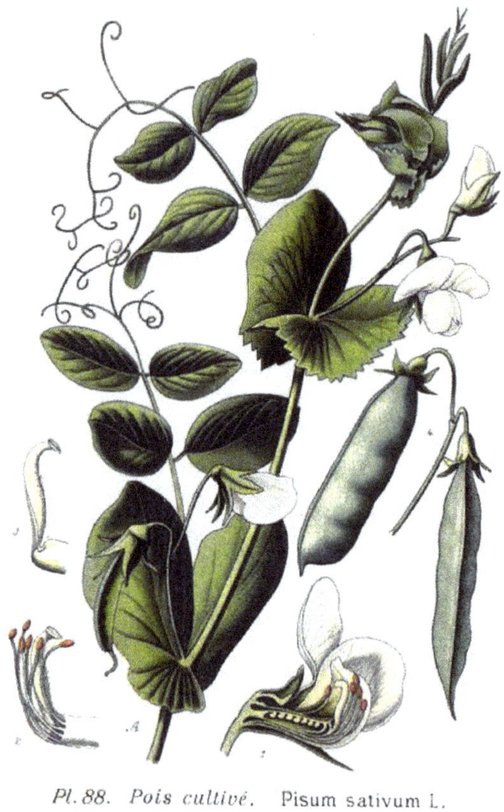

dominant both exhibit the dominant phenotype, even though the latter only has dominant alleles to pass on to offspring, whereas the former contains one dominant allele and one recessive allele. As Mendel demonstrated with his experiment, all true-breeding individuals are homozygous.

One trait of the garden pea that Mendel examined was cotyledon color. The cotyledon in a seed is the first pair of leaves that will eventually grow up and out of the soil. In peas, it is actually what is commonly referred to as the "pea". The plants Mendel selected produced peas in two distinct colors, yellow and green. To start his experiment, Mendel selected one true-breeding plant with yellow peas and one true-breeding plant with green peas. These plants became known as the parental, or **P generation** (Figure 8.3). Mendel transferred pollen from green pea plants to the carpels of green pea plants to mate the plants together. This process is known as *crossing*. Because true-breeding individuals with two different phenotypes were crossed, this process was also known as **hybridization,** and the resulting offspring were known as hybrids. The first generation of offspring produced through crossing individuals from the P generation was known as the **F1 generation**, which is short for first filial (Latin for "son") generation. Mendel observed that all plants in the F1 generation had yellow peas, and that they were exactly the same shade of yellow as the parent in the P generation (Figure 8.3). Later, individuals from the F1 generation were allowed to self-pollinate or were crossed with other individuals from the F1 generation. The offspring resulting from this cross were known as the **F2 generation**, or second filial generation. Mendel carried out these crosses until he had collected approximately 8,000 individuals from F2 generation. Of all of the F2 individuals Mendel observed, approximately 3/4 (6,022) had yellow peas and approximately 1/4 (2,001) had green peas (Figure 8.3).

Mendel noted the trait of green peas was not present in the F1 generation but appeared again in the F2 generation. It appeared as if the trait of yellow peas sometimes hides or masks the trait of green peas but does not alter or destroy it. The observation that the yellow and green peas present in the F2 generation looked exactly like the yellow and green peas in the P generation, rather than appearing as an intermediate form of yellow-green, disproved the previously prevailing hypothesis that traits of offspring develop as a blend of the traits in their parents. Furthermore, it is based on this observation that Mendel proposed the term "dominant" to describe a trait capable of masking another trait and recessive to describe the trait hidden by the dominant trait. The ratio of dominant to recessive traits observed in the F1 and F2 generations provided the information from which Mendel deduced the law of segregation. Mendel reasoned that the only scenario mathematically consistent with his observations was the separation of two alleles in the parents so that each gamete receives only one copy of each gene to be combined with a gene copy from the other parent during fertilization. If Mendel had only followed the plants through the F1 generation, he likely would not have been able to deduce how the traits he observed were passed from generation to generation.

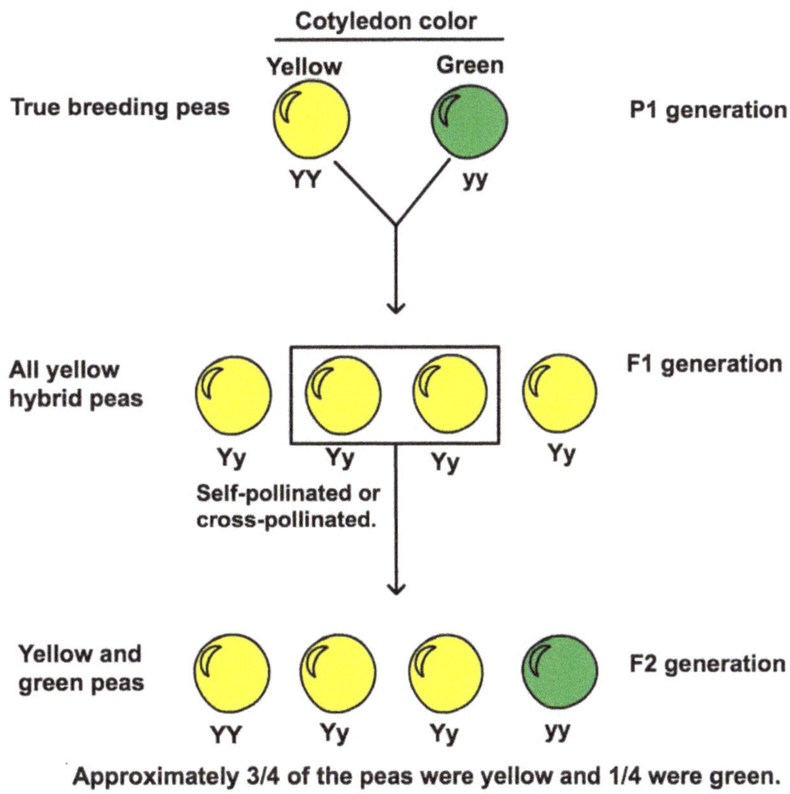

FIGURE 8.3: Law of dominance—The yellow trait masks the expression of the green trait in the hybrids of the F1 generation. The green trait reappears in the F2 generation.

Mendel went on to observe and analyze the inheritance of seven different traits in pea plants, and noted the same patterns for each trait (Table 8.1). Each trait was observed in two distinct forms; only one form as present in the F1 generation, and the other form reappeared in approximately one-fourth of the F2 individuals. This provided Mendel confidence in his conclusions and also led to his proposal of the law of independent assortment. Among the seven traits he examined, Mendel noticed that the combinations of traits that appeared in offspring were random. That is, the segregation of each allele into a gamete occurred independently of the segregation of all other alleles. No two alleles were more or less likely to be inherited together. Mendel's model is useful not only for explaining how traits were inherited in previous generations; this knowledge can be used to predict the probability of genetic outcomes in future generations.

When calculating the probability of genetic outcomes for Mendelian traits, one tool that is particularly helpful is the Punnett square. A **Punnett square** is a diagram used to predict the

TABLE 8.1: Mendelian Traits in Pea Plants (Phenotypes)

	Flower color	Flower position	Pod color	Pod shape	Cotyledon color	Seed shape	Stem length
P	Purple x White	Axial x Terminal	Green x Yellow	Inflated x Constricted	Yellow x Green	Round x Wrinkled	Tall x Short
F1	Purple	Axial	Green	Inflated	Yellow	Round	Tall

allele compositions of offspring resulting from a cross or mating. A Punnett square contains one box to account for each different combination of alleles possible in the offspring. Both parents have two copies of each gene; two multiplied by two yields four possible outcomes for each gene. The two alleles from one parent are indicated along the top of the Punnett square, and the two alleles from the other parent are indicated along the left side of the Punnett square. It does not matter which parent's alleles are listed in either location. However, there is one important convention that must be adhered to when constructing a Punnett square or identifying a genotype for any other reason. A capital letter should always be used to represent the dominant allele, and a lower case letter of the same type should be used to represent the recessive allele. Each gene is symbolized by its own letter, which often represents the dominant phenotype unless that letter has already been used to symbolize another trait under consideration.

For example, the letter Y is used to represent the cotyledon or seed (pea) color trait (Y is for yellow). As revealed by Mendel's observations, yellow is the dominant trait, so capital Y is used to represent the yellow allele; lower-case y is used to represent the recessive allele associated with the recessive trait of green peas. Therefore, the genotype of the yellow pea, true-breeding parent from the P generation is written as YY, and the genotype of the green pea, true-breeding parent from the P generation is written as yy. All of the offspring in the F1 generation received one allele from each parent and have the genotype Yy (Figure 8.4A). A Punnett square showing the crossing of two F1 individuals to produce the F2 generation has Yy written across the top to represent the genotype of one parent, and Yy written along the side to represent the genotype of the other parent (Figure 8.4B). With each parent contributing one of their alleles to each offspring, the possible genotype combinations in the F2 generation are YY, Yy, Yy, and yy (Figure 8.4B). The expected phenotypes and genotypes of a generation can also be expressed as ratios to indicate how likely each combination is to appear. In the previous example, the genotypic ratio of F2 generation is written as 1:2:1 to signify that for every one offspring expected to have the genotype YY, two are expected to be Yy, and one to be yy. The phenotypic ratio is 3:1; for every three yellow pea plants in the F2 generation, one green pea plant is expected. If a sufficiently large number of offspring were examined, the true ratios of observed outcomes would likely be very close to these ratios. However, if only a small number of offspring are examined, the ratio may not necessarily be close to the expected ratio caused by the effect of random chance.

FIGURE 8.4A: **Punnett square method**—Crossing plants with hybrid genotypes using a Punnett square helps keep track of alleles from the parents and possible offspring.

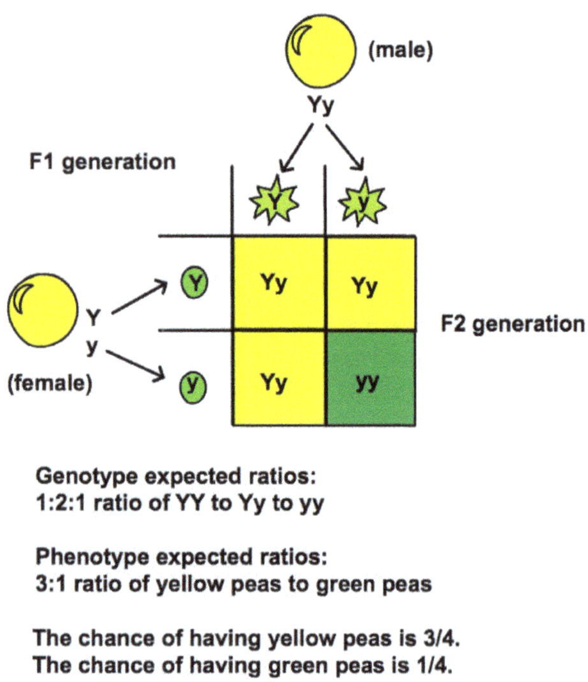

Punnett squares are useful for determining the expected genotypic and phenotypic ratios for a single Mendelian trait, and even for two traits at the same time. However, to determine the probability that offspring will inherit a particular phenotype or genotype for two or more Mendelian traits at the same time, a more useful tool is the multiplication rule. The

multiplication rule states that the probability of two events occurring can be determined by multiplying the probability of each individual event. For example, the probability that one coin toss will result in heads landing face up is 1/2. The probability that a second coin toss will result in heads landing face up is also 1/2. According to the multiplication rule, the probability of a coin toss resulting heads up twice in a row is 1/2 × 1/2, which is 1/4. Similarly, in the F2 generation of Mendel's experiment, the probability of an individual having yellow peas was 3/4 (Figure 8.4B). The probability of an individual from the F2 generation having green peas is 1/4 because green is the recessive trait for pea color. The multiplication rule can be applied to determine the probability of any number of outcomes occurring. Another interesting example is that of male and female sex chromosomes in humans. Each offspring has roughly a 1/2 chance of being genetically female and the same chance of being male. To determine the probability that a couple would have four genetically female children in a row, one would multiply 1/2 by 1/2 by 1/2 by 1/2, which yields a result of 1/16. Therefore, if a large number of families with four children were examined, approximately 1/16 would be expected to have four genetically female children.

FIGURE 8.4B: **Multiplication rule method**—Crossing plants with hybrid genotypes using the multiplication rule uses the probability to determine the chance of certain phenotypes in the offspring.

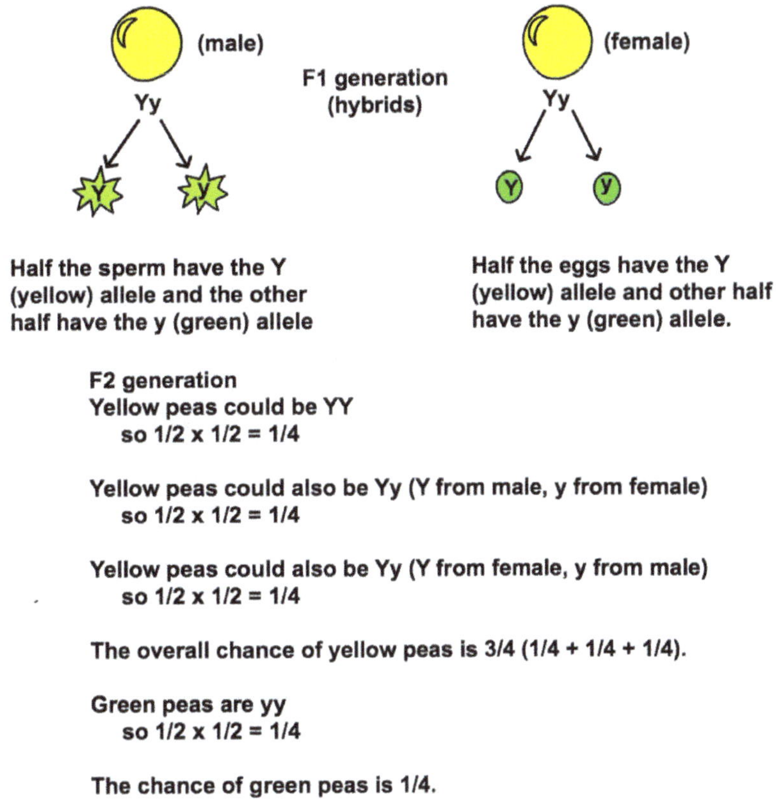

Monohybrid Cross, Dihybrid Cross, and Test Cross

To develop the law of segregation, Mendel needed only to follow the inheritance of one trait from the F1 generation to the F2 generation. When only one trait is followed, the genetically hybrid (heterozygote) individuals in the F1 generation are referred to as monohybrids. The mating of two individuals who are each heterozygous for one Mendelian trait is known as a **monohybrid cross** (Figure 8.4). To develop the law of independent assortment, which describes inheritance patterns of multiple genes, Mendel used a dihybrid cross. A **dihybrid cross** is the mating of individuals who are heterozygous for two Mendelian traits. A dihybrid cross for the traits of pea color and pea pod color is illustrated in Figure 8.5. The odds of predicted genotypes and phenotypes resulting from a dihybrid cross can be determined in three different ways. The method for predicting the outcome of a dihybrid cross is to construct a Punnett square that includes both alleles for both genes in both parents (Figure 8.5A). This Punnett square contains a total of 16 possible outcomes. For any dihybrid cross, it is estimated that 9/16 of the

FIGURE 8.5A: Dihybrid cross using a Punnett square

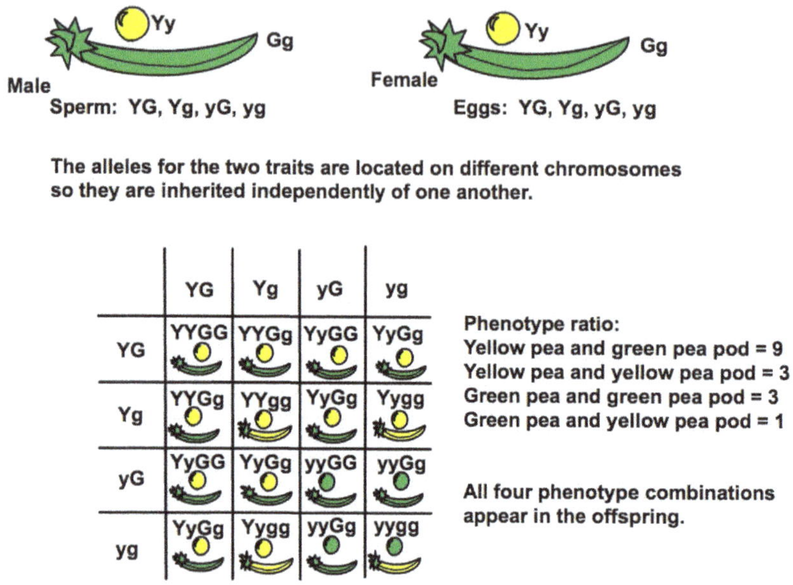

FIGURE 8.5B: Dihybrid cross using the multiplication method

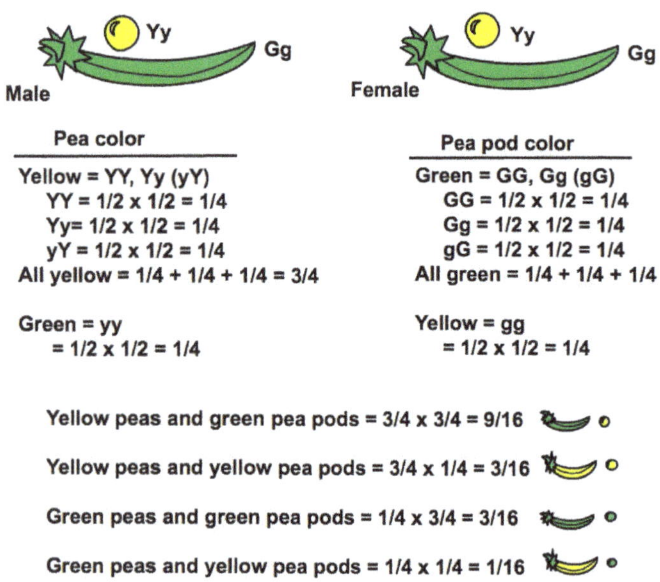

offspring will inherit the dominant phenotype for both traits, 3/16 will inherit the dominant phenotype for the first trait and the recessive phenotype for the second trait, 3/16 will inherit the recessive phenotype for the first trait and the dominant phenotype for the second trait, and 1/16 will inherit the recessive phenotype for both traits. In this example, it is predicted that 9/16 of the offspring will have yellow peas and green pea pods, 3/16 will have yellow peas and yellow pea pods, 3/16 will have green peas and green pea pods, and 1/16 will have green peas and yellow pea pods (Figure 8.5A). These outcomes can also be determined using the multiplication rule (Figure 8.5B). A third method for predicting the phenotypes of the offspring produced by a dihybrid cross is to memorize the ratio 9:3:3:1, then remember which combination of phenotypes each number presents. Of these methods, memorizing the ratio 9:3:3:1 is the least applicable to other situations, because it only provides information about the phenotypes of a dihybrid cross. The other two methods can be applied for determining both the genotypic and phenotypic ratios of a dihybrid cross, as well as for calculating the probability of inheritance two Mendelian traits when one or both parents is homozygous rather than heterozygous for one or both traits. Further, the multiplication rule can be applied to virtually any scenario that involves calculating the predicted frequency of inheritance outcomes for multiple Mendelian traits.

In addition to predicting the genotypes and phenotypes of offspring with parents of known genotype, Mendel developed a method for determining whether an organism that expresses the dominant phenotype for a Mendelian trait is homozygous dominant or heterozygous. This technique, called a **test cross,** involves mating an individual of unknown genotype that expresses a dominant phenotype with an individual known to be homozygous recessive for that trait. If all of the offspring in the F1 generation express the dominant trait, the individual in question can be identified as homozygous dominant. If some of the offspring express the recessive trait, the unknown individual must be heterozygous. Mating a heterozygous individual with a homozygous recessive individual is expected to produce heterozygotes and recessive homozygotes at 1:1 ratio. Although it is now possible, through the use of modern technology, to determine an organism's genotype by sequencing its genome, the test cross is still used by botanists and genetics because of its simplicity and efficiency.

Mendelian Traits in Humans

Mendelian patterns of inheritance are observed in a number of physical traits in humans and other organisms. Several examples of human traits often described as being inherited in a Mendelian manner include eye color, facial dimples, earlobe structure, freckles, hand-clasping tendency, mid-digital hair, PTC tasting ability, "hitchhiker's thumb," tongue rolling, and widow's peak hairline (Table 8.2). For each of these traits, if the parents' genotypes are known, the probability that their offspring will have a certain genotype or phenotype can be predicted using a Punnett square. The probability that a collection of two or more of these traits will be inherited in the same offspring can be further predicted using the multiplication rule. The majority of observable human characteristics follow more complex patterns of inheritance than Mendelian traits. It has even been suggested that some of the traits listed in Table 8.2, which were identified through observational studies as following Mendelian patterns of inheritance, may also be affected by other genetic influences, such as continuous variation.

TABLE 8.2: Mendelian Traits in Humans

Trait description	Dominant trait	Recessive trait
Eye (iris) pigment	Darkly pigmented irises (eyes any color other than blue or gray)	Lightly pigmented irises (blue or gray eyes)
Dimples	Presence of facial dimples	Absence of facial dimples
Earlobes	Free earlobes	Attached earlobes
Freckles	Presence of freckles	Absence of freckles
Hand clasping	Clasping hands together with left thumb on top	Clasping hands together with right thumb on top
Mid-digital hair	Hair growth on middle segments of fingers	No hair growth on fingers
PTC tasting	Ability to taste the bitter compound PTC	Inability to taste the bitterness of PTC
Hitchhiker's thumb	Straight thumb	Bent ("hitchhiker's") thumb
Tongue rolling	Ability to physically roll the tongue	Inability to physically roll the tongue
Widow's peak	Widow's peak present in hairline	Widow's peak absent

Dominant and Recessive Genetic Diseases and Disorders

In addition to a number of physical features, several human genetic disorders and diseases are inherited in a Mendelian manner (Table 8.3). The mutations that cause each of these conditions is present in approximately 1 in 10,000 to 1 in 40,000 people in the general population, with the exception of the mutation that causes sickle cell disease, which is significantly more common, especially among people descendent from sub-Saharan Africa, India, and parts of the Middle East. In the United States, approximately 1 in 365 Black or African-Americans have inherited two copies of the sickle cell mutation, causing them to develop sickle cell disease. Approximately 8%, or 1 in 13, of people in this population inherit only one copy of the mutated beta hemoglobin gene and are referred to as having sickle cell trait. Although sickle cell disease often causes serious and potential fatal health problems, sickle cell trait is characterized by much milder symptoms and is not fatal.

Although achondroplasia dwarfism is a dominantly inherited disorder, the majority of the population does not exhibit this condition, which is characterized by an abnormally short stature resulting from an impeded ability to convert cartilage into bone. Similarly, most people do not have the fatal neurodegenerative disease known as Huntington's or chronically high cholesterol resulting from a defect in the proteins required for shuttling cholesterol between the bloodstream and cells of the body.

The rarity of dominant genetic disorders highlights the misconception that dominant traits are more common than recessive. Recall that the frequency of each allele in a population is primarily determined by other factors, such as the length of time that each allele has existed in the population and the effect of each allele on fitness (the ability to survive and reproduce successfully). The mutation that causes sickle cell disease is one example of these forces at work. Because sickle cell disease significantly reduces fitness, it would be logical to expect this mutation to be increasingly rare. However, this allele has persisted for thousands of years, and is one of the most common recessive genetic disorders among certain populations, such as central Africans and those with African ancestors. The uneven distribution of this mutation across the globe, and the particularly high prevalence of heterozygotes with sickle cell trait, provides a clue to why natural selection and evolution have not caused this mutation to become scarcer or disappear altogether. The explanation is thought to involve human evolution influenced by one of the most common infectious diseases in the world: malaria. Sickle cell disease and malaria are connected by one important characteristic, which is that both affect red blood cells. The mutation that causes sickle cell disease leaves red blood cells misshapen and unable to transport oxygen properly, whereas malaria parasites require red blood cells to complete their life cycle. In areas where the *Plasmodium falciparum* parasite that causes malaria is common, such as central Africa and other equatorial regions, individuals with this mutation experience reduced frequency and severity of malaria attacks. For individuals whose ancestors lived in malaria-prone regions, this infectious disease has long been one of the leading causes of mortality. Therefore, it is thought that a genetic mutation that confers resistance to malaria would experience strong positive selection through many generations. Protection against malaria is conferred whether an individual has two copies of the mutation or one. This means that the heterozygous condition is highly preferred evolutionarily; these individuals are more likely to survive malaria outbreaks, and also they would not experience decreased fitness because of sickle cell disease.

For many mutations that cause disorders or diseases that decrease fitness, there is no explanation for why they continue to persist in the population that involves protection against an unrelated disease, similar to that of sickle cell trait and malaria. However, the presence of heterozygotes is still key to the persistence of many recessively inherited disorders. People who inherit two copies of the mutation that causes cystic fibrosis experience mucus accumulation in the lungs, pancreas, and digestive tract leading to chronic bronchitis, recurrent infections, poor absorption of nutrients, and reduced life expectancy, yet people with only one copy of the mutation do not experience any of these effects (Figure 8.6). Similarly, individuals who are homozygous recessive for the oculocutaneous albinism II (OCA2) mutation completely lack pigment in their skin, hair, and eyes, and those homozygous for a mutation in phenylalanine hydroxylase experience a potentially severe metabolic disorder that may disrupt brain functioning, yet heterozygous individuals, in both instances, are unaffected. In fact, for the majority of autosomal recessive disorders and diseases, only individuals who inherit two copies of the mutated gene display symptoms. In these cases, heterozygous individuals who do not exhibit the disorder or disease but carry one copy of the mutated gene are known as **carriers**. However, if two individuals who are carriers of the same recessive mutation have offspring, each offspring has approximately a 25% chance of inheriting two copies of the mutated gene and developing the disease (Figure 8.6). Therefore, individuals who may be carriers of serious genetic diseases often want to know whether they possess the disease-causing mutation.

Determining one's status as a carrier for a recessive disorder can be easily determined through analysis of family history. If one biological parent has a recessive genetic disease or disorder and the other does not, all of their offspring who do not exhibit the disease or disorder must be carriers. For cases where determining carrier status is not as straightforward, routine genetic testing can provide a definitive result for the majority of conditions of a known genetic cause. Individuals considering having children who are concerned about potential genetic disorders are encouraged to seek the help of genetic counselors. Modern biotechnology provides a multitude

FIGURE 8.6: Autosomal recessive inheritance

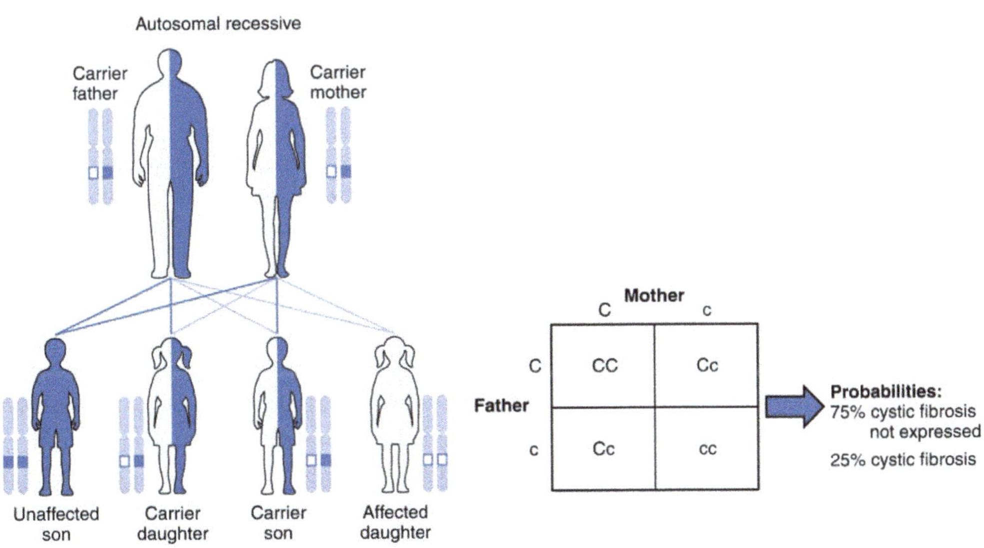

TABLE 8.3: Mendelian Genetic Disorders and Diseases

Disorder or disease	Gene responsible	Type of inheritance
Achondroplasia dwarfism	Fibroblast growth factor receptor 3 (FGFR3)	Autosomal dominant
Albinism	Oculocutaneous albinism II (OCA2)	Autosomal recessive
Cystic fibrosis	Cystic fibrosis conductance transmembrane regulator (CFTR)	Autosomal recessive
Huntington's disease	Huntingtin (HTT)	Autosomal dominant
Hypercholesterolemia, type B	Low-density lipoprotein receptor (LDLR); apolipoprotein B (APOB)	Autosomal dominant
Phenylketonuria (PKU)	Phenylalanine hydroxylase (PAH)	Autosomal recessive
Sickle-cell anemia	Beta hemoglobin (HBB)	Autosomal recessive

of options for addressing these concerns, including newborn screening for genetic mutations, fetal genetic testing, assisted reproductive technology that allows for the production or selection of embryos unaffected by mutations, and many approaches for preventing, treating, and accommodating harmful or undesired symptoms of these conditions in children and adults.

All of the examples listed in Table 8.3 are the result of mutations in genes located on autosomes. The term **autosome** refers to all chromosomes that are not sex chromosomes. Normally each individual has two copies of each autosome. A number of other disorders and diseases, such as Duchenne muscular dystrophy and hemophilia A, result from mutations in a single gene located on either the X or Y sex chromosome. **Sex chromosomes** differ between males and females, and contain genes involved in the differentiation of male and female sex characteristics (as well as some genes not related to sexual differentiation). These are inherited according to all of the principles described by Mendel, except that their representation in offspring is modulated by the fact that some offspring inherit two X chromosomes, whereas others inherit one X chromosome and one Y chromosome. As with physical traits, most diseases that have a genetic basis, such as most types of cancer, result from mutations in more than one gene, and thus do not follow Mendelian patterns of inheritance.

Exceptions to Mendel's Rules

Although many species have some traits inherited in a Mendelian manner, expression of the majority of traits involves a more complex interplay between genes and other factors. When inheritance of characteristics is influenced by more than one gene or by one gene for which there are more than two possible alleles, Mendel's rules do not apply. If the locus of a gene is on one of the X or Y sex chromosomes, Mendelian inheritance patterns do not apply because those chromosomes, unlike all others, are not always paired with another chromosome of the same type. Another type of scenario in which Mendel's model does not accurately predict phenotypic and genotypic outcomes is when inheritance does not follow the law of dominance. For example, sometimes heterozygous individuals, rather than exhibiting the dominant phenotype, display a phenotype that is an intermediate or combination of the dominant and recessive phenotypes. Principles and examples of non-Mendelian genetics are discussed in more detail in another chapter of this book.

8.2 SUMMARY

- Gregor Mendel discovered several fundamental principles of genetics by observing the inheritance of traits in pea plants
- Mendelian traits are those controlled by a single gene with two different alleles, one dominant and one recessive, that assort independently during gamete formation
- In Mendel's experiments, the P generation was true-breeding and homozygous; all individuals in F1 generation were heterozygous and expressed the dominant phenotype; the F2 generation had a 3:1 phenotypic ratio and a 1:2:1 genotypic ratio
- Alternative versions of a gene are called alleles; the location of a gene on a chromosome is called a locus
- The frequency of Mendelian traits in future generations can be predicted using Punnett squares and the multiplication rule
- A number of traits and diseases in humans are inherited in a Mendelian manner, though most human characteristics follow a more complex pattern of inheritance
- A carrier is an individual who is heterozygous for a recessive disorder or disease who has a normal phenotype but may pass on the genetic condition to offspring
- If traits are influenced by a gene with more than two alleles, if one allele is not completely dominant or if one than one gene influences a trait, the rules of Mendelian inheritance do not apply

8.3 KNOWLEDGE CHECK

1. What is the name for alternative versions of the same gene?
   a. alleles
   b. carriers
   c. hybrids
   d. locus
   e. multiplication rule

2. In Gregor Mendel's pea plant experiment, what was the genotypic ratio of the F2 generation?
   a. 1:1
   b. 3:1
   c. 4:0
   d. 1:2:1
   e. 1:1:1:1

3. Yellow-colored peas express the dominant trait and green-colored peas express the recessive trait. What fraction of the offspring is expected to have yellow peas if two heterozygous individuals have offspring?
    a. 0/4
    b. 1/4
    c. 2/4
    d. 3/4
    e. 4/4

4. Conducting a dihybrid cross would be most helpful for predicting inheritance patterns in the offspring of which of the following?
    a. two individuals who are heterozygous for one trait
    b. two individuals who are heterozygous for two traits
    c. two individuals who are homozygous for two traits
    d. two individuals who have an unknown genotype for two traits
    e. two individuals who are homozygous for four traits

5. Hypercholesterolemia, type B is an autosomal dominant disorder. If two individuals who have hypercholesterolemia, type B have a child who does not have hypercholesterolemia, type B, which of the following would best describe the child in terms of the gene that causes hypercholesterolemia, type b?
    a. carrier
    b. heterozygous
    c. homozygous dominant
    d. homozygous recessive
    e. hybrid

6. Approximately 1/2 of all human children conceived are genetically female and the other 1/2 are genetically male. Each conception occurs independently of the others. If a couple conceives five children, what is the likelihood that all five will be genetically female?

7. Describe three different ways to determine the predicted phenotype frequencies among the offspring produced by a dihybrid cross.

8. Identify one Mendelian genetic disorder or disease present in humans. Describe the pattern of inheritance for this disease or disorder, and list the symptoms or characteristics of individuals who have this condition.

8.4 ACTIVITY

Probability and Multiplication Rule

This activity is designed to demonstrate and examine core concepts of Mendelian genetics by observing and analyzing the outcome of coin tosses. This is a useful demonstration for understanding Mendelian genetics because of the similarities between tossing a coin and the sorting of alleles for Mendelian traits that occurs during meiosis. A coin has two sides that are equally likely to land face up each time a coin is tossed, and the outcome of one coin toss does not influence and is not dependent upon the outcome of any other coin toss. Similarly, for each Mendelian trait, there are two different gene alleles. According to the law of segregation, the pair of alleles for each gene is separated during meiosis, and each is equally likely to appear in any given gamete. Each pair of alleles segregates independently and is not affected by the segregation of other allele pairs, according to the law of independent assortment. It is important to note that this is true only for Mendelian traits.

In this activity, each coin represents one parent. Each parent is heterozygous for the trait, with one copy of the dominant allele (heads) and one copy of the recessive allele (tails).

The gene represented here is responsible for the trait of seed shape. The dominant trait is round (R); the recessive trait is wrinkled (r).

Materials

For this activity, you will need two coins of the same denomination and a method of recording results.

Procedure

One member of your group should toss the coin, while another records the results. If your group has a third member, their role can be keeping track of the total number of coin tosses. All members of the group should participate in analyzing data and discussing conclusions.

1. Toss both coins at the same time to represent the combination of alleles received from both parents.
2. Record the outcome as RR (heads/heads), Rr (heads/tails), or rr (tails/tails).
3. Repeat Steps 1 and 2 for a total of 50 times.
4. Complete the data table with your group data and the cumulative data of the entire class.
5. Answer the questions that accompany the data table.

8.5 ACTIVITY

Analyzing Mendel's Experiment

Gregor Mendel discovered seven observable traits in pea plants that are inherited according to the principles described in this chapter. These are referred to as Mendelian traits. The purpose of this activity is to apply the law of independent assortment, the law of segregation, and the multiplication rule to determine inheritance patterns for these traits.

TABLE 8.4: Mendelian Traits in Pea Plants (Genotypes)

	Flower color	Flower position	Pod color	Pod shape	Seed color	Seed shape	Stem length
Dominant	purple (P)	axial (F)	green (G)	inflated (V)	yellow (Y)	round (R)	tall (L)
Recessive	white (p)	terminal (f)	yellow (g)	constricted (v)	green (y)	wrinkled (r)	short (l)

Use Table 8.4 to answer questions about the knowledge learned from analyzing Gregor Mendel's experiments.

8.6 ACTIVITY

Mendelian Traits in Humans

Some human traits follow the laws of Mendelian inheritance. Examples of these traits are listed in the following text. Although there are reports that some of these traits exhibit continuous variation, each of these traits are commonly thought to be controlled by a single gene with two distinct alleles. According to convention, a different letter of the alphabet symbolizes each gene, as indicated; the allele for the dominant trait is symbolized by a capital letter, and the recessive allele is identified with a lower-case letter.

- **Dimples**—Having facial dimples is a dominant trait (D), whereas the absence of dimples is recessive (d).
- **Earlobes**—Having free earlobes (E) is a dominant trait, whereas attached earlobes are recessive (e).
- **Eye (iris) pigment**—Having darkly pigmented irises (producing eyes any color other than blue or gray) is a dominant trait (P), whereas having no pigment in the front of the iris (producing blue or gray eyes) is recessive (p).
- **Freckles**—Having freckles (F) is dominant, whereas not having freckles is recessive (f).
- **Hand clasping**—When you clasp your hands together, if the left thumb is on top, that is the dominant trait (C), whereas having the right thumb on top is recessive (c).
- **Hitchhiker's thumb**—A straight thumb is dominant (S), whereas a bent or hitchhiker's thumb is recessive (s)
- **Mid-digital hair**—If any hair grows on the middle segments of your fingers, that is the dominant trait (H), whereas having no hair is recessive (h).
- **PTC tasting**—The ability to taste the bitter compound PTC is dominant (T), whereas an inability to taste the bitterness of PTC is recessive (t).
- **Tongue rolling**—The ability to physically roll the tongue is a dominant trait (R), whereas non-rolling is recessive (r).
- **Widow's peak**—Having a distinct peak in the center of the hairline on the forehead is a dominant trait (W), whereas having a straight hairline is recessive (w).

The purpose of this activity is to develop and demonstrate an understanding of Mendelian inheritance as it applies to human traits. For each trait described, record your phenotype and potential genotype(s) in the data table, then record and analyze data for the entire class as indicated. Use the letter symbols previously given for each genotype.

ACTIVITY QUESTIONS

8.4

TABLE 8.5: Activity 8.4 Data Table

	RR	Rr	rr
Tally marks for 50 tosses (of both coins)			
Total times outcome observed (Group data)			
Total times outcome observed (Class data)			

1. Complete Table 8.5.
2. In the data collected by your group, what was the percent of the total outcomes for each of the following: RR, Rr, and rr?
3. In the data collected by the entire class, what percent of the total outcomes was each of the following: RR, Rr, and rr?
4. Express the class data as a genotypic ratio.
5. Draw a Punnett square to show the predicated outcomes of this cross.
6. What percent of the predicted outcomes was each of the following: RR, Rr, and rr?
7. Express the predicted outcomes of this cross as a genotypic ratio.
8. Discuss how your group data and the class data compared to the expected results.
9. In his experiments, Gregor Mendel examined thousands of individuals to determine the frequency of each allele pair. Describe why it is important to examine a large number of individuals when analyzing probability.

8.5

1. Write the seed shape phenotypes and corresponding genotypes of the individuals in the P generation in Mendel's experiment.
2. Draw a Punnett square to show the potential outcomes of the F1 generation in terms of seed shape in Mendel's experiment.
3. Draw a Punnett square to show the potential outcomes of the F2 generation in terms of seed shape in Mendel's experiment.

4. Indicate the phenotypic ratio for seed shape of the F2 generation.
5. Indicate the genotypic ratio for seed shape of the F2 generation.
6. In Mendel's experiment, what fraction of the offspring in the F2 generation had wrinkled peas and green pods? Show the work or describe the reasoning used for determining this answer.
7. In Mendel's experiment, what fraction of the offspring had smooth peas, yellow pods, and purple flowers? Show the work or describe the reasoning used for determining this answer.
8. Describe how each of the following was demonstrated in the work you completed to answer the previous question: law of independent assortment, law of segregation, and multiplication rule.

8.6

1. Complete Table 8.6.
2. Describe how your personal phenotype compares with the class data.

TABLE 8.6: Activity 8.6 Data Table

Trait	Your Phenotype	Your Possible Genotype(s)	Class Data	
			Recessive	Dominant
Dark eye pigment or Light eye pigment				
Have dimples or Do not have dimples				
Free ear lobes or Attached ear lobes				
Hand clasp left over right or Hand clasp right over left				
Have freckles or Do not have freckles				
Mid-digital hair or No mid-digital hair				
PTC taster or non-PTC taster				
Straight thumb or Bent thumb				
Tongue roller or Non tongue roller				
Widow's peak or Straight hairline				

3. Summarize the class data. Propose an explanation for the frequency of dominant/recessive traits observed.
4. True Mendelian traits exhibit discontinuous variation (they exist as two distinct forms rather than a spectrum of intermediate forms). Based on discussions with your classmates, do all of the traits examine appear to be true Mendelian traits? Did any of the traits examined appear to display continuous variation? How could you tell?
5. Indicate the possible phenotype(s) of someone who has the genotype FfhhRrWW.
6. Indicate the possible genotype(s) and phenotype(s) for someone who is heterozygous for both the widow's peak gene and the tongue-rolling gene.
7. Draw a Punnett square to show the possible genotypes of offspring from one individual who is heterozygous for the gene associated with widow's peak and one individual who is homozygous dominant for the gene associated with widow's peak.
8. What fraction of offspring would have dimples and attached earlobes if their parents were both heterozygous for both traits? Show your work or reasoning.
9. If two parents are heterozygous for three genes, what is probability that their first child will have the dominant phenotype for all three genes? Show the work or describe the reasoning used for determining this answer.
10. If two parents are heterozygous for four genes, what is probability that their first child will have the dominant phenotype for two genes and the recessive phenotype for the other two genes? Show the work or describe the reasoning used for determining this answer.

8.7 REFERENCES

Centers for Disease Control. (2016). Sickle cell disease (SCD). [On-line] Last modified August 31, 2016. https://www.cdc.gov/ncbddd/sicklecell/data.html.

Chial, H. (2008). Rare genetic disorders: Learning about genetic disease through gene mapping, SNPs, and microarray data. *Nature Education, 1*(1), 192.

Gayon, J. (2016). From Mendel to epigenetics: History of genetics. *Comptes Rendus Biologies, 339*(7–8), 225–30.

Genetic Science Learning Center. (2017). The basics and beyond: An introduction to heredity. [On-line] *University of Utah*. Access date: July 7, 2017. http://learn.genetics.utah.edu.

Germain, D. P. (2006). General aspects of X-linked diseases. In A. Mehta, M. Beck, G. SunderPlassmann (Eds). *Fabry disease: Perspectives from 5 years of FOS*. Oxford: Oxford PharmaGenesis.

Mendel, J. G. (1865). Versuche über Pflanzen-Hybriden. *Verhandlungen des Naturforschenden Vereines in Brünn, IV*, 3–47.

Monaghan, F. and Corcos, A. (1984). On the origins of the Mendelian laws. *Journal of Heredity, 75*(1), 67–69.

FIGURE CREDITS

FIG. 8.1: William Bateson, "Gregor Mendel," Mendel's Principles of Heredity: A Defence. Copyright in the Public Domain.

FIG. 8.2: Amédée Masclef, "Garden pea (Pisum sativum)," Atlas des plantes de France. Copyright in the Public Domain.

FIG. 8.6: Copyright © OpenStax College (CC by 3.0) at https://commons.wikimedia.org/wiki/File:2926_Autosomal_Recessive_Inheritance-new.jpg.

Non-Mendelian Genetics

LEARNING OBJECTIVES

- Describe and identify examples of incomplete dominance, complete dominance, codominance, epistasis, pleiotropy, and gene linkage
- Describe the concept of polygenic inheritance using human skin color as an example
- Identify the inheritance pattern that produces ABO blood groups
- Apply knowledge of ABO blood groups to determine which blood donors are compatible with which recipients, and which possible blood types may appear in the offspring of parents with a known blood type
- Explain why X-linked traits are more common in males than females
- Discuss examples of epigenetic inheritance, including genomic imprinting

9.1 INTRODUCTION

The principles of genetic inheritance proposed by Gregor Mendel are useful for understanding how certain traits are passed from parents to offspring. Although some human traits follow all the rules of Mendelian genetics, many traits observed in humans and other organisms that reproduce sexually are inherited in a different manner. Alternative modes of inheritance include incomplete dominance, codominance, multiple alleles, polygenic inheritance, epistasis, gene linkage, and inheritance of sex-linked traits. This chapter describes each of these inheritance patterns and provides examples. Also described in this chapter is the concept of epigenetic inheritance, which involves regulation of genes independent of the nucleotide sequence of the gene.

Incomplete Dominance

Mendel's rules are applicable only to traits controlled by a single gene for which each individual inherits one allele from each parent. According to Mendel, these alleles can be either dominant or recessive, and must follow the law of independent assortment. One additional characteristic shared by all Mendelian traits is that they exhibit **complete dominance**. This means that heterozygous individuals who have one dominant allele and one recessive allele for the same gene express only the dominant phenotype, whereas the recessive phenotype is completely masked. One alternative mode of genetic inheritance is known as **incomplete dominance**. Incomplete dominance is a form of inheritance in which the dominant allele only partially masks the recessive allele, resulting in expression of an intermediate phenotype in heterozygotes.

One example of incomplete dominance is found in the carnation. Some individual carnation plants produce red flowers, and some produce white flowers. Red flowers are an expression of the dominant allele, so homozygous dominant carnation plants produce red flowers. White flowers are an expression of the recessive allele, so carnations that are homozygous recessive produce white flowers. However, heterozygous carnations do not produce red flowers, as would be expected based on the complete dominance observed for Mendelian traits. Instead, heterozygous carnations

FIGURE 9.1: Punnett square showing incomplete dominance—Incomplete dominance produces intermediate phenotypes in heterozygotes, such as the pink flowers of carnations with one allele for red flower color and one allele for white flower color.

	C^R	C^W
C^R	C^RC^R	C^RC^W
C^W	C^RC^W	C^WC^W

produce pink flowers, which is an intermediate phenotype that represents a mixing of the dominant and recessive alleles. Inheritance of traits that exhibit incomplete dominance can be predicted using a Punnett square, similar to the Punnett squares used to track Mendelian inheritance. However, the rules for how genotypes are written for incompletely dominant traits are different than those for Mendelian traits. When a Punnett square is generated to illustrate the previous example, C^R is used to represent the dominant allele and C^W is used for the recessive allele. Therefore, a homozygous dominant individual would have the genotype C^RC^R, a heterozygous individual would have the genotype C^RC^W, and a homozygous recessive genotype would be written C^WC^W. As shown in Figure 9.1, the genotypic ratio resulting from a cross between two heterozygous carnations would be 1:2:1 ($1C^RC^R$:$2C^RC^W$:$1C^RC^W$), and the phenotypic would also be 1:2:1 (1 red: 2 pink: 1 white).

Codominance and Multiple Alleles

Another alternative mode of inheritance, known as **codominance,** occurs when both alleles for a trait are simultaneously expressed in heterozygotes. Codominance can also be observed in some species of flowers that possess red and white alleles for the trait of flower color. In those instances, rather than being pink, as is the case with incomplete dominance, the flowers of heterozygotes have patches of red and patches of white. One of the most well-known examples of codominance is the ABO blood type in humans. This trait is also an example of there being more than two alleles for the same gene, which causes a non-Mendelian pattern of inheritance in itself. For this trait, there is one gene for which there are three possible alleles instead of two. Type A blood is derived from the allele I^A, type B is derived from the I^B allele, and type O is derived from the allele written as i. The letters A and B refer to two different types of carbohydrates that may be expressed on the surface of red blood cells, and the letter O indicates that neither carbohydrate is present. Even though there are three possible alleles for the gene that controls ABO blood type, each individual still only inherits one allele from each parent. Although the rules of Mendelian inheritance do not apply in the case of multiple alleles, a Punnett square can still be used to predict the frequency of genotypes and phenotypes among offspring. For ABO blood types, the expected outcome of mating between one heterozygous individual with type A blood and one heterozygous individual with type B blood has a genotypic ratio of 1:1:1:1 and a phenotypic ratio of 1:1:1:1 as well (Figure 9.2).

FIGURE 9.2: Punnett square showing codominance—ABO blood types are an example of codominance because two alleles may be expressed simultaneously in one phenotype.

	I^A	i
I^B	$I^A I^B$	$I^B i$
i	$I^A i$	$i i$

The possible ABO blood type genotypes and their corresponding phenotypes are indicated in Figure 9.3. Although once used as a crude measure for identifying individuals involved in crimes or paternity cases, the significance of ABO blood types is mostly a result of their influence on the success of blood transfusions. Blood transfusions are a potentially life-saving method of transferring blood cells from one individual to another individual who has experienced significant blood loss or insufficiency. It is critical that blood transfusions only be given from donors who have a blood type compatible with the recipient. This is because the A and B carbohydrate molecules are antigens, which are molecules that the immune system recognizes

FIGURE 9.3: ABO blood types—The four ABO blood groups result from inheritance of different pairings of three alleles.

Genotype	I^AI^A or I^Ai	I^BI^B or I^Bi	I^AI^B	ii
Phenotype (blood type)	A	B	AB	O
Can donate blood to:	A, AB	B, AB	AB	A, B, AB, O
Can receive blood from:	A, O	B, O	A, B, AB, O	O

and will attack if they are determined to be from a foreign source rather than from one's self. An individual's immune system usually will not attack the antigens naturally present on its own cells but will attack antigens not usually present. Therefore, a person with type A blood can safely receive a blood transfusion from another person with type A blood but not from a person with type B blood. Because type O blood does not contain any of these carbohydrate antigens, every person can safely receive type O blood. Conversely, people with type O blood can only receive blood transfusions that are type O because their immune system will attack any red blood cells expressing A or B antigens.

In addition to the ABO blood types, people also have "positive" and "negative" blood types. For example, an individual may have type O negative blood or type O positive blood. This secondary blood typing refers to expression of a protein known as Rh factor. Rh factor is the product of a different gene than the A and B carbohydrates, and is inherited independently of the genes that produce the ABO blood type. Similar to the A and B antigens, individuals who are Rh negative cannot receive blood transfusions from people who are Rh positive because their immune system will attack red blood cells expressing the foreign Rh factor. People who are Rh positive can tolerate blood that is Rh positive or Rh negative because their immune system is already familiar with the antigen, and if the antigen is not present, there is no cause for the immune system to attack. Rh factor also plays a significant role during pregnancy. The immune system of an Rh negative mother may attack the red blood cells of an Rh positive baby, causing a potentially fatal condition known as hemolytic disease of the newborn. The term hemolytic is derived from "hemo" meaning blood and "lysis," which means bursting. Fortunately, hemolytic disease of the newborn can be effectively prevented by administration of a common drug called Rho(D) immune globulin or RhoGam.

Although the normal role of the carbohydrates that comprise the A and B antigens is not well known, evidence suggests their expression contributes to the regulation of blood clotting. In addition to this, these antigens may serve as cellular entry points for certain infectious microorganisms. For example, people with type O blood are more likely to be infected with norovirus, which is the most common cause of gastroenteritis or "stomach flu", than people who have one of the other blood types. However, people with type O blood are also somewhat protected against the parasite that causes malaria. Interestingly, these antigens have also been linked to variations in the risk of a number of other diseases such as coronary heart disease and certain types of cancer.

Polygenic Inheritance

Although Mendelian traits and the aforementioned examples of incomplete dominance and codominance all result from the effects of a single gene, some traits are derived from the combined effects of multiple genes. **Polygenic inheritance** is the process by which the phenotype for a single trait is determined by the additive effects of two or more genes. Traits for which there are multiple variations along a spectrum of possible outcomes, such as height and weight, are often the result of polygenic inheritance. Another example of polygenic inheritance observed in humans is skin color. The many shades of skin pigmentation observed in humans result from the additive effects of multiple gene products as well as environmental and behavioral factors. Much of the variation in human skin color can be accounted for by examining the contribution of three genes, often termed A, B, and C for the purposes of simplification. Each of these genes exists in two discrete forms: light and dark. For each of these three genes, an individual will inherit either one light or one dark allele, providing a total of six alleles. Individuals who inherit six copies of the light allele will have very pale skin, whereas individuals who inherit five light alleles and one dark allele will have slightly darker skin. Each additional dark allele has an additive effect, with inheritance of two to five dark alleles producing a range of intermediate skin tones. If two individuals who are heterozygous for genes A, B, and C produce an unlimited number of offspring together, 1/64 of the offspring would be expected to inherit six light alleles, 6/64 would inherit five light and one dark allele, 15/64 would inherit two dark alleles and four light alleles, 20/64 would inherit three of each allele, 15/64 would inherit four dark and three light alleles, 6/64 would inherit five dark and one light allele, and 1/64 would inherit six alleles that are all of the dark variety. This would produce seven different gradations of skin color. To reiterate, development of human skin color is even more complex than what is represented by this example, yet this description effectively captures the polygenic nature of this trait.

The contribution of each dark allele to human skin is production of a dark colored pigment called **melanin**, expression of more melanin-producing genes results in development of a darker skin color. The number of light and dark alleles that an individual may inherit is limited by the combination of alleles possessed by that individual's parents. For example, a parent who is heterozygous for genes A and B, and homozygous for the dark allele for gene C can only pass on zero, one, or two light alleles to each gamete, along with four to six dark alleles. The evolution of human skin color can be explained by a fascinating story that revolves around the fact that melanin blocks ultraviolet radiation from the sun from entering skin cells.

Epistasis

A second type of inheritance that involves interaction between two or more genes is epistasis. **Epistasis** occurs when expression of one gene alters or interferes with the expression of one or more other genes. Epistasis often manifests when the effect of one gene product silences or masks expression of a phenotype produced by the product of another gene. Pigment expression in a variety of animals including mice, dogs, and humans is controlled by epistasis. For example, the coat color in Labrador retrievers, a type of dog commonly referred to as labs, results from epistasis between two genes. One gene, customarily denoted by the letter B, determines whether Labrador retrievers will have black or brown fur. Black fur is the dominant trait so it is designated with the capital letter B, whereas brown fur, the recessive trait, is designated by the lower-case letter b. Labrador retrievers that inherit the genotype BB or

FIGURE 9.4: Epistasis in Labrador retrievers—The coat color of Labrador retrievers exemplifies the effect of epistasis, where one gene controls expression of another gene.

	BE	bE	Be	be
BE	BBEE (black)	BbEE (black)	BBEe (black)	BbEe (black)
bE	BbEE (black)	bbEE (brown)	BbEe (black)	bbEe (brown)
Be	BBEe (black)	BbEe (black)	BBee (yellow)	Bbee (yellow)
be	BbEe (black)	bbEe (brown)	Bbee (yellow)	Bbee (yellow)

Bb are capable of producing black fur, whereas Labrador retrievers with the genotype bb are capable of producing brown fur. These are commonly known as black labs and chocolate labs, respectively. The word "capable" is used here because whether or not an individual Labrador retriever produces black or brown fur is dependent upon the effect of a second gene. This second gene, which controls coat pigment expression, is represented by the letter E. Expression of coat pigment is the dominant trait, and non-expression of coat pigment is the recessive trait. Therefore, Labrador retrievers with the genotype EE or Ee will produce either black or brown fur, depending on their genotype for the B gene. Labrador retrievers with the genotype ee will not produce pigment in their fur regardless of the genotype they have for the B gene and instead will have light-colored fur. These individuals are known as yellow labs.

Figure 9.4 shows the expected outcomes of offspring produced by two Labrador retrievers that are each heterozygous for both of these genes (BbEe × BbEe). In this type of cross, 9/16 offspring are expected to have black fur, 3/16 are expected to have brown fur, and 4/16 are expected to have yellow fur. Other types of epistasis may produce different phenotypic ratios depending on the number of phenotypes affected and whether each phenotype is derived from a dominant or recessive allele. Similar to the Labrador retriever's example, the gene that causes albinism in humans is an epistatic gene; individuals with the genotype AA or Aa express pigment-producing gene products, yet individuals with the genotype aa do not produce pigment in their hair, skin, or eyes no matter what their genotype is for their pigment-producing genes. However, albinism in humans results from epistatic interactions involving a number of different genes expressed in different cell types.

Pleiotropy

Genes are often discussed as if each gene only affects a single trait. However, the products of most genes are known to affect multiple different, and often unrelated, characteristics. The ability of genes to affect two or more independent characteristics is known as **pleiotropy**. One example of pleiotropy in humans is sickle-cell disease, which involves a mutation in a single gene leading to a number of different effects throughout the body. Many other inherited diseases are also the result of pleiotropy.

Gene Linkage

When Gregor Mendel reported his observations in pea plants, he had chosen, through either great luck or great care, a set of traits controlled by genes that were not all located on the same chromosome. The independent assortment that Mendel observed, and which formed the basis of his laws of genetics, only occurs between genes located on different chromosomes. This is because individual chromosomes are segregated and independently shuffled into newly forming gametes during meiosis, yet the genes on each chromosome travel as a set. Gene alleles inherited together more often than independent assortment would predict are said to be **linked genes**. Gene linkage is not an all or nothing event. Whereas independent assortment predicts an even 50% that each allele from a gene pair will end up in the same gamete as any other randomly chosen allele, gene linkage results in a spectrum of probabilities that ranges from less than 50% to near 100% concordance.

The near infinite set of possible outcomes in allele combinations is the result of a process that occurs in meiosis known as genetic recombination. **Genetic recombination** produces offspring with combinations of alleles and traits that are different than those present in either parent. The mechanism that causes genetic recombination, called crossing-over, is described in Chapter 11. Because Gregor Mendel was presumably not aware of the concept of gene linkage, he likely would have had difficulty making sense of the phenotypic ratios he would have observed had he chosen to study any traits controlled by linked genes.

Sex-Linked Traits

Sex-linked traits are phenotypes that result from expression of genes located on the sex chromosomes. More specifically, traits derived from genes on the X chromosome are known as X-linked traits, and traits derived from genes on the Y chromosome are known as Y-linked traits. In humans, the complement of sex chromosomes that produces genetic males is one X chromosome and one Y chromosome, whereas genetic females have two X chromosomes. The sex chromosomes are the only one of the 23 pairs of chromosomes for which humans do not possess two homologous chromosomes. In other words, sex chromosomes are the only ones for which we do not have a pair of chromosomes that contain all of the same genes. As a result of this, inheritance of sex-linked traits does not follow Mendelian patterns of inheritance. Certain sex-linked traits are more common in one sex or the other, whereas some sex-linked traits, which are derived from genes located on the Y chromosome, are only observed in males.

The X and Y chromosome differ not only in which genes they contain but also in the number of genes found on each. The X chromosome contains approximately 1,100 genes, whereas the Y chromosome contains only 78 total genes, some of which are duplicates of the same gene. Because of this difference in size, there are many more X-linked traits than Y-linked traits. Most of the traits controlled by genes located on the Y chromosome are related to male sex differentiation, such as testes formation. The much larger X chromosome includes a number of genes not related to sexual characteristics, such as a gene that contributes to the development of color vision. Because females inherit an X chromosome from each parent, they can inherit an X-linked trait from either parent. However, fathers cannot pass on X-linked traits to their sons because males receive their only X chromosome from their mothers. Y-linked traits, on the other hand, are only passed from father to son.

Because females inherit twice as many X chromosomes as males, it might seem that X-linked traits, including X-linked disorders, would be more common in females. However, X-linked traits are actually much more common in males. One reason is that males express all of the alleles on their X chromosome, regardless of whether those alleles act in a dominant or recessive (or other) manner. Conversely, because females possess two copies of the X chromosome, some alleles on one X chromosome will be masked by alleles on the other

X chromosome. This is especially important in the case of recessive X-linked disorders. One example of such a disorder is Duchenne muscular dystrophy, which is caused by a mutation in the gene responsible for production of an important muscle protein known as dystrophin. People with Duchenne muscular dystrophy experience progressive deterioration of muscle function and coordination, and rarely survive to the age of 30. Women who are heterozygous for this gene will not develop Duchenne muscular dystrophy because their normal allele will produce dystrophin, effectively masking the mutated allele. These women will, however, be carriers of the disease-causing allele and will pass on the mutated allele to approximately half of their gametes. When these gametes are combined with those of a male who does not carry the Duchenne muscular dystrophy allele, all of the resulting female children will avoid developing the disease because of the presence of the normal X chromosome from their father. However, approximately 50% of the resulting male children will develop the disease because they received the mutation copy of the X chromosome from their mother. Because the male children receive the Y chromosome from their father and not his normal X chromosome, they do not benefit from the protection of the dominant allele. The 50% of male children who receive the normal X chromosome from their mother will not develop the disease and will not even be carriers.

As mentioned earlier, few genes are located on the Y chromosome, and most of them are involved in determination of sexual characteristics. Interestingly, genetic control of physical sexual characteristics is largely controlled by a single gene on the Y chromosome and its interaction with a single gene located on the X chromosome. The Y chromosome gene primarily responsible for triggering male sexual development is known as *SRY*, and the X chromosome gene primarily responsible for triggering female sexual development is known as *DAX*. The presence of SRY, as in someone with XY chromosomes, results in development of male sexual characteristics, whereas the absence of SRY, as in someone with XX chromosomes, results in development of female sexual characteristics. Occasionally, embryos will form containing an abnormal number of sex chromosomes. Some of these individuals inherit only one copy of the X chromosome and no Y chromosome. These individuals are genetically female and have Turner syndrome. Turner syndrome is characterized by infertility, heart defects, and physical abnormalities. It is also possible for individuals to inherit XXY for their sex chromosomes. These individuals are genetically male and have Klinefelter syndrome, which is characterized by low testosterone production, inability to make sperm, above average height, and increased risk of type 2 diabetes and heart defects. Even more rarely, some individuals with an XY chromosome complement carry an extra copy of DAX on their X chromosome resulting from a duplication mutation. These individuals, though genetically male, physically develop more like females.

Epigenetic Inheritance

The alternative modes of inheritance described in this chapter represent only a partial list of how inheritance and expression of genetic traits are modulated. It is important that discussion of genetic traits does not conclude without describing one more method of gene control that is of broad and rapidly growing interest to scientists today: epigenetics. In short, **epigenetic inheritance** is the passage of traits from parents to offspring that does not rely on the genetic code. In other words, epigenetics involves transmission of observable characteristics not encoded by the DNA sequence but instead result from modifications to DNA molecules involving the addition or removal of other molecules that modify gene expression. One of the most common examples of epigenetic modification is DNA methylation. Methylation involves the addition of methyl groups (CH_3) to certain nucleotide bases within genes. The presence of a large number of methyl groups on the nucleotides of a gene reduces how often that gene is transcribed to mRNA, effectively reducing the amount that gene product is expressed. For example, if a gene that encodes the sequence for producing an enzyme involved in pigment production is highly methylated, less pigment will be produced from that gene than if it was not highly methylated. Epigenetic inheritance has also been linked to more complex and consequential traits such as learning, memory, and intelligence; mood disorders such as depression and anxiety; and development of serious diseases such as cancer.

Each individual inherits a unique collection of epigenetic modifications. This is true even for identical twins, which helps explain why identical twins do not always appear completely identical even though they share the same genetic code. To understand why differences in epigenetic modifications occur even in twins, it may help to think of epigenetics as a tool cells use to mark which parts of the DNA sequence should receive the most attention and which parts can be ignored. Imagine identical twins each reading this textbook and highlighting excerpts they want to study later. Even though they have the same DNA, and each is reading the same information, they will likely choose to highlight different parts of the text. Later, when they return to study the parts of the book they highlighted, one might become better prepared for an exam over the material than the other if that twin chose to highlight information that is more useful for answering exam questions.

There are many types of epigenetic inheritance, and many different purposes and consequences for these modifications. One type of epigenetic inheritance that affects all of us is genomic imprinting. **Genomic imprinting** is a type of epigenetic modification in which the expression of certain genes is dependent upon which parent the gene was inherited from. Although each individual inherits two copies of each gene, one from each parent, expression of only one copy of some genes is necessary to carry out the functions of life. For these genes, one copy of the gene is silenced, or turned off, by epigenetic modifications. Genomic imprinting has some surprising implications. One example of the effects of genomic imprinting is the case of Angelman syndrome and Prader-Willi syndrome. Angelman syndrome is a nervous system disorder characterized by delayed development, intellectual disability, hyperactivity, difficulty sleeping, and a happy, excitable demeanor. Individuals with Prader-Willi syndrome experience much different symptoms including poor muscle tone, difficulty waking up, obesity, short stature, and behavioral and psychiatric disturbances. Both of these syndromes are caused by deletion of the same part of the same chromosome, chromosome 15. The sole factor that determines which syndrome someone with this mutation will develop is genomic imprinting; if the mutation occurs on the maternally derived copy of the chromosome 15, the individual will develop Prader-Willi syndrome, and if it occurs on the copy of chromosome 15 from the father, the individual will develop Angelman syndrome.

That babies are born at all is the result of a type of competition between maternally and paternally imprinted genes. This finding was demonstrated by an interesting set of experiments with mice conducted in the 1980s. In these experiments, scientists attempted to produce baby mice from two parents of the same sex. In one experiment, the DNA was removed from an unfertilized egg and replaced with the combined DNA from two females. In a second experiment, the DNA was removed from a different unfertilized egg and replaced with the combined

DNA from two males. Both eggs were stimulated to divide as embryos and implanted into the uterus of a surrogate mouse. Neither embryo successfully made it through the stages of fetal development to have a chance at producing a viable mouse pup. Each embryo failed to develop for a different and very telling reason. The embryo generated from two mothers appeared to have formed normally but failed to produce a placenta capable of capturing the nutrients and oxygen supply from the mother necessary to sustain life. The embryo generated from two fathers, on the other hand, produced a large and healthy looking placenta, but the embryo itself was nothing more than a disorganized blob of cells. In light of the phenomenon of genomic imprinting, these findings led to the conclusion that genes responsible for directing much of embryonic development are active only in the maternally derived DNA, whereas genes responsible for producing the placenta are active only in paternally derived DNA. The copies of the same genes in the opposite sex parent are somehow silenced through imprinting.

Evolutionarily, this arrangement can be explained by considering the competing interests of males and females in reproduction. It is in the interest of females to produce healthy offspring that will carry their genetic material into future generations, while not sacrificing so many resources to the developing fetus that the body of the mother is deprived of nutrients or harmed in any way. For much of human history, birth of babies that were too large was a leading cause of death of women. Thus, maternal genes are the subject of a conflict of interest in terms of balancing the needs of the mother and the fetus. However, it is in the interest of males that the developing fetus obtains as many nutrients as possible, growing big and healthy to increase chances of survival after birth, leading to continuation of the genetic lineage. Therefore, maternal genes have been imprinted to focus on maintaining maternal health while supporting fetal development, and male genes have been imprinted to focus on maximizing the ability of the placenta to extract nutrients from the mother.

Genomic imprinting is also one of the main roadblocks to the cloning of animals. To clone an animal, the nucleus of a cell from the body of a living adult animal is removed and fused into an egg cell from an individual of the same species that has had its nucleus removed. If the procedure works, the egg cell will start dividing and progress along the normal path of embryonic development. The embryo will be implanted into the uterus of another adult animal where it will grow and be born like a normal baby. The cloned baby contains the exact same DNA as the adult animal that donated the nucleus. There is no mixing of DNA from two individuals as there is during sexual reproduction. The first successfully cloned mammal, a sheep named Dolly, was born in 1996 and went on to live a relatively normal life, biologically speaking. Since then, approximately 20 species of animals, including fish, rabbits, pigs, and dogs have been successfully cloned. However, focusing on these successes obscures the fact that years of countless unsuccessful attempts and experimental trials had to be worked through to produce each successful clone. Many times, the fused egg would fail to start developing, would stop developing shortly after fusion, or the embryo would fail to thrive at some point during gestation. Although scientists verified the DNA sequence of the embryo contained no potentially fatal mutations, and no chromosomes were lost or abnormally structured, problems still arose with the development of hereditary traits. The root of these problems was that many genes, approximately 100 to 200 genes in a sheep, are affected by genomic imprinting. Because the entire set of DNA in a clone is derived from one individual, and that individual can be only genetically male or genetically female, some of the imprinted genes will be inactivated. If the donor of the DNA is female, the genes normally "turned off" in females will be off, and there will be no male complement of DNA to provide copies of the genes that are "turned on." The reverse would happen if the clone were derived from a male DNA donor. To produce a viable clone, scientists must find a way to circumvent genomic imprinting so that the clone receives an active copy of all necessary genes. We know from the fact that cloning has been successful that this has been done, however, the precise way that genomic imprinting can be overcome remains somewhat of a mystery. Despite the growing history of successes, it is still quite difficult to produce cloned animals. Unlike mutations that affect the genetic code, there is still no effective way to detect potentially harmful epigenetic changes in developing embryos.

Although genomic imprinting is an inherit method of selecting alleles based on the parental origin of the chromosome on which they are located, many other factors also influence different types of epigenetic inheritance. These include diet, exercise, health status, and drug use by parents before a child is conceived or during pregnancy. Not all epigenetic modifications occur before birth. Exposure to certain chemicals, high levels of stress, emotional trauma, and exercise can all cause epigenetic changes to an individual's DNA during their lifetime. These changes can affect the health of the individual later in life and can even be passed on to their offspring. Despite all that is already known about epigenetics, it clear that much remains to be discovered. Based on preliminary studies conducted within the last few years, it is likely that epigenetics will become a key factor in the development of individually customized treatments for diseases such as cancer and therapies for a wide array of other disorders and modifiable characteristics.

9.2 SUMMARY

- Not all traits follow Mendelian patterns of inheritance
- Incomplete dominance results in heterozygotes that express a phenotype that is intermediate to the dominant and recessive phenotypes
- Codominance results in heterozygotes that express both alleles for a trait
- Human skin color results from the additive effects of several genes through a process called polygenic inheritance
- Epistasis occurs when the effect of one gene product silences or masks expression of a phenotype produced by the product of another gene
- ABO blood types are an example of multiple alleles and codominance
- X-linked traits and genetic disorders are more common in males than females
- Alleles for genes located close together on the same chromosome are more likely to be inherited together than genes located far apart or on different chromosomes because of the phenomenon of gene linkage
- Epigenetic inheritance results from hereditary factors that are not encoded in the DNA that account for a wide range of variability in physical traits and disease susceptibility

9.3 KNOWLEDGE CHECK

1. All of the following modes of inheritance are inconsistent with Gregor Mendel's laws of genetics, EXCEPT:
   a. complete dominance
   b. incomplete dominance
   c. gene linkage
   d. polygenic inheritance
   e. sex-linked traits

2. ABO blood types are an example of which of the following?
   a. gene linkage
   b. incomplete dominance
   c. multiple alleles
   d. X-linked trait
   e. Y-linked trait

3. What are the possible blood types of the offspring of someone with type A blood and someone with type B blood?
   a. AB only
   b. A and B only
   c. A, B, and AB only
   d. A, B, and O only
   e. A, B, AB, and O

4. Which of the following occurs when expression of one gene alters or interferes with the expression of one or more other genes?
   a. codominance
   b. complete dominance
   c. epistasis
   d. incomplete dominance
   e. polygenic inheritance

5. Which of the following statements about X-linked traits is true?
   a. X-linked traits are expressed more often in females because they inherit twice as many copies of X chromosome genes
   b. X-linked traits are expressed more often in females because females can be carriers of X-linked recessive traits
   c. X-linked traits are expressed more often in females because they always receive an X chromosome from both parents
   d. X-linked traits are expressed more often in males because they do not have a second copy of the gene involved that could mask expression of a recessive trait
   e. X-linked traits are expressed more often in males because the presence of the SRY gene determines expression of sex-linked traits

6. Identify the inheritance pattern that produces pink carnations.

7. People with blonde hair are more likely to have blue eyes than people who do not have blonde hair, whereas people with brown hair are more likely to have brown eyes than people who do not have brown hair. Explain the pattern of inheritance responsible for this observation.

8. Explain how genomic imprinting affects the success of cloning.

9.4 ACTIVITY

Genetics of ABO Blood Types Case Study

ABO blood groups have important implications for blood transfusions. Receiving blood from a non-compatible donor may cause a potentially life-threatening immune system reaction that destroys donor red blood cells. ABO blood typing also provides some information that can be used to help track inheritance of traits from parents to offspring. Although ABO blood types do not provide enough unique information to confirm whether two individuals are related, some purported relationships can be ruled out if there is no possibility that a given blood type could have been inherited from a given pair of individuals. The purpose of this activity is to investigate inheritance of ABO blood types and the significance of ABO blood types to blood transfusions. This will be accomplished by working individually or in small groups, as directed, to complete the case study titled "Why Is He Different from Both Parents?: The Genetics of ABO Blood Types" published by the National Center for Case Study Teaching in Science (NCCSTS).

"Why Is He Different from Both Parents?: The Genetics of ABO Blood Types" http://sciencecases.lib.buffalo.edu/cs/files/blood_type_genetics.pdf

9.5 ACTIVITY

ABO Blood Types and Pedigree Analysis Case Study

The purpose of this activity is to use ABO blood groups as an example for analyzing inheritance patterns with Punnett squares and pedigree analysis. A pedigree is a tool used to trace the passage of traits through multiple generations involving multiple family members. This activity will also provide an opportunity to apply knowledge of autosomal and sex-linked traits to predict inheritance patterns. The learning objectives for this activity will be accomplished by working in small groups to complete the case study titled "Bloodline: A Human Genetics Case" published by the National Center For Case Study Teaching In Science (NCCSTS).

Bloodline: A Human Genetics Case http://sciencecases.lib.buffalo.edu/cs/collection/detail.asp?case_id=572&id=572

9.6 ACTIVITY

Epigenetics Case Study

The purpose of this activity is to examine how epigenetic modifications lead to changes in gene expression that may affect development of complex traits such as mental illness. This activity provides an opportunity to investigate and explain how epigenetic factors may produce different phenotypes in identical twins and how drugs that target epigenetic modifications such as methylation could potentially be used to treat disorders such as schizophrenia. The learning objectives for this activity will be accomplished by working individually or in small groups to complete the case study titled "Identical Twins, Identical Fates?: An Introduction to Epigenetics" published by the National Center For Case Study Teaching In Science (NCCSTS).

Identical Twins, Identical Fates?: An Introduction to Epigenetics http://sciencecases.lib.buffalo.edu/cs/files/epigenetics.pdf

9.7 ACTIVITY

Cloning Case Study

The purpose of this activity is to discuss the concept of human cloning and promote understanding of the process of cloning. This activity provides an opportunity to evaluate the ethical and social implications of a scientific technology. This activity involves working in small groups to complete the case study titled "Bringing Back Baby Jason: To Clone or Not to Clone" published by the National Center For Case Study Teaching In Science (NCCSTS).

"Bringing Back Baby Jason: To Clone or Not to Clone" http://sciencecases.lib.buffalo.edu/cs/files/cloning.pdf

ACTIVITY QUESTIONS

9.4

1. Follow instructions to work individually or form small groups for completing the case study and answering the indicated questions from the case study.

9.5

1. Follow instructions to work individually or form small groups for completing the case study and answering the indicated questions from the case study.

9.6

1. Follow instructions to work individually or form small groups for completing the case study and answering the indicated questions from the case study.

9.7

1. Follow instructions to work individually or form small groups for completing the case study and answering the indicated questions from the case study.

9.8 REFERENCES

Cassidy, S. B. and Schwartz, S. (1998). Prader-Willi and Angelman syndromes. Disorders of genomic imprinting. *Medicine (Baltimore)* , *77*(2), 140–151.

Contu, L. and Hawkes, C. A. (2017). A review of the impact of maternal obesity on the cognitive function and mental health of the offspring. *International Journal of Molecular Sciences, 18*(5), 1093.

Dean, L. (2005). *Blood groups and red cell antigens.* Bethesda, MD: National Center for Biotechnology Information (US), Chapter 5, The ABO blood group. PDF e-book.

Freda, V., Gorman, J. G., and Pollack, W. (1966). Rh Factor: Prevention of isoimmunization and clinical trial on mothers. *Science, 151*(3712), 828–830.

McGreevy, J. W., Hakim, C. H., McIntosh, and Duan, D. (2015). Animal models of Duchenne muscular dystrophy: From basic mechanisms to gene therapy. *Disease Models & Mechanisms, 8*, 195–213.

McPherson, S., McMullin, M. F., and Mills, K. (2017). Epigenetics in myeloproliferative neoplasms. [On-line] *Journal of Cellular and Molecular Medicine.* doi: 10.1111/jcmm.13095.

Miller, W. J. (1997). Dominance, codominance and epistasis. *Brazilian Journal of Genetics, 20*(4), 663–665.

Pulst, S. M. (1999). Genetic linkage analysis. *Archives of Neurology, 56*(6), 667–672.

Ridley, M. (1999). *Genome: The autobiography of a species in 23 chapters.* New York: HarperCollins Publishers.

Swain, A., Narvaez, V., Burgoyne, P., Camerino, G., and Lovell-Badge, R. (1998). Dax1 antagonizes SRY action in mammalian sex determination. *Nature, 391*, 761–767.

Whitfield, F. (2001). Imprinting marks clones for death. [On-line] *Nature*. Last modified July 6, 2001. doi:10.1038/news010712-1.

FURTHER READING

Darwin, C. (1859). *The origin of species by means of natural selection or the preservation of favoured races in the struggle for life.* 1st ed. London: Penguin.

Federoff, N. V., and Brown, N. M. (2006). *Mendel in the kitchen: A scientist's view of genetically modified food.* Washington, D.C.: Joseph Henry Press.

Johnson, S. (2006). *The ghost map: The story of London's most terrifying epidemic – and how it changed science, cities and the modern world.* New York, NY: Riverhead Books.

Judson, O. (2002). *Dr. Tatiana's sex advice to all creation.* New York, NY: Henry Holt and Company.

Kean, S. (2012). *The violinist's thumb: And other lost tales of love, war, and genius, as written by our genetic code.* New York, NY: Back Bay Books.

Kolata, G. (2017). *Mercies in disguise: A story of hope, a family's genetic destiny and the science that rescued them.* New York, NY: St. Martin's Press.

Maddox, B. (2002). *Rosalind franklin: The dark lady of DNA.* New York, NY: Harper Collins.

Marantz Henig, R. (2000). *The monk in the garden: The lost and found genius of Gregor Mendel, the father of genetics.* 1st ed. Boston: Houghton Mifflin Company.

Monaghan, P., and Birkhead, T. (2010). Dad V Mom: The ultimate battle of the sexes. *New Scientist,* 205(2749), 36–39. Web.

Mukherjee, S. (2011). *The emperor of all maladies: A biography of cancer.* New York, NY: Scribner.

Preston, R. M. (1995). *The hot zone: The terrifying true story of the origins of the Ebola virus.* New York, NY: Anchor Books.

Sachs, J. S. (2007). *Good germs, bad germs: Health and survival in a bacterial world.* New York, NY: Hill and Wang.

Shubin, N. (2008). *Your inner fish: A journey into the 3.5-billion-year history of the human body.* New York, NY: Vintage Books.

Skloot, R. (2010). *The immortal life of Henrietta Lacks.* 1st ed. New York, NY: Broadway paperbacks.

Watson, J. (2001). *The double helix: A personal account of the discovery of the structure of DNA.* New York, NY: Touchstone.

Cell Division and Cancer

10.1 INTRODUCTION

The **cell cycle** is the ordered sequence of events in the life of a cell, from its formation until it divides to make two new cells. The cell cycle consists of an interphase, during which the cell grows and replicates its genetic material and cellular contents, as well as the mitotic phase, in which replicated DNA is separated, along with the cytoplasm and organelles, into two daughter cells. The main steps of this process are similar for most of the somatic (non-reproductive) cells of eukaryotic organisms, including those belonging to animals, plants, and fungi. Cells contain a multitude of regulatory mechanisms to ensure progression through the cell cycle occurs properly, and that cessation of the cell cycle occurs when appropriate. However, sometimes damage to DNA causes these regulatory mechanisms to fail, which may have disastrous consequences. One example of such a consequence is cancer. Cancer is sometimes called "a disease of the cell cycle" because it results from loss of cell cycle control leading to unregulated cell division. This chapter describes the cell cycle and discusses key components of cell cycle regulation, cancer development, and cancer treatment.

Alternatives to Mitosis

The part of the cell cycle in which the genetic material in the nucleus is divided is called **mitosis**. Each round of mitosis begins with one somatic cell and results in two genetically identical somatic cells. This process is necessary for replacing worn out or damaged cells throughout the body, as well as for growth of tissues and body structures during development. Mitosis occurs only in eukaryotic somatic cells. This is in contrast to the process of meiosis used for production of reproductive gametes of eukaryotic organisms, which results in the formation of four genetically different daughter cells. Prokaryotic cells replicate through yet a different process known as binary fission. During **binary fission**, prokaryotic cells reproduce asexually by replicating their chromosome, then splitting their cell in half so that two cells are formed, each containing all of the necessary components of that cell. Binary fission is a much simpler process than the mitotic phase of eukaryotic cells because prokaryotes do not have a

LEARNING OBJECTIVES

- Synthesize an ordered list of the steps in the cell cycle
- Summarize what occurs in each stage of interphase (G1, S, G2, and G0 phase)
- Describe what occurs in each stage of the mitotic phase (prophase, metaphase, anaphase, telophase, and cytokinesis)
- Identify the role of cyclins and cyclin-dependent kinases (CDKs) in cell cycle regulation
- Describe how cancer cells differ from normal cells in terms of their cell cycle
- Discuss how mutations in BCL2, BRCA1/BRCA2, P53, and telomerase may affect cell cycle regulation and contribute to the development of cancer
- Discuss the purpose of apoptosis
- Describe how metastasis and density-dependent inhibition affect cancer development
- Define the term carcinogen and provide an example
- Describe two different ways that cancers may be classified
- List three types of cancer treatments and describe when each is most appropriate

nuclear membrane or membrane-bound organelles, and the majority of their DNA is located on a single chromosome.

Mitotic Phase

During the mitotic phase of the cell cycle, replicated DNA condenses into chromosomes and is evenly divided into two identical nuclei during the formation of new daughter cells. Mitosis can be divided into four main phases: prophase, metaphase, anaphase, and telophase. Sometimes mitosis is described as including another phase between prophase and metaphase, known as prometaphase; however, for the purposes of this discussion, the events that are sometimes ascribed to prometaphase will be discussed as part of prophase.

Prophase is the first phase of mitosis. During prophase, four major events occur (Figure 10.1). First, long strands of DNA in the form of **chromatin** condense around histones and other associated proteins into compact chromosomes. Prophase occurs after DNA replication has occurred, so at this point there are four copies of each chromosome. The two copies of each chromosome that were initially present in the parent cell are called **homologous chromosomes**. The two identical copies of each of the homologous chromosomes that were produced during DNA replication are referred to as **sister chromatids**. Prior to mitosis, the chromatin was loosely organized in the cell, similar to an unwound spool of yarn. This facilitated access of enzymes to the DNA for the purposes of transcription and DNA replication. However, during mitosis, sister chromatids must be paired together and evenly divided, so it is important that the chromatin is condensed into compact, well-organized structures, much like yarn that has been neatly wound around a spool.

Next, the nuclear envelope dissolves. This allows the chromosomes to be pulled from the area of the original nucleus toward opposite ends of the cell. To accomplish this, assemblages of microtubule proteins called **mitotic spindle fibers** attach to each chromosome and pull them toward one of the poles of the cell. The third major event that occurs in prophase is movement of centrioles toward opposite ends of the cell. **Centrioles** are cellular structures made of a protein called tubulin. Centrioles form in pairs called **centrosomes,** which function to organize the mitotic spindle fibers responsible for separating homologous chromosomes pairs during mitosis. Formation of mitotic spindle fibers and attachment of these fibers to the sister chromatids is the fourth major event of prophase. To use an analogy, centrosomes can be thought of as similar to the reel of a fishing pole, whereas mitotic spindle fibers are like the fishing line, and sister chromatids (chromosomes) represent the fish. The region where spindle fibers attach to the chromosome is known as the **centromere**. The connection of the centromere with the spindle fiber is strengthened by association with a complex of proteins called a **kinetochore**. Thus, centromeres are like the hook attached to a fishing line and the kinetochore is similar to the mouth of the fish (although it is located in the center of the chromosome rather than at one end).

The second phase of mitosis is **metaphase** (Figure 10.2). During metaphase, sister chromatids line up at an imaginary structure in the center of the cell called the **metaphase plate**. Sister chromatids are moved into position by the pull of mitotic spindle fibers and remain attached to each other until after metaphase. Chromosomes are highly condensed and neatly organized during metaphase, making this an ideal time for chromosomal analysis. It is during metaphase that cells are often checked to determine whether they contain the appropriate number of chromosomes or if they are the product of aneuploidy. This determination can be accomplished using a karyotype, which is a visual display of all the chromosome pairs ordered in by size.

During **anaphase**, which is the third phase of mitosis, sister chromatids separate from each other and move toward opposite ends of the cell as they are pulled apart by the mitotic spindle fibers (Figure 10.3). The point at which sister chromatids separate from each other is technically when they become known as chromosomes (even though the term chromosome is often used to refer to sister chromatids prior to this). At the same time as the chromosomes move toward the poles of the cell, components of the cytoskeleton begin to elongate the cell in preparation for forming two separate cells.

Telophase is the last phase of mitosis (Figure 10.4). During **telophase,** mitotic spindle fibers disassemble and a new nuclear membrane forms around each genetically identical daughter nuclei. This process is essentially the reverse of what occurs during prophase. Elongation of the cell, which began in anaphase, continues to occur throughout telophase. By the end of telophase, the cell appears as if it is two complete cells joined together by a shared plasma membrane.

During the four phases of mitosis, only the nuclear contents are divided. However, the mitotic phase is not complete until the cytoplasm and other cellular contents have also been separate into two distinct cells. This occurs immediately after telophase through a process known as **cytokinesis**. Cytokinesis

FIGURE 10.1: Prophase—During prophase, the nuclear membrane dissolves, chromatin condenses, and mitotic spindles form.

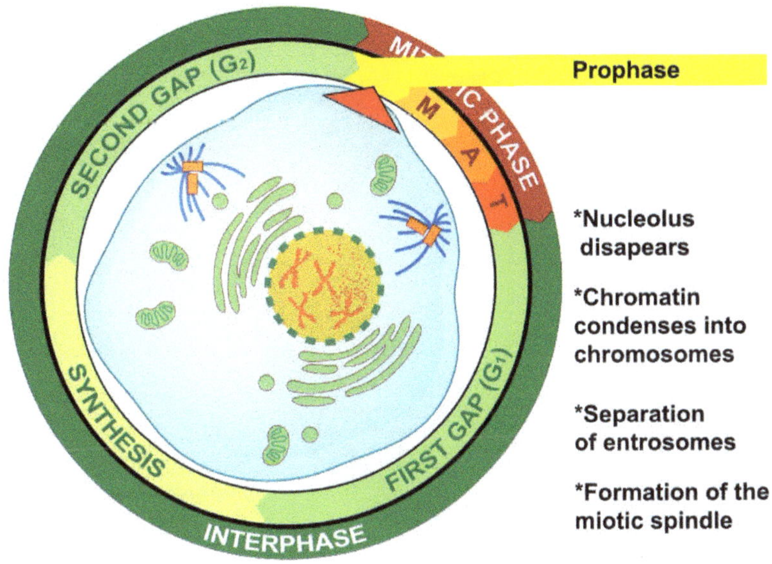

FIGURE 10.2: Metaphase—During metaphase chromosomes line up in the center of the cell along an imaginary line called the metaphase plate.

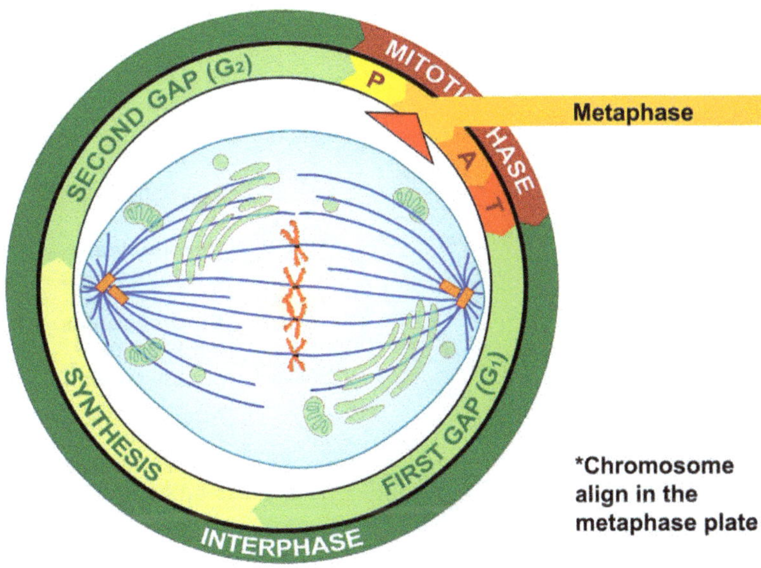

FIGURE 10.3: **Anaphase**—During anaphase, spindle fibers pull chromosomes to opposite ends of the cell.

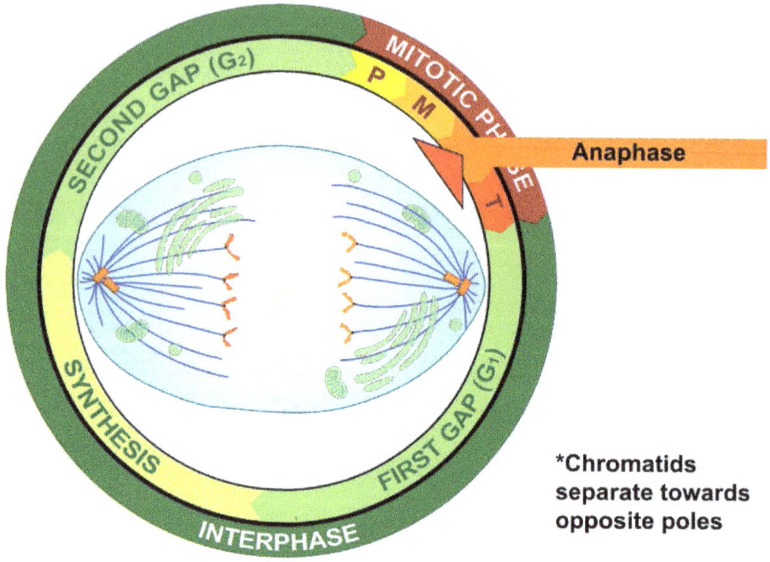

FIGURE 10.4: **Telophase**—During telophase, nuclear membranes reform around each set of chromosomes, spindle fibers dissolve, and chromatin unwinds. Telophase is immediately followed by cytokinesis, which is when the cytoplasm and plasma membrane split to form two separate cells.

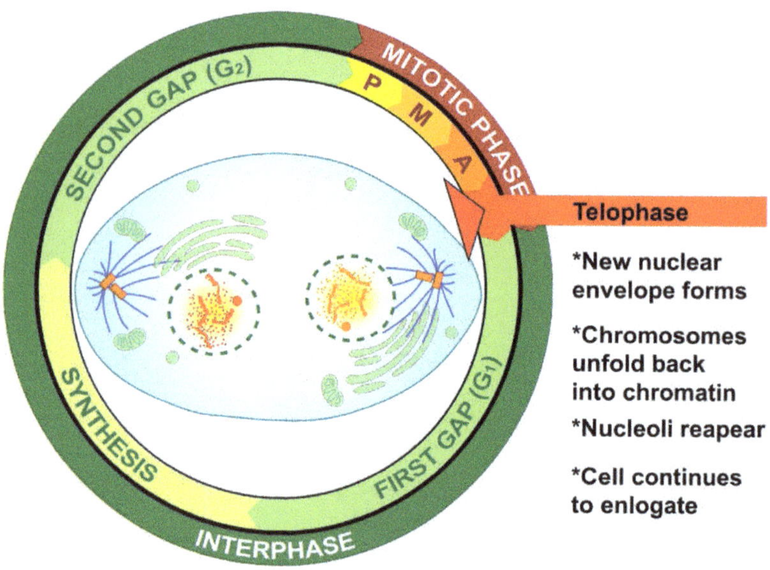

actually begins some time during anaphase or telophase but cannot be completed until after the end of telophase, so it is often listed as occurring after mitosis. In animal cells, cytokinesis occurs by a process known as cleavage in which the cellular membrane gradually pinches together while forming a structure known as a cleavage furrow. In plant cells, a structure known as a cell plate forms along the center of the cell to separate the two daughter cells. The cell plate forms a connection between the cell wall on either end of the cell and provides a foundation for formation of a new cell wall between the daughter cells.

Human cells complete the entire mitotic phase in approximately 24 hours. Some types of eukaryotic cells, such as yeast, complete their mitotic phase in as little as 90 minutes. The frequency with which a cell undergoes mitosis varies widely among cell types, even within the same organism. For example, the epithelial cells that comprise the outer layer of human skin divide much more rapidly than the nerve cells found in our brains. Although a single cell may divide dozens of times during its lifespan, the process of cell division accounts for a relatively short amount of time.

Interphase

When not actively dividing, cells spend the majority of their time—up to 90% or more—in another part of the cell cycle called interphase. During **interphase,** cells grow, replicate their DNA, and complete all of their necessary cellular functions. Interphase is subdivided into three stages based on the different activities the cell undergoes in each stage.

The first stage of interphase is known as the first gap (G1) phase (Figure 10.5). The term "gap" is a misnomer that stems from early incorrect assumptions that this period of the cell cycle was merely a gap in cellular activity. Although few observable structural changes occur within a cell during this phase, it is now known that cells are highly active on a chemical level during **G1 phase**. In addition to constantly completing normal cellular functions, cells must begin to prepare for DNA replication during G1 phase. This involves accumulation of energy and DNA building blocks, such as nucleotides, as well as production of proteins, such as DNA polymerase, that will be used during DNA replication. Cells also grow in size during gap phases, so a more appropriate term for them might be "growth" phases. Depending on the type of cell and the signals it is receiving from other cells, a single cycle through the G1 phase may last from hours to years. Once the cell is signaled to prepare for another round of division, it will begin to produce an identical copy of all of its genetic material.

DNA is replicated during the second stage of interphase, known as the synthesis (S) phase. Cells usually contain two copies of each chromosome; one inherited from each parent. As mentioned earlier, these two copies of each chromosome are known as homologous chromosomes. During **S phase**, DNA replication produces one identical new copy of each homologous chromosome (Figure 10.6). The two identical copies of each homologous chromosome are called sister chromatids. From S phase until telophase in mitosis, the cell will contain four complete sets of DNA. This means that a human somatic cell, which usually contains one pair of 23 chromosomes for a total of 46 chromosomes, will contain two pairs of 23 chromosomes for a total of 92 chromosomes at the completion of S phase. Given the size of the human genome and the rate at which our cellular machinery is capable of replicating DNA, most of our cells must spend approximately 10–12 hours in S phase to complete the task of DNA replication.

S phase is quickly followed by the final stage of interphase, the second gap (G2) phase (Figure 10.7). The main outcomes of the **G2 phase** are cellular growth, duplication of some organelles, and energy storage. This prepares the cell to become two complete daughter cells during the mitotic phase. Cells that have not received signals to progress into S phase may enter a fourth stage of interphase, known as **G0 phase**. This is known as cell cycle arrest because the cell stops moving through the stages of the cell cycle while in G0 and instead enters a resting state called quiescence. The G0 phase

FIGURE 10.5: **First gap phase**—During G1 phase, the cell grows and prepares for DNA replication.

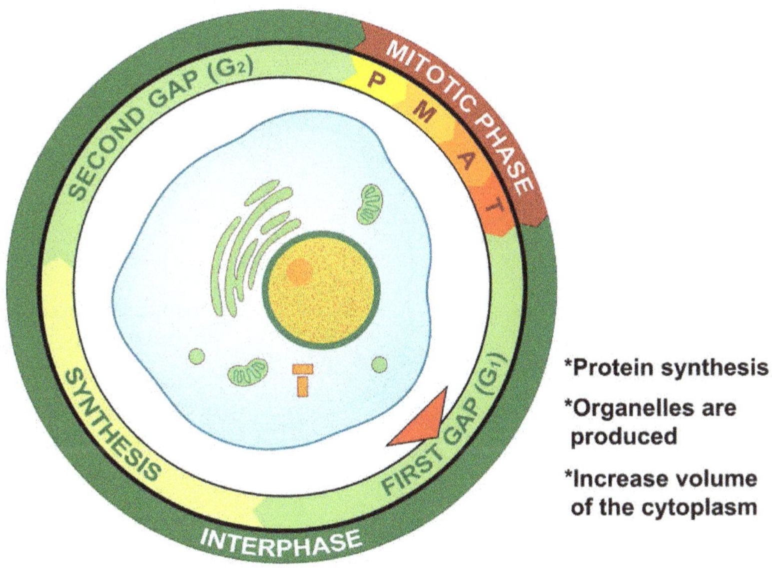

FIGURE 10.6: Synthesis phase—During S phase, DNA is replicated.

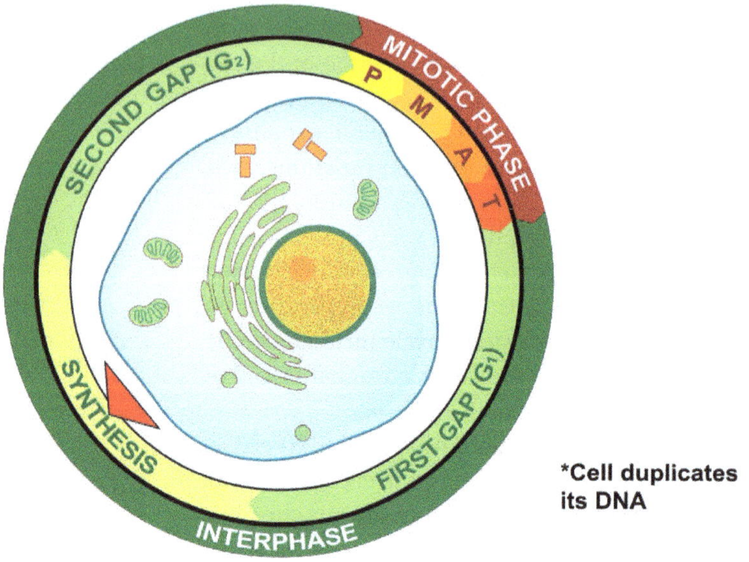

FIGURE 10.7: Second gap phase—During G2 phase, organelles are produced and the cell grows in size.

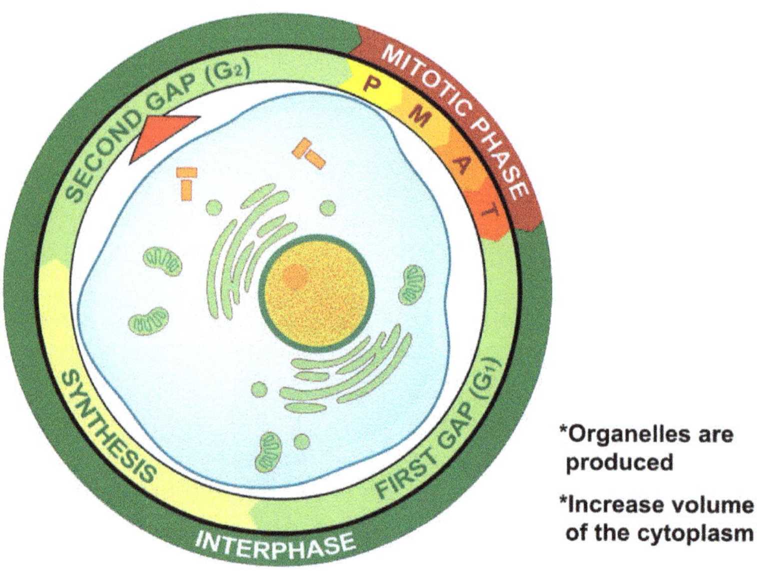

was once thought of as a type of cellular retirement from which cells never returned to the work force of the body; however, it is now known that some cells will reenter the cell cycle from G0 if they later receive appropriate signals.

Cell Cycle Checkpoints

To maintain the health and integrity of an organism, it is crucial for cell division to occur only when it is needed. Because of this, the cell cycle is highly regulated. Throughout interphase are a number of cell cycle checkpoints at which regulatory mechanisms within the cell make sure there are no serious issues within the cell's DNA or other structures. The cell also checks for signals from other cells to indicate whether it is appropriate for it to continue on to the next stage of the cell cycle or to wait until given the proper signals to do so. Depending on which stage of the cell cycle the cell is in, this could mean verifying that sufficient growth has occurred, checking newly synthesized DNA for mistakes and completeness, or waiting for exterior conditions to be present that are favorable for cell division.

To prevent abnormal proliferation, cells possess an entire network of internal regulatory mechanisms that monitor conditions within the cell. Most normal cells will not proceed to the next phase of the cell cycle until the appropriate signals are received. The G1 checkpoint can be considered the most important cell cycle regulator, because once the cell completes the early part of

G1 phase, it will almost certainly continue all the way through S phase and G2 phase and enter mitosis. Between G1 and S phase is another checkpoint called the G1/S checkpoint. At this point, the cell verifies the integrity of the DNA before DNA replication commences. Near the end of S phase, there is another checkpoint responsible for ensuring that the DNA has been replicated without error. If mutations are detected, the cell will attempt to repair them. When there are no errors detected in the DNA, the cell proceeds to G2 phase. However, the next main checkpoint does not occur until the middle of mitosis. This checkpoint, known as the M checkpoint or the spindle checkpoint, ensures that spindle fibers are properly attached to each of the sister chromatids. This is crucial because separation of sister chromatids during anaphase is an irreversible process that may result in the generation of a potential fatal chromosomal abnormality if carried out improperly. Once a cell clears the M checkpoint, it will complete mitosis and continue on to cytokinesis.

If, during any of the checkpoints, a cell determines that it contains errors that cannot be fixed, it may undergo programmed cell death called **apoptosis**. Cells of the immune system also engage in routine surveillance of the cells in the body to determine if they contain abnormalities, and if so, may signal the abnormal cell to undergo apoptosis. This process prevents abnormal cells from producing daughter cells containing errors and also ensures that cells that are not functioning normally do not continue to use the body's resources or otherwise cause damage to the organism. Approximately 50 million cells are eliminated from the adult human body every day through the process of apoptosis! For the majority of these cells, apoptosis is triggered as a result of the limitations of normal cellular aging; however, many cells that are on their way to forming cancerous tumors are destroyed in this manner. Although this is a very large number of cells for one body to be continually losing and replacing, it represents only a fraction of 1% of the 37 trillion cells estimated to be comprise an adult human body.

The intrinsic methods of cell cycle regulation are complemented by exterior cell cycle control mechanisms. One way that cells signal to each other when it is time to divide is by releasing growth factors. **Growth factors** are proteins or lipid hormones that stimulate cellular growth, differentiation, and replication. One example of the effect of external growth factors is a phenomenon known as **density-dependent inhibition** that prevents overcrowding by stopping cell division in cells that are closely touching neighboring cells. This type of regulation is necessary for maintaining proper tissue structure and function. Cells that ignore density-dependent inhibition may form abnormal clumps of cells called tumors. If the mass of abnormal cells is not capable of infiltrating nearby tissue or spreading to other parts of the body, it is known as a benign tumor. **Benign tumors** are not cancerous and are generally not harmful, although they may sometimes cause discomfort or other problems if they compress a nearby nerve or obstruct a blood vessel or duct. In contrast, **malignant tumors** are cancerous groups of abnormally proliferating cells capable of damaging nearby tissues, possibly spreading to other parts of the body.

Cyclins and Cyclin-Dependent Kinases (CDKs)

A second important mechanism for regulating cell cycle progression involves two complementary sets of proteins called cyclins and cyclin-dependent kinases (CDKs) (Figure 10.8). **Cyclins** are proteins that accumulate throughout different phases of the cell cycle until they reach a level that signals the cell to advance to the next phase of the cell cycle. At that point, the level of that cyclin will drop and the level of a different cyclin may start to rise to control movement

FIGURE 10.8: Cyclin-dependent cell cycle regulation—Cyclins and cyclin-dependent kinases control movement through the stages of the cell cycle.

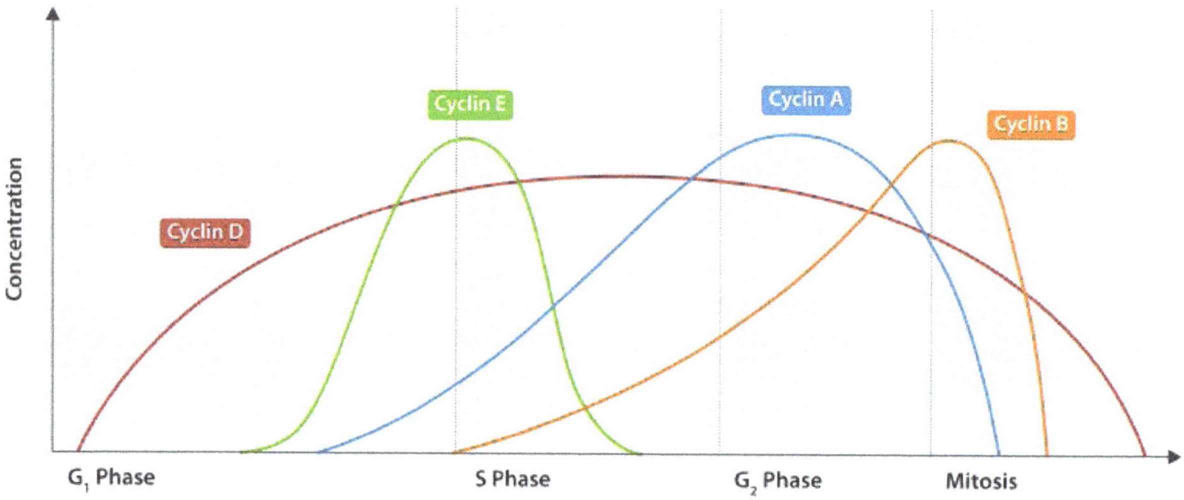

through the next phase of the cycle. This process repeats over and over again as the cell completes each cell cycle phase. It is not the cyclins themselves that control the actions of the cell but rather their effect on CDKs that is important. As cyclins build up in the cell, they bind to and activate CDKs. **CDKs** are enzymes that speed up various chemical processes important for cell division. For example, during G2 phase, levels of cyclin B steadily rise and bind to a type of CDK known as Cdk1. Together, cyclin B and Cdk1 form a functional protein complex called mitosis-promoting factor (MPF) that helps break apart the nucleus and build spindle fibers out of microtubules. These steps are required for the cell to leave interphase and enter the first phase of mitosis (prophase). By the end of mitosis, levels of a different cyclin known as cyclin D will build up and start initiating the steps required to progress from mitosis into G1 phase. Different cyclin/CDK pairs are responsible for regulating passage from G1 phase to S phase, and from S phase to G2 phase. If a cell acquires a mutation that inactivates one of the genes that encodes a cyclin or CDK, it may become stuck in one of the phases of the cell cycle. Alternatively, a mutation that results in overproduction of a cyclin/CDK pair may push the cell into the next phase of the cell cycle prematurely and may even cause the cell to divide when it should not. It has already been demonstrated experimentally that fusing a cell in mitotic phase to a cell that is in G1 phase will cause the latter cell to immediately begin mitosis even though the DNA has not been duplicated.

Cancer

Cancer accounts for more deaths worldwide than any other cause except cardiovascular diseases. Although cancer can be thought of as a single entity, the term **cancer** actually does not refer to a single disease but rather describes a group of highly variable diseases. Cancers vary greatly in terms of their cause, site of origin, biochemical characteristics, and effect on the body. However, all cancers are united with one commonality, which is that they all result from abnormal cell division. Because of this, cancer is sometimes referred to as a "disease of the cell cycle." There are a number of reasons why cancerous cells may stop adhering to cues of normal cell cycle regulation. One reason is that cancer cells may undergo changes that allow them to escape the need for external growth factors to grow and divide. This may be because they have improperly started to produce their own growth factors or because they overexpress growth factor receptors causing them to overreact to normal growth stimuli. A second explanation for the development of cancerous growth is that CDKs may become overactive in cancer cells, and proteins that regulate the function of CDKs may stop working. Cancerous cells also ignore internal and external signals to stop growing and dividing, and continue to proliferate abnormally.

Some cancer cells acquire the ability to leave their site of origin and start growing and dividing in another part of the body. The development of a malignant tumor at a secondary site in the body is known as **metastasis**. Figure 10.9 shows tumors of pancreatic cancer that have grown in the liver. Although this cancer is now growing in the liver, it is still considered pancreatic cancer because the cells it is made from are derived from the pancreas. To most effectively treat this cancer, physicians must recognize that it is metastatic pancreatic cancer, rather than liver cancer, and treat it as such. Metastatic cancers are usually much more difficult to treat than non-metastatic cancers and generally have a poorer prognosis. To be effective, treatments for metastatic cancers must penetrate the entire body rather than just a localized area.

Another trait of cancer cells that allow them to grow in an uncontrolled manner is the ability to evade destruction by the immune system. Cells throughout the body are constantly undergoing changes that lead them to grow or divide abnormally, but the immune system is often able to recognize and eliminate potentially problematic cells before they start to cause damage. However, cancer cells may be able to evade the immune system by altering the expression of molecules on their cell surface to signal immune cells not to destroy them, thus allowing them to escape the body's natural defense system.

Each of the characteristics of cancer cells described in this chapter is the result of at least one individual genetic mutation that causes the loss or gain of expression of a gene product that affects cell cycle regulation. A common misconception about cancer is that it results from a single mutation, when, in fact, most cancers do not start to develop until multiple mutations in key genes have accumulated within the same cell. This process often occurs over a period of years, during which time there are usually no noticeable signs that any potentially cancer-causing changes have occurred. Other misconceptions about cancer are that the same thing causes all cancers and that one day it may be possible to develop a single cure that prevents or treats all cancers. As mentioned earlier, cancer is not one disease but many different diseases. The constellation of genetic mutations and other factors that lead to the development of cancer differs between types of cancer, between individuals with the same type of cancer, and even among different cancerous cells within the same individual. Therefore, a panacea for cancer would only be effective if it was able to target something common to cancers with unrelated origins.

Cancer patients often undergo genetic testing as part of their diagnosis to determine which genetic mutations are present in the cells of their tumor(s). This often helps oncologists determine how to best treat the cancer. For example, breast cancer is characterized based on the presence or absence of three cell surface receptors: estrogen receptor (ER), progesterone receptor (PR), and human epidermal growth factor receptor 2 (HER2). Estrogen and progesterone are hormones that regulate development and function of the female reproductive system and some secondary sexual characteristics. These hormones are produced by female reproductive organs, as well as by some other tissues such as the adrenal gland and the liver. Some breast cancer cells express high numbers of one or both of these receptors, allowing them to react to these hormones as strong growth factors. This essentially means that estrogen fuels the growth of ER-positive breast cancer cells, and progesterone fuels the growth of PR-positive breast cancer cells. Therefore, medicines that block these hormones from binding to their receptors can be used to help slow or stop the growth of cancerous tumors that rely on these receptors to fuel their growth. This type of treatment is called hormone therapy. Hormone therapy can be highly effective for treating some cancers; however, most cancers do not rely on these receptors to fuel their growth and thus will not respond to this type of treatment. Likewise, a drug called Herceptin that blocks HER2 from binding to the protein it recognizes is highly effective at treating HER2-positive breast cancers but has no effect on HER2-negative cancers. Breast cancer tumors in which less than 1% of cells express any of these three receptors are known as triple-negative breast cancers. These cancers are often the most difficult types of breast cancer to treat because treatment options are limited.

Several examples of how mutations in genes involved in cell cycle regulation may lead to the development of cancer are described in Table 10.1. These genes are divided into two main categories called proto-oncogenes (oncogenes) and tumor suppressor genes. Proto-oncogenes usually produce products that facilitate normal cell growth, but if these genes acquire a mutation that causes them to become overactive or permanently active, they become **oncogenes**. Expression of oncogenes promotes cancer development and/or growth. One example of an oncogene is BCL2 (Table 10.1). The normal function of BCL2 is to suppress apoptosis, which prevents normal cells from unnecessarily undergoing apoptosis because of an oversensitive response to pro-apoptotic signals. If a genetic mutation leads to overexpression of BCL2, this will prevent an abnormal cell from undergoing apoptosis even when it should. This may lead to unregulated cell division and development of a malignant tumor.

FIGURE 10.9: **Liver cancer**—This images shows a human liver filled with a type of cancer known as adenocarcinoma. This cancer metastasized to the liver from a primary tumor in the pancreas of this individual.

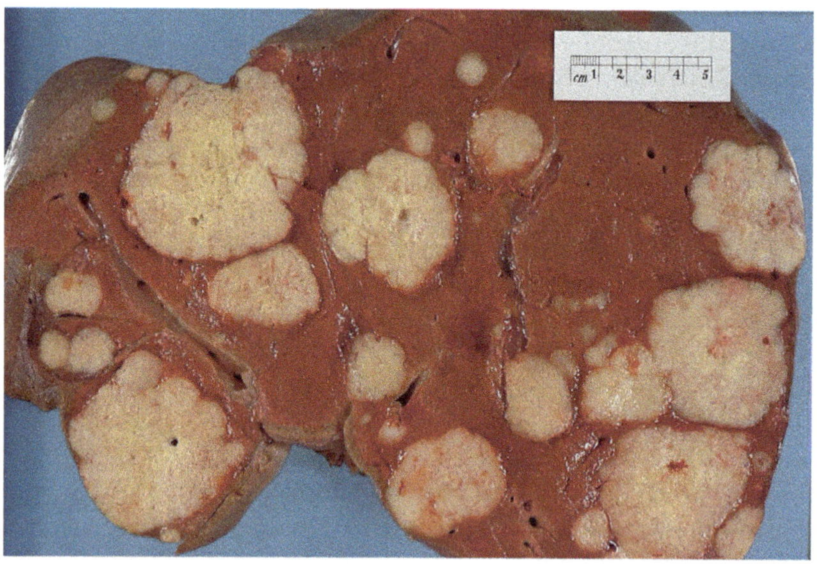

Telomerase genes are proto-oncogenes that are normally only active during embryonic development and in stem cells. In adults, telomerase genes are normally inactive, unless a mutation is incurred that reactivates their expression. Collectively, telomerase genes produce an enzyme called telomerase, which lengthens the protective telomeres on the ends of DNA strands

TABLE 10.1: Examples of Genes Responsible for Cell Cycle Regulation

Gene (protein)	Activity in Normal Cells	Activity in Cancer Cells
BCL2	BCL2 suppresses apoptosis and is involved in other cellular functions	Known as an oncogene; cancer cells produce elevated levels of BCL2 protein, which can prevent abnormal cells from undergoing apoptosis when they should
BRCA1/BRCA2	BRCA genes produce enzymes that repair damaged DNA	Known as a tumor suppressor gene; when mutated in cancer cells they no longer repair DNA damage, increasing the likelihood that harmful mutations will accumulate
P53	P53 is an important regulator at the G1/S and G2/M checkpoints; when cells experience DNA damage or other stressors, levels of p53 protein increase, preventing progression through the cell cycle and/or inducing apoptosis	Known as a tumor suppressor gene; when mutated in cancer cells they will continue to grow and divide even when they should undergo apoptosis
Telomerase genes	Telomerase genes are normally only active in stem cells; they produce telomerase, an enzyme that lengthens the telomeres on the ends of DNA strands after each round of DNA replication	In cancer cells, these genes become active allowing cells to continue dividing after they normally would stop

after each round of replication. This is important because the cellular machinery responsible for DNA replication invariably cuts off a few nucleotide bases from the end of sequence each time DNA is replicated. If the coding sequence of genes extended all the way to the end of the DNA strand, important parts of genes would be lost every time a cell divides. Telomeres provide a buffer zone to prevent that from happening, and when part of a telomere is cut off, there are no negative consequences. That is, until the cell has divided so many times that it runs out of telomeres. At this point, the cell must stop dividing and must inevitably die. The alternative is to risk accumulating potentially deleterious mutations. Therefore, the length of one's telomeres places a limit on the number of times an individual's cells can divide and theoretically determines the individual's maximum lifespan.

Indeed, analysis of the DNA of the oldest surviving humans in recent history has shown they tend to have abnormally long telomeres. A potential solution to this problem is the restorative telomerase that our genomes are capable of producing, yet cease to do so long before we reach old age. This observation has led to speculation that finding a way to reactivate telomerase could be a veritable fountain of youth, allowing us to live longer than currently possible. The problem with this is that our cells are programmed to stop dividing and die at a certain point for a very important reason. Cells that live too long acquire genetic mutations and other problems that increase the risk of cancer development and other abnormal cellular functions. When telomerase genes get switched back on, cells that should be entering G0 phase and/or undergoing apoptosis may continue to divide uncontrollably. This is, in fact, a key to the success of cancerous tumors. Cancer cells are so adept at dividing and dividing and dividing that they are sometimes referred to as "immortalized." There are, right now, cancerous cells growing in laboratories that were isolated from patients whose bodies died many decades ago. One famous example is Henrietta Lacks who died in 1951, and whose ovarian cancer cells continue to live on not only in laboratories around the world, but also in a 2010 novel and 2017 movie, both titled *The Immortal Life of Henrietta Lacks.*

Expression of **tumor suppressor genes**, on the other hand, normally limits progression through the cell cycle. Mutations that reduce or prevent expression of tumor suppressor genes lead to unregulated cell division. For example, P53 is an important regulator at the G1/S and G2/M checkpoints. When cells experience DNA damage or other stressors, levels of P53 protein increase, preventing damaged or abnormal cells from progressing through the cell cycle. If a mutation inactivates P53, it will no longer stop abnormal cells from dividing, which may also lead to development of a malignant tumor. P53 is one of the most common mutations linked to cancer, found in approximately 50% of all cancerous tumors. Mutations in two other tumor suppressor genes, known as BRCA1 and BRCA2, account for 20%–25% of hereditary breast cancers, as well as significant number of ovarian cancers. Approximately 70% of women who inherit both of these mutations will develop breast cancer in their lifetime, compared with about 12% of all women. This effect is so strong that many women who discover they have these mutations seek drastic treatments such as mastectomy and hysterectomy to prevent the development of cancer before it even develops.

Mutagens and Carcinogens

Substances and exposures that can lead to the development of cancer are called **carcinogens**. Many carcinogens are **mutagens**, which are physical or chemical agents that cause mutations in DNA. Some carcinogens do not directly affect the DNA sequence but instead disrupt biochemical processes in a way that promotes uncontrolled cell division. Fortunately, most genetic mutations and metabolic disruptions will not lead to cancer; only those that adversely affect genes or cellular processes involved in cell cycle regulation are potentially carcinogenic. Despite this criterion, many diverse carcinogens exist. Examples of carcinogens include chemicals such as asbestos and formaldehyde, and certain types of radiation, such as gamma rays, x-rays, and ultraviolet radiation. Exposure to physical and chemical carcinogens may be passively encountered in the environment, such as by breathing polluted air, or caused by engaging in certain behaviors such as smoking or firefighting. A number of other behavioral factors have also been linked to the development of cancer as a result of their ability to damage tissues or induce a chronic inflammatory state, which itself has been shown to increase the risk of cancer. These include eating low-fiber, high-fat diets; obesity; and living a sedentary lifestyle. Infectious agents can also be carcinogenic because of the way they interact with host cell DNA or because they cause chronic inflammation, which leads to cancer-promoting changes in cells. A number of viruses, such as human papilloma virus, Rous sarcoma virus, and hepatitis B virus have all been shown to cause cancer, as have some bacteria such as *Helicobacter pylori.*

Types and Stages of Cancer

Cancer is often described as not just one single disease but many different diseases. Just as there any many different causes of cancer, there are many different types of cancer. There are also many ways to classify different cancers. One common method of classifying cancer is based on the type of cell in which the cancer originated. Based on this characteristic, cancers can be divided into three main categories: (i) carcinomas are cancers that arise in the epithelial cells that line the airways, digestive tract, and compose the outer layers of the skin; (ii) leukemia, lymphoma, and myeloma, also known as hematopoietic cancers, are cancers that originate in different types white blood cells found in the bone marrow, blood, and lymphatic system; and (iii) sarcomas are cancers that affect connective tissues, such as fat, muscle, and bone. Sometimes cancers are identified by referring to the region of the body in which they originated; for example, cancers of the mouth, tongue, throat, and nasal passages are all referred to as head and neck cancers, whereas cancers of the bladder, kidney, prostate, and testicles are referred to as genitourinary cancers. Cancer may also be identified based on the specific organ in which it originated. In men, the organs in which cancer most often originates are the prostate, lungs, colon and rectum, urinary bladder, and the skin; in women, cancer most commonly originates in the breast, followed by the lungs, colon and rectum, uterus, and thyroid. In both men and women, lung cancer is responsible for more deaths than any other type of cancer.

The type of cancer is just one of many factors that affects the prognosis and outcome of a cancer diagnosis. Another important factor affecting disease outcome, as well as treatment, is the stage of progression of the cancer in the body. Stage 0 cancers consist only of small tumors that have not invaded surrounding tissues. These early stage cancers are also known as "in situ" cancer, which means "cancer in place." Stage I cancers are relatively small primary tumors that have begun to invade surrounding tissue but are still confined to the localized region. Stage II cancers are also still confined to the localized region but are characterized by a larger, more advanced primary tumor. When the cancer has spread to nearby regional lymph nodes, it becomes known as a Stage III cancer, and when a cancer has metastasized to lymph nodes or tissues in distant parts of the body, it becomes a Stage IV cancer. Early stage cancers make be difficult to detect because the primary tumors are small and often do not cause recognizable symptoms. However, late stage cancers generally have poorer prognosis. The type of treatment that is appropriate, and the expected outcome of that treatment, varies greatly depending on the type of cancer and other factors that may vary between individuals, such as which genetic mutations contributed to development of the cancer and the age of the person with the cancer.

Cancer Treatment

Although there are many different types of cancer, there is also a wide array of treatments available for treating cancer. Oncologists choose which type of treatment to use for each specific case of cancer based on a number of factors including the type of cancer; how far the cancer has progressed and whether it has metastasized; the genetic mutations associated with the cancer; the age, gender, and ethnic background of the patient; the patient's overall health; and how well the patient will tolerate the side effects of a particular treatment. The main types of cancer treatment are chemotherapy, radiation therapy, hormone therapy, immunotherapy, surgery, and adjuvant therapy.

Chemotherapy involves administration of chemical compounds that have the potential to inhibit cancerous growth and shrink the size of malignant tumors. In simplistic terms, chemotherapy works by disrupting the replication process of dividing cells and is therefore most effective when used to treat rapidly growing cancers. Chemotherapy is often the preferred treatment for metastatic cancer because the chemical compounds circulate throughout the body and can target cancer cells at multiple sites simultaneously. However, chemotherapy often causes undesirable effects such as hair loss that many patients would prefer to avoid, as well as harsh side effects such as nausea and constipation that is sometimes so severe it cannot be tolerated. Radiation therapy can also shrink tumors and slow cancerous growth by disrupting the division of cells that divide abnormally quickly. However, radiation must be delivered from a static source and directed at a particular area of the body, such as the brain or the breast. Because of this, radiation is not as effective at treating metastatic cancer as chemotherapy is. Surgery involves the physical removal of a malignant tumor. This approach is most effective at treating tumors easily accessible by a surgeon, have not thoroughly invaded surrounding tissue, and have not metastasized. Surgery may also be preferred for treating malignant tumors that are slow growing, because chemotherapy and radiation are not optimally effective for treating cancers that do not divide rapidly. Sometimes chemotherapy is combined with radiation and/or surgery. Adjuvant therapy is the use of chemotherapy after surgery. This type of treatment is particularly effective for treating cancers largely confined to one or a small number of easily accessible sites when there is a high risk that a small number of cells have started to metastasize. Hormone therapy, as discussed earlier, is useful for targeting cancers dependent on hormones such as estrogen to promote their growth. One final type of cancer treatment is immunotherapy. This treatment relies on administration of biological agents to mimic or augment the natural immune response the body uses to control tumor growth. In addition to use of these therapies, doctors may suggest cancer patients adopt lifestyle modifications to increase their chances of survival and good health. These include change of diet, smoking cessation, abstinence from alcohol, increased physical activity, and weight loss.

10.2 SUMMARY

- The cell cycle is the ordered sequence of events in the life of a cell, which can be divided into mitosis (prophase, metaphase, anaphase, and telophase), cytokinesis, and interphase (G1, S, and G2)
- During prophase, chromosomes condense, the nuclear membrane dissolves, and mitotic spindle fibers appear
- During metaphase, chromosomes line up along the center of the cell
- During anaphase, sister chromatids move toward opposite ends of the cell
- During telophase, mitotic spindle disassemble and nuclear membranes form around daughter nuclei
- During interphase, the cell grows, replicates organelles and DNA, and moves through a series of cell cycle checkpoints before proceeding to divide
- Cancer cells escape normal cell cycle control by improperly producing or responding to growth factors, resisting apoptosis, and evading immune system surveillance
- Oncogenes and tumor suppressor genes involved in regulating the cell cycle are often mutated in cancerous cells
- Carcinogens are physical and chemical factors that cause cancer by creating genetic mutations or interfering with important metabolic processes

- There are many different types of cancer, which may be classified based on the type of cell, body region, or organ in which they originated
- Cancer is classified into five stages, with Stage 0 being the smallest and most confined, and Stage 4 being the most widespread with the poorest prognosis
- Chemotherapy, radiation therapy, surgery, and a number of other treatments are available to treat cancer patients, depending on the characteristics of their tumor, health, and lifestyle

10.3 KNOWLEDGE CHECK

1. If an otherwise normal cell does not receive growth factors, what is most likely to occur?
   a. it will be destroyed by the immune system
   b. it will continually repeat the S phase until growth factors are received
   c. it will enter G0 phase until growth factors are received
   d. it will immediately undergo apoptosis
   e. it will start to produce its own growth factors

2. During which phase of the cell cycle is DNA replicated?
   a. G0 phase
   b. G1 phase
   c. G2 phase
   d. M phase
   e. S phase

3. During which part of the mitotic phase are chromosomes pulled toward opposite ends of the cell?
   a. anaphase
   b. cytokinesis
   c. metaphase
   d. prophase
   e. telophase

4. Which of the following occurs the soonest after interphase?
   a. telophase
   b. prophase
   c. metaphase
   d. cytokinesis
   e. anaphase

5. Increased activity of which of the following is most likely to cause cancer to form?
   a. apoptosis
   b. density-dependent inhibition
   c. kinetochores
   d. oncogenes
   e. tumor suppressor genes

6. Describe how the cell cycle of cancer cells differs from the cell cycle of normal cells.

7. Describe how a mutation in BCL2 might lead to cancer progression, then compare this to the effect that a mutation in BRCA1 might have on cancer progression.

8. List four different types of cancer treatment and describe when each is most effective.

10.4 ACTIVITY

Mitosis Worksheet

The goals of this activity are to identify the stages of mitosis, describe what occurs during each part of interphase, and discuss various ways that genes and proteins regulate the cell cycle.

FIGURE 10.10: Cells in various stages of the cell cycle

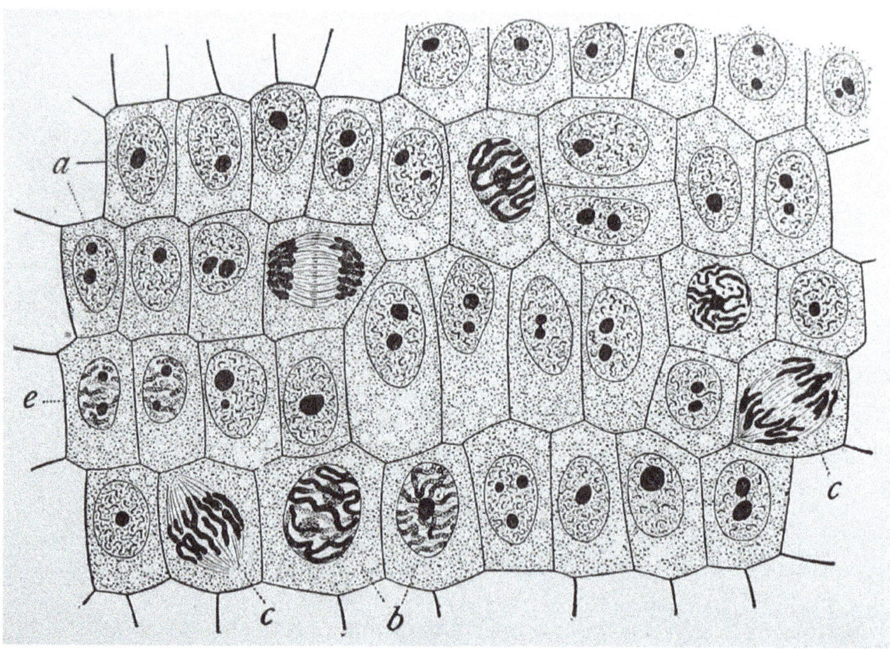

This activity can be completed in small groups, as instructed, after reviewing the relevant course materials. Use Figure 10.10 to answer Questions 1–3.

10.5 ACTIVITY

Cell Cycle and Cancer Virtual Lab

The main objectives of this activity are to describe how the cell cycle of cancer cells is different than the cell cycle of normal cells, and to discuss different types of cancer treatment. This activity also provides an opportunity to analyze cancerous and non-cancerous cells from different tissues as part of virtual laboratory simulation titled "Virtual Lab: The Cell Cycle and Cancer" which is provided by Glencoe. This virtual lab also reiterates the events that occur in each stage of mitosis. Virtual Lab: The Cell.

For this activity, work in groups of two unless otherwise instructed. First, open "Virtual Lab: The Cell Cycle and Cancer" by entering the following link into a web browser:

http://www.glencoe.com/sites/common_assets/advanced_placement/mader10e/virtual_labs_2K8/labs/BL_03/.

Then, once the virtual lab is open, follow the Procedure on the left hand side of the screen. Answer the questions provided for this activity instead of those in the virtual lab.

ACTIVITY QUESTIONS

10.4

1. In Figure 10.10, identify whether each labeled cell is in anaphase, interphase, metaphase, prophase, or telophase.

   a ____________________ b ____________________

   c ____________________ e ____________________
2. For the phase not identified in Question 1, draw a sketch of a cell that is in that phase.
3. Do the cells in Figure 10.10 appear to be animal cells or plant cells? Explain your reasoning.
4. Describe what occurs in G0 phase, G1 phase, G2 phase, and S phase.
5. In most normal adult cells, telomerase genes are not actively expressed. Propose a hypothesis that might explain the evolutionary advantage of this.
6. Describe how a mutation in BRCA1/BRCA2 might lead to the development of cancer.

TABLE 10.2: Cancer Virtual Lab Data Table

	Number of cells in each phase of the cell cycle*					% cells dividing
	Anaphase	Interphase	Metaphase	Prophase	Telophase	
Normal lung						
Cancerous lung						
Normal Stomach						
Cancerous Stomach						
Normal Ovary						
Cancerous Ovary						

10.5

1. Complete Table 10.2 with the data you collected from the virtual lab. Note: Count all cells on the screen, not just the ones you labeled.
2. List the following phases of the cell cycle in order, and briefly describe what occurs during each phase: anaphase, interphase, metaphase, prophase, telophase.
3. Describe how the cell cycle of cancer cells differs from the cell cycle of normal cells.
4. Is the cell cycle the same for all normal cells? Explain.
5. List three different types of treatment for cancer and describe the factors that would determine which one would be most appropriate for treating a particular case of cancer.
6. Common side effects of chemotherapy treatment include hair loss and gastrointestinal problems, but it is rare for chemotherapy to damage the lungs or brain cells. Propose an explanation for this observation, using what you now know about the cell cycle and chemotherapy to support your claim.

10.6 REFERENCES

Adams, J. M. and Cory, S. (2007). The Bcl-2 apoptotic switch in cancer development and therapy. *Oncogene, 26*, 1324–1337.

Collignon, J., Lousberg, L., Schroeder, H., and Jerusalem, G. (2016). Triple-negative breast cancer: Treatment challenges and solutions. *Breast Cancer* , *8*, 93–107.

Cong, Y. S., Wright, W. E., and Shay, J. W. (2002). Human telomerase and its regulation. *Microbiology and Molecular Biology Reviews* , *66*(3), 407–425.

Contu, L. and Hawkes, C. A. (2003). Breast and ovarian cancer risks due to inherited mutations in BRCA1 and BRCA2. *Science, 302*(5645), 643–646.

Iyengar, N. M., Gucalp, A., Dannenberg, A. J., and Hudis, C. A. (2016). Obesity and cancer mechanisms: Tumor microenvironment and inflammation. *Journal of Clinical Oncology, 34*(35), 4270–4276.

Iyengar, N. M. and Morris, P. G. (2013). Identifying the challenges of biomarkers that predict response to HER2-directed therapies. *Expert Opinion on Pharmacotherapy, 14*(11), 1437–1440.

Kumar, M., Lechel, A., and Günes, C. (2016). Telomerase: The devil inside. *Genes (Basel), 7*(8), doi: E43.

Luo, Q., Beaver, J. M., Liu, Y., and Zhang, Z. (2017). Dynamics of p53: A master decider of cell fate. *Genes (Basel), 9*(8), E66.

Mikhailov, A. and Rieder, C. L. (2002). Cell cycle: Stressed out of mitosis. *Current Biology, 12*(9), R331–R333.

Müller, I., Boyle, S., Singer, R. H., Bickmore, W. A., and Chubb, J. R. (2010). Stable morphology, but dynamic internal reorganisation, of interphase human chromosomes in living cells. *PloSOne, 5*(7), e11560.

Sudhakar, A. (2010). History of cancer, ancient and modern treatment methods. *Journal of Cancer Science and Therapy* , *1*(2), 1–4.

Surbone, A. (2011). Social and ethical implications of BRCA testing. *Annals of Oncology, 22*, i60–i66.

FIGURE CREDITS

FIG. 10.1: LadyofHats, "Prophase," https://commons.wikimedia.org/wiki/File:Prophase_eukaryotic_mitosis.svg. Copyright in the Public Domain.

FIG. 10.2: LadyofHats, "Metaphase," https://commons.wikimedia.org/wiki/File:Metaphase_eukaryotic_mitosis.svg. Copyright in the Public Domain.

FIG. 10.3: LadyofHats, "Anaphase," https://commons.wikimedia.org/wiki/File:Anaphase_eukaryotic_mitosis.svg. Copyright in the Public Domain.

FIG. 10.4: LadyofHats, "Telophase," https://commons.wikimedia.org/wiki/File:Telophase_eukaryotic_mitosis.svg. Copyright in the Public Domain.

FIG. 10.5: LadyofHats, "First Gap phase," https://commons.wikimedia.org/wiki/File:First_gap_cell_lifecycle.svg. Copyright in the Public Domain.

FIG. 10.6: LadyofHats, "Synthesis phase," https://commons.wikimedia.org/wiki/File:Synthesis_cell_lifecycle.svg. Copyright in the Public Domain.

FIG. 10.7: LadyofHats, "Second Gap phase," https://commons.wikimedia.org/wiki/File:Second_gap_cell_lifecycle.svg. Copyright in the Public Domain.

FIG. 10.8: WikiMiMa, "Cyclin-dependent cell cycle regulation," https://commons.wikimedia.org/wiki/File:Cyclin_Expression.svg. Copyright in the Public Domain.

FIG. 10.9: Haymanj, "Liver cancer," https://commons.wikimedia.org/wiki/File:Secondary_tumor_deposits_in_the_liver_from_a_primary_cancer_of_the_pancreas.jpg. Copyright in the Public Domain.

FIG. 10.10: Edmund Beecher Wilson, "Cells in various stages of the cell cycle," The Cell in Development and Inheritance (2nd ed.), New York: The Macmillan Company, 1900. Copyright in the Public Domain.

Meiosis and Reproduction

LEARNING OBJECTIVES

- Describe differences between asexual and sexual reproduction
- Explain the difference in chromosome number in somatic cells and gametes
- Discuss the role of meiosis in the life cycle of humans
- Describe the structure of chromosomes, including homologous pairs
- Summarize the steps of meiosis
- Discuss types and causes of chromosomal abnormalities
- Compare and contrast the two types of cell division, mitosis and meiosis

11.1 INTRODUCTION

Meiosis is a type of cell division used by organisms for **sexual reproduction**. Sexual reproduction occurs when sex cells from two parents are combined to produce a new unique offspring different from both parents. This differs from **asexual reproduction**, which results in offspring genetically identical to the one parent. Some organisms produce offspring only through sexual reproduction, some use only asexual reproduction, and some use both. For example, humans reproduce exclusively through sexual reproduction, whereas bacteria only reproduce asexually. Most fungi and many types of plants are capable of undergoing both asexual and sexual reproduction. Even some animals, such as the sea star, are capable of both types of reproduction. Usually the sea star produces sperm and eggs to reproduce sexually; however, if one of their legs is cut off, it can regrow an entire body, forming a new organism through asexual reproduction. Other examples of animals that reproduce asexually include invertebrates like coral and hydra, which can produce a whole new adult organism by budding off of the parent organism. Plants utilize asexual reproduction in many different ways. One way is by sending out a stem above and along the ground that takes root and produces another plant. Look at Figure 11.1 to see different examples of asexual reproduction. Both types of reproduction have advantages and disadvantages. In fact, some organisms may even switch from one kind to another depending on the environmental conditions. Table 11.1 summarizes the advantages and disadvantages of asexual and sexual reproduction.

The haploid cells directly produced as a result of meiosis are called **gametes** or sex cells in animals or spores in plants. Gametes can be egg and sperm, like those in animals. In plants the spores will undergo mitosis to produce the gametes, ovules and pollen. During the process of **fertilization,** the two sex cells come together to produce a zygote. In animals, the zygote undergoes mitosis to produce more cells, eventually becoming an embryo. In the early stages of development, the embryo is referred to as a morula, and later it is known as a blastocyst. During this time, the cells begin to specialize, turning on certain genes in some cells and different ones in others. This specialization process lays the foundation for the development

FIGURE 11.1A: Sea star regenerating legs

FIGURE 11.1B: Stolon

FIGURE 11.1C: *Hydra oligactis*

of special tissues, including the ones that will undergo meiosis and produce gametes in the adult. At about 8 weeks post fertilization, when all of the major body structures have formed, the embryo becomes known as a fetus. An overview of the human life cycle, including the processes of meiosis and fertilization, is shown in Figure 11.2.

Most cells of organisms that undergo sexual reproduction contain two sets of each chromosome. Cells that contain this full set of chromosome pairs are referred to as **diploid,** which is sometimes represented as 2n because the number of chromosomes varies between species. In Table 11.2, there are some examples of the number of chromosomes for several different organisms. Notice that the number of chromosomes does not reflect how "complex" an organism is. Because the genetic material of two cells is combined during fertilization, the cells produced by meiosis must have half the number of chromosomes normally present in the **somatic** cells of that type of organism. Somatic cells are all the cells that are not gametes. A cell that contains only one of each chromosome is called **haploid.** This is generally represented as n. If gametes were diploid instead of haploid, newly forming organisms would have twice as many chromosomes as they should.

TABLE 11.1: Asexual/Sexual Reproduction Comparison

	Advantages	Disadvantages
Asexual reproduction	• Can occur rapidly • Uses less energy (don't need to find a mate or produce special cells) • All advantageous traits passed on	• Disadvantageous traits are passed on • Loss of variation is unlikely to result in adaptations to changing environmental conditions • Risk of extinction is increased
Sexual reproduction	• Increases variation • Some advantageous traits are passed on • More likely to result in adaptations to changing environmental conditions	• Takes longer since there are many steps • Some advantageous traits are not passed on • Uses more energy (looking for a mate, producing special cells)

Chromosome Structure

During meiosis, the duplicated chromosomes form **homologous pairs**, which are chromosomes similar in size and genetic information. In humans, there are 22 homologous pairs of autosomes, or non-sex chromosomes, and one pair of sex chromosomes. One member of the pair is inherited from the mother and the other from the father. Each member of the pair is made of sister chromatids. **Sister chromatids** are identical chromosomes joined together at their centromeres. The **centromere** of a chromosome is a pinched-in area that separates the two "arms" of a chromosome. There is also a structure made of proteins called a **kinetochore**, which will attach to the spindle that pulls apart the chromosomes during meiosis. The ends of chromosomes contain stretches of noncoding DNA that help to protect the coding sections of the chromosome called **telomeres**. Figure 11.3 illustrates the main chromosomal structures.

During prophase I of meiosis, the homologous chromosomes pair up together during a process called **synapsis**. It is during this process that the homologous chromosomes undergo the process of crossing-over or recombination. During **crossing-over**, segments of homologous chromosomes are swapped, resulting in a new combination of maternal and paternal alleles. This formation of new combinations of alleles during meiosis is also called **recombination**. Crossing-over and genetic recombination only occurs between homologous chromosomes. It would be disastrous for genetic recombination to involve non-homologous chromosomes because it would result in duplication of some genes and loss of others. Crossing-over happens randomly within chromosomes and occurs at different locations within

FIGURE 11.2: **Human life cycle**—The stages of the human life cycle from diploid adults to haploid gametes via meiosis and back to diploid via fertilization.

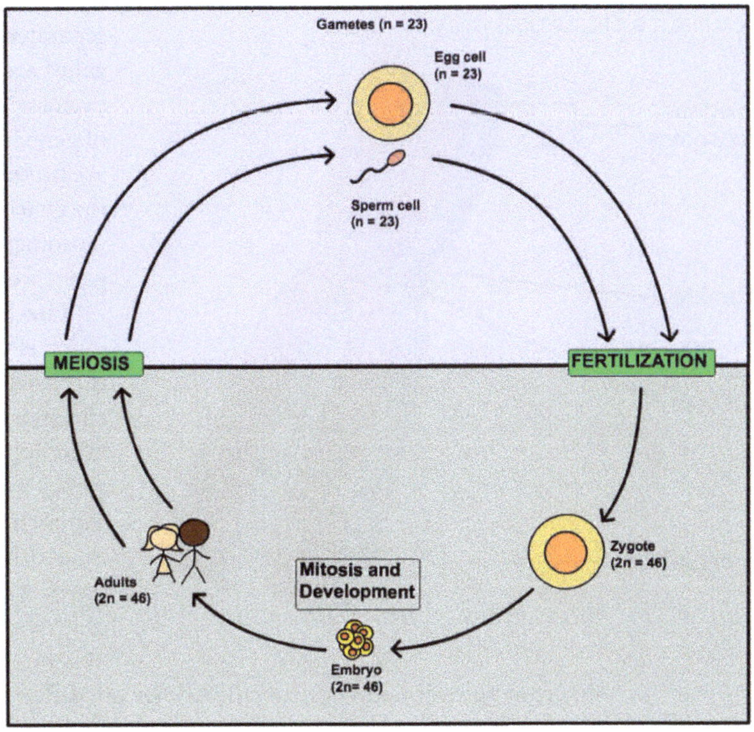

TABLE 11.2: Chromosome Counts for Various Organisms

Organism	Scientific name	Diploid number of chromosomes
Great white shark	*Carcharodon carcharias*	82
Cotton	*Gossypium hirsutum*	52
Chimpanzee	*Pan troglodyte*	48
Human	*Homo sapiens*	46
Earthworm	*Lumbricus terrestris*	36
Rice	*Oryza sativa*	24

FIGURE 11.3: Homologous chromosome pair—The two red chromosomes are identical so they are called sister chromatids. They are joined together by the kinetochore. The red and blue chromosomes are a homologous pair.

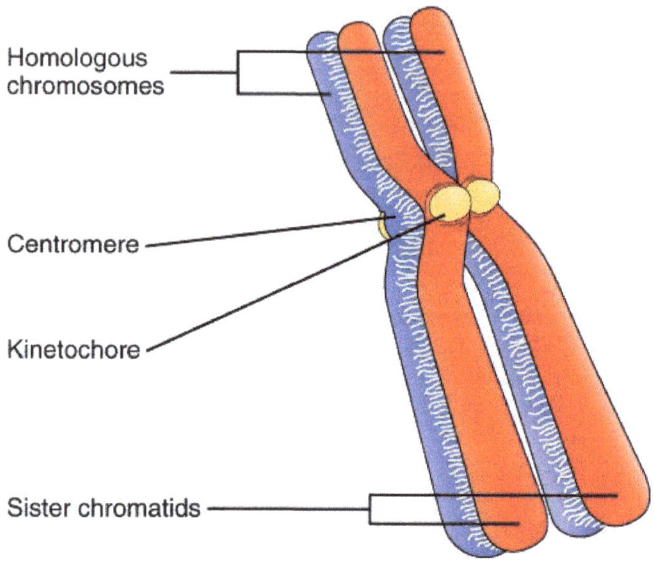

the chromosome during each round of meiosis. The farther apart two genes are on a chromosome, the more likely the alleles for those genes are to be separated by crossing-over. The closer together two genes are, the less likely they are to be separated by crossing-over. As a result, there are gradations of gene linkage. By observing how strongly traits are linked, one can determine how close together the genes that control those traits are on the chromosome. This is, in fact, how the locations of many genes were first mapped.

The homologous chromosomes physically touch each other in areas called **chiasmata,** and this is where they exchange pieces. A piece of one chromosome will detach and reattach to the other homologous chromosome. Recall that chromosomes are long pieces of DNA wrapped around supporting proteins, and some sequences of the nucleotide base along this DNA molecule are genes. A chromosome can have a few to thousands of genes. The alleles located on these pieces will now be inherited in different combinations than if crossing-over hadn't occurred. **Alleles** are different versions of genes. They differ in their nucleotide base sequence so they code for slightly different proteins. In a homologous pair, you have one allele that codes for a trait from your mother and one allele that codes for a trait from your father. Crossing-over creates new combinations of alleles that you can inherit from your mother and father. As a result, crossing-over increases the variety of alleles in the gametes produced by meiosis and in the possible offspring (Figure 11.4).

Phases of Meiosis

After DNA is replicated during interphase, the cell can begin dividing through the process of meiosis. Meiosis is divided into meiosis I and meiosis II, which contain all of the same phases but have some important differences. The names of the phase are the same as in mitosis but are designated I and II; however, what happens during the phases in meiosis I differs significantly from what happens in mitosis. During prophase I, the nuclear envelope breaks down, and the chromosomes condense and form homologous chromosome pairs. This is also when crossing-over occurs (Figure 11.4). The microtubule spindle begins to form during prophase I as well. In metaphase I, the homologous chromosome pairs move to the center of the cell and line up along the equator of the cell. The chromosome pairs attach to the spindle fibers by their kinetochore. Next the homologous pairs get pulled apart and move toward opposite ends of the cell, getting pulled by the spindle during anaphase I. The last step of meiosis I is telophase I. During this phase, the chromosomes reach the opposite ends of the cell, and the spindle disassembles. Cytokinesis then occurs, producing two cells with one-half of each homologous pair of chromosomes. The nuclear envelope doesn't reform and, depending on the organism, the cells may rest for some time or they may go right into the phases of meiosis II.

FIGURE 11.4: Crossing-over—The red and blue chromosomes match up as a homologous pair and undergo crossing-over. Some of the resulting chromosomes are recombinant, and some are not.

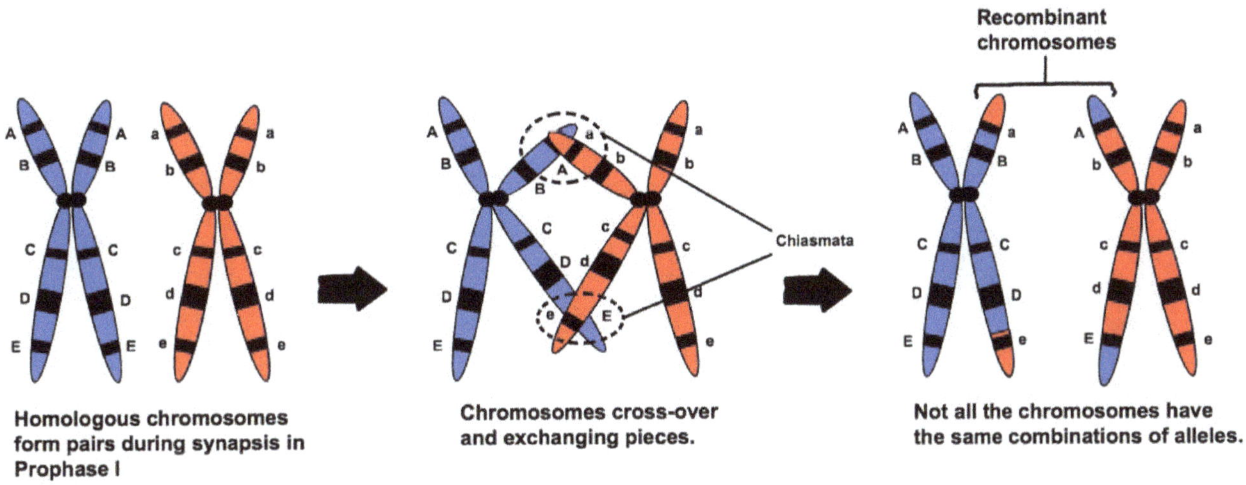

The movement of the chromosomes in the phases of meiosis II resemble mitosis as this is when the sister chromatids separate. During prophase II, the spindle reforms and attaches to the chromosomes. This time, it is the sister chromatids that line up along the center of the cell during metaphase II. In anaphase II, the sister chromatids are pulled apart by the spindle and are pulled to the opposite ends of the cells. Lastly, during telophase II, the spindle disassembles, the nuclear envelope reforms, and the cell splits in two. There are now four haploid cells. Figure 11.5 illustrates the movement of the chromosomes during the phases of meiosis I and II.

Increasing Variation

The process of sexual reproduction produces a large amount of variation in the genetic material of offspring. The combination of different chromosomes from both parents ensures that the offspring will have some of the genes and characteristics of the parents but will have a unique combination for each child.

Independent assortment of the chromosomes also increases variation in the gametes and possible offspring. During prophase I, the homologous pairs of chromosomes line up. Remember that one of the pair of chromosomes is from the mother and one from the father. When the chromosomes line up, whether the mother's chromosomes end up on the right or left, for example, is random for each pair. Each of our 23 homologous pairs of chromosomes can pair up in one of two ways. This results in a unique combination of alleles for each gamete (Figure 11.6). In humans, that means 223 or more than 8 million genetically unique gametes may be produced through independent assortment. Genetic diversity is further increased through a process called crossing-over. Crossing-over, or the exchange of chromosome pieces that takes place during prophase I, further increases the uniqueness of each gamete. The portions of the chromosome where crossing-over occurs are called chiasmata.

FIGURE 11.5: **Stages of meiosis**—Observe what happens to the chromosomes during each stage.

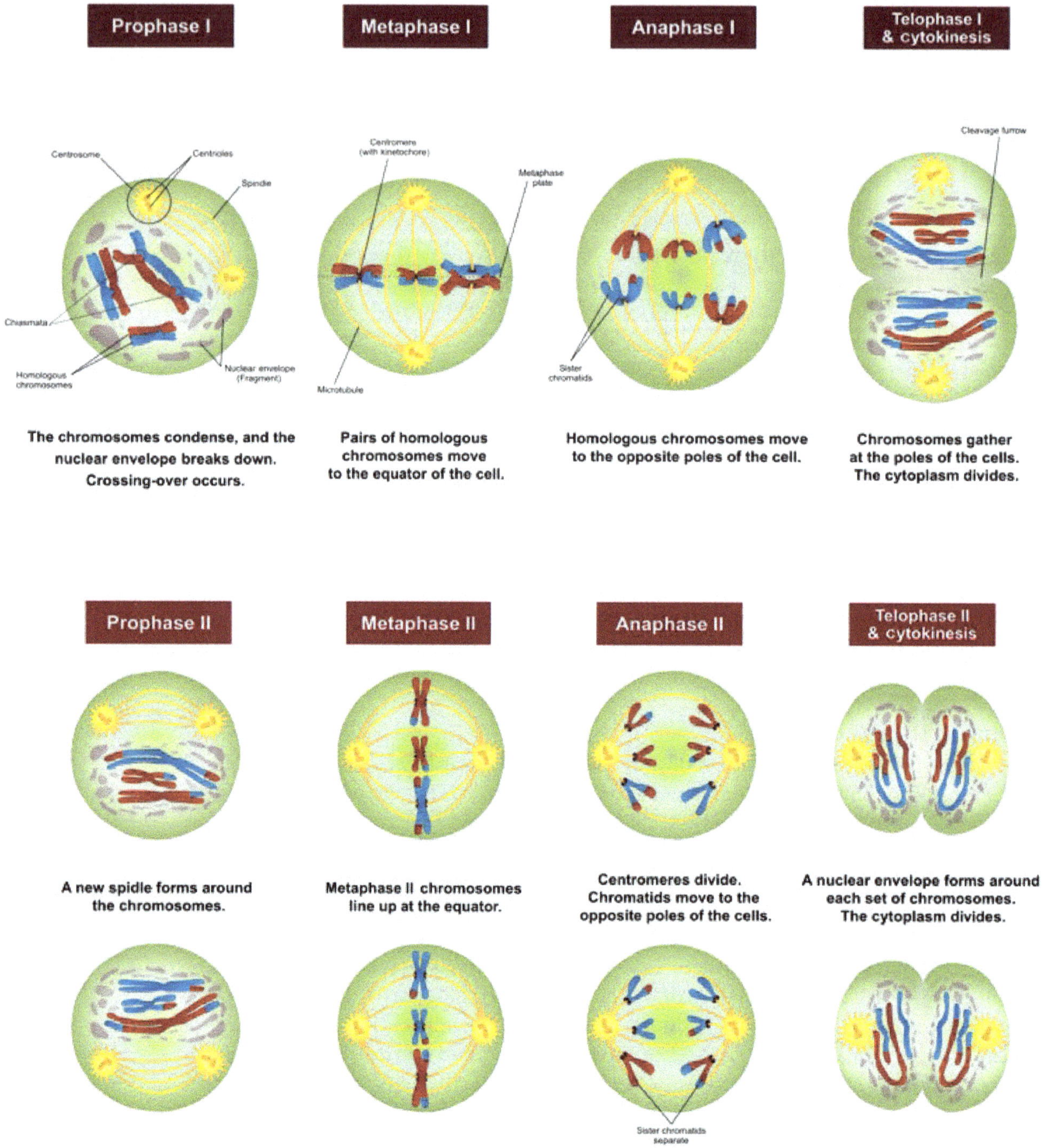

FIGURE 11.6: **Independent assortment**—Homologous chromosomes will pair up randomly. There are many different combinations possible, producing many unique gametes.

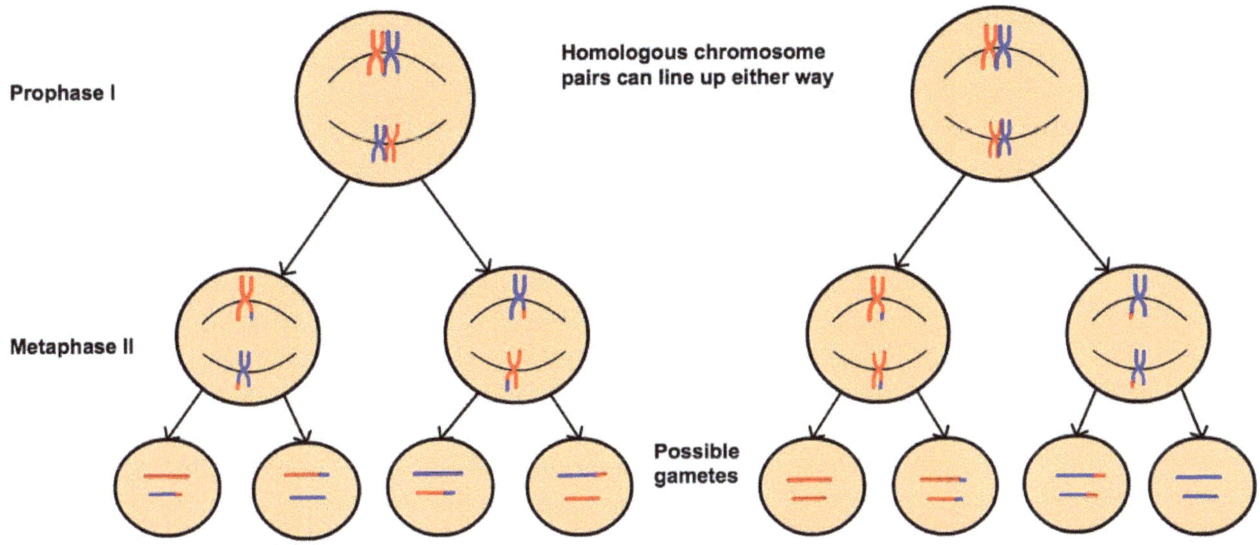

Chromosomal Alterations That Can Occur During Meiosis

In addition to mutations that involve changes in a small number of nucleotide bases, which were discussed earlier in this textbook, some genetic mutations involve large parts of chromosomes containing multiple genes, or even entire chromosomes or sets of chromosomes. Although it may seem that such large scale mutations would be exceedingly rare, given all of the mechanisms cells have for regulating and repairing genetic material, chromosomal abnormalities occur with surprising frequency. It is estimated that approximately one-third of all conceptions (when an egg is fertilized by a sperm) in humans involve a large-scale chromosomal alteration. However, this does not result in the same proportion of adult humans containing such genetic defects. This is because changes to the genetic code that involve the deletion or duplication of more than one gene are usually so damaging to development that the affected individual cannot make it to adulthood. Oftentimes, chromosomal abnormalities result in a miscarriage in which the developing embryo does not survive to birth as a result of overwhelming problems with development.

The structure of individual chromosomes can be altered by errors that occur during meiosis or damaged by ultraviolet radiation and other mutagens. Four types of alterations that may occur are deletions, duplications, inversions, and translocations. A chromosomal deletion is similar to a deletion point mutation in that it results in the absence of part of the genetic code; however, chromosomal deletions usually involve the loss of one or more entire genes. Chromosomes can also be affected by **duplication**, which is when a large segment of the chromosome is repeated. Inversions and translocations both involve breakage and misalignment of a chromosome rather than a gain or loss of gene segments. **Inversion** occurs when a fragment of a chromosome breaks off and reattaches at the same place but in the reverse orientation. Although this preserves the correct set of genes on the chromosomes, this type of misalignment can have disastrous consequences as the two copies of each gene inherited from each parent no longer correspond to each other. If part of a chromosome breaks off and attaches to a different, non-homologous chromosome, that is known as **translocation**. Translocation events may also result in serious consequences including miscarriage or development of a genetic disorder. Examples of human genetic disorders caused by each of these types of chromosomal abnormalities are identified in Table 11.3.

Sometimes errors that occur during meiosis lead to a fertilized egg having the wrong number of chromosomes. During meiosis, sets of homologous chromosomes (copies of the same chromosome) are normally separated during anaphase I, and pairs of sister chromatids separate during anaphase II. This results in each gamete (sperm or egg) produced during meiosis containing one of each chromosome. However, sometimes one pair of chromosomes or sister chromatids does not separate properly. This is called **nondisjunction**. After nondisjunction, gametes may retain both copies of that chromosome and another gamete will receive no copies of the chromosome. If either of these gametes is combined with a normal gamete through fertilization, the resulting zygote will contain an abnormal number of chromosomes. The condition of having an abnormal number of chromosomes is known as **aneuploidy**. **Monosomy** occurs when a zygote only contains one copy of a chromosome, and **trisomy** occurs when a zygote contains three copies of the same chromosome (Table 11.3).

As mentioned earlier, substantial chromosomal alternations, such as monosomy and trisomy, often result in nonviable zygotes. However, there are a few types of aneuploidy that do often result in zygotes that develop into adult individuals. The most well-known example of aneuploidy in adult humans is Down syndrome, which affects approximately 1 in 700

TABLE 11.3: Genetic Disorders Resulting from Chromosomal Alterations

Type of Mutation	Affected Chromosome	Name of Disorder	Characteristics
Deletion	Chromosome 5	Cri-du-chat syndrome	High pitched (catlike) cry, microcephaly, development delays, abnormal growth
Duplication	Chromosome 17	Charcot-Marie-Tooth disease	Degeneration of nerves, muscle weakness
Monosomy	X Chromosome	Turner Syndrome	Sterility, abnormal physical features, heart defects
Inversion	Chromosome 9	Chromosome 9 inversion	Usually no symptoms; may cause infertility
Translocation	Chromosome 8	Burkitt's lymphoma	Cancer affecting immune cells (B cells)
Trisomy	Chromosome 21	Down Syndrome	Developmental delays, abnormal facial features, short stature, heart defects

children born in the United States. Down syndrome is most often the result of trisomy involving chromosome 21, although it can also be caused by a translocation involving part of this chromosome. Individuals with Down syndrome exhibit a number of noticeable characteristics, including developmental delays, heart defects, short stature, and specific facial features. Although the life expectancy for those with Down syndrome is lower than normal, people with this disorder can live relatively healthy and independent lives well into middle age and even beyond. The next most common examples of aneuploidy in adult humans involve the X and Y sex chromosomes. Klinefelter syndrome is the result of an extra X chromosome, producing males who are XXY instead of XY, and occurs in about 1 in 500 males. Turner syndrome occurs when females only have one X chromosome, which is denoted as X0 instead of the normal XX, and occurs in about 1 in 2,000 females. A few other types of trisomy have been documented in human children, but Turner syndrome is the only known viable monosomy in humans.

FIGURE 11.7A: **Down syndrome**—This individual has Down syndrome, illustrated by the three copies of chromosome 21.

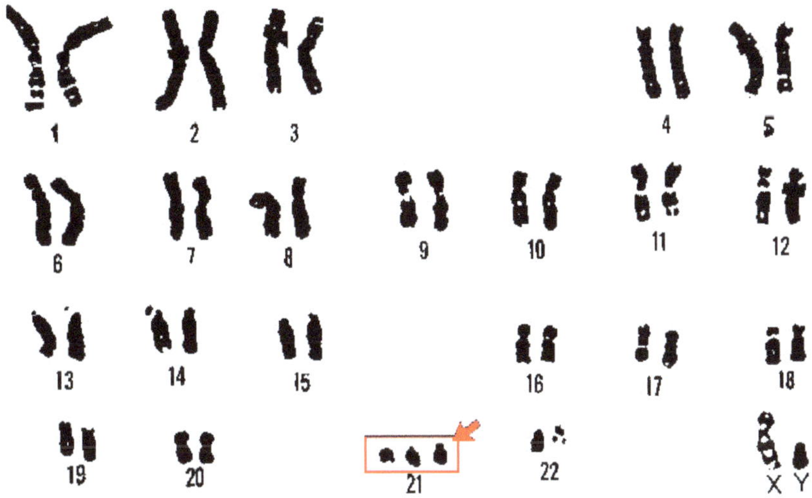

FIGURE 11.7B: **Klinefelter syndrome**—This individual has Klinefelter syndrome because they have two X chromosomes and one Y chromosome.

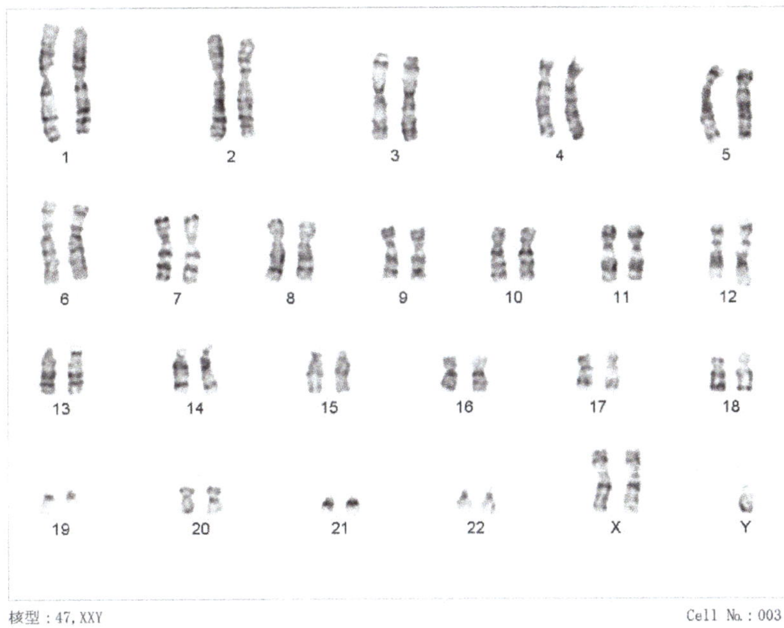

The reason aneuploidy involving these particular chromosomes leads to live births more often than aneuploidy involving other chromosomes is not because these chromosomes are more likely to experience nondisjunction during meiosis. Instead, the reason is that these chromosomes contain fewer genes than most of the other chromosomes, so the effect of their alteration is less severe. Figure 11.7 shows a picture of all the chromosomes in a person with Down syndrome and Klinefelter syndrome. This picture is called a **karyotype** and can be used to identify chromosomal abnormalities. The chromosomes are lined up from the largest, number 1, to the smallest, number 22, and then the sex chromosomes are last.

It is even possible for organisms to have additional complete sets of chromosomes in their cells. This condition is known as **polyploidy**. More specifically, when an organism's cells contain three full sets of chromosomes, it is known as triploidy; four sets of chromosomes is tetraploidy. Polyploidy is rare in animals but is fairly common among plants, especially those that have been cultivated by humans. Although it may seem counterintuitive, polyploid organisms, in the species in which they exist, generally appear much closer to normal than individuals affected by monosomy or trisomy.

Comparing Mitosis and Meiosis

Meiosis is similar to mitosis in several ways, including the fact that it is divided into several phases. One other similarity between mitosis and meiosis is that the cell goes through interphase before dividing. During interphase, the DNA replicates. This is the only time the DNA will replicate, but the cell and chromosomes will divide twice, which is how we end up with haploid gametes.

There are several important differences between the two types of cell division, especially concerning how the chromosomes line up and divide. The two types of cell division are part of the life cycle of organisms and share many similarities but also have many differences (Table 11.4).

TABLE 11.4: Comparing and Contrasting Mitosis and Meiosis

	Mitosis	Meiosis
End result	Two identical diploid cells	Four unique haploid cells
Number of cell divisions	One	Two
Type of reproduction	Asexual reproduction	Sexual reproduction
Type of cells produced	Somatic cells	Gametes or spores
Homologous chromosomes?	Homologous chromosomes do not line up and crossing over does not occur	Homologous chromosomes line up and crossing over occurs

11.2 SUMMARY

- Organisms can use asexual or sexual reproduction to produce offspring, each of which has advantages and disadvantages
- Each type of organism has a specific number of chromosomes normal for that type of organism
- Meiosis produces special cells that are used for sexual reproduction called gametes
- During meiosis, the chromosomes form homologous chromosomes and sometimes exchange pieces during a process called crossing-over
- Diploid cells are somatic cells that have 2n chromosomes, and haploid cells are gametes that have n chromosomes
- The cell goes through two cycles of division during meiosis, the DNA only replicates one time
- The phases of meiosis include: prophase I, metaphase I, anaphase I, telophase I, cytokinesis, prophase II, metaphase II, anaphase II, telophase II, and cytokinesis
- Independent assortment and crossing-over help to increase the variations of genetic information in the gametes
- Alterations to the structure and number of the chromosomes can occur during meiosis, leading to genetic disorders

11.3 KNOWLEDGE CHECK

1. If a somatic cell of an organism usually has 32 chromosomes, how many chromosomes should each daughter cell have at the end of meiosis?
   a. 64
   b. 32
   c. 16
   d. 8
   e. 4

2. Which one of the following best describes what occurs during independent assortment?
   a. only one allele actually makes protein while the other is inactive
   b. the segments of homologous chromosomes that are exchanged are selected randomly and independently of other segments
   c. only some chromosomes are duplicated
   d. homologous chromosomes are distributed to daughter cells randomly and independently of one another
   e. the chromosomes are sorted into several independent groups

3. Which of the following occurs during crossing-over?
    a. homologous chromosomes line up and separate
    b. sister chromatids are pulled apart
    c. homologous chromosomes line up and exchange pieces
    d. sister chromatids exchange pieces
    e. non-homologous chromosomes line up and exchange pieces

4. Which of the following does NOT result from nondisjunction?
    a. Turner syndrome, where an individual only has one X-chromosome
    b. Cri-du-chat syndrome, where an individual is missing part of chromosome number 5
    c. Edwards syndrome, where an individual has three copies of chromosome number 18
    d. Klinefelter syndrome, where an individual has two X chromosomes and one Y chromosome
    e. Patau syndrome, where a person has three copies of chromosome number 13

5. Which of the following best describes what happens during an inversion mutation?
    a. Part of a chromosome breaks off and attaches to another chromosome
    b. Part of a chromosome breaks off and attaches in the reverse direction
    c. Part of a chromosome is repeated
    d. The way the nucleotide code is grouped into codons is shifted
    e. Two chromosomes fail to separate properly during meiosis

6. Explain several evolutionary advantages to asexual reproduction in contrast to sexual reproduction.

7. Discuss the two processes that occur during meiosis that help to increase genetic diversity in the gametes.

8. Compare and contrast the phases and outcomes of mitosis and meiosis.

11.4 ACTIVITY

Simulating Meiosis

The purpose of this activity is to simulate the movement of the chromosomes during the different stages of meiosis I and II. The structure of the chromosomes will be represented by four different colors of pop beads, which lock together and can be used to build models of chromosomes. The centromere section of the chromosomes can be modeled using magnetic connectors. It is also possible to use pipe cleaners and pony beads to model chromosomes and alleles. You will also use three pieces of paper, one with a large circle to simulate a single cell, another with two small circles to simulate two cells, and a third with four cells representing four gametes.

1. Constructing the chromosomes: You are making four single chromosomes using the pop beads, each with a magnetic centromere placed somewhere near the middle. Two of the chromosomes should be the same length, one is one color and the other one is a different color. The other two should be a different length from the first two, also using two different colors. Take a picture.
2. Interphase: Simulate DNA replication by making copies of each of your chromosomes so that you have eight total chromosomes. Now join them together to make sister chromatids using the magnetic centromeres. Remember that sister chromatids are identical to each other. In real cells, the chromosomes wouldn't yet be visible, but we are showing them here as part of our simulation. Take a picture.
3. Prophase I: Put all the chromosomes in your single cell and pair up homologous chromosomes. Remember that homologous pairs are similar in size and genetic information. Then simulate crossing-over by exchanging chromosome pieces between homologous chromosomes by switching beads. Take a picture.
4. Metaphase I: Line up the homologous chromosomes pairs along the center of the cell. Remember you can arrange them in any way. Take a picture.
5. Anaphase I: Pull apart the homologous pairs and move the sister chromatids toward opposite ends of the cell. Take a picture.
6. Telophase I and cytokinesis: Transfer your chromosomes to the paper with two cells. This is the end of meiosis I. Take a picture.
7. Prophase II: Chromosomes are visible in each cell. Take a picture.
8. Metaphase II: Line up the sister chromatids along the center of the cell. Take a picture.
9. Anaphase II: Pull apart the sister chromatids and move the single chromosomes toward the opposite ends of the cell. Take a picture.
10. Telophase II and cytokinesis: Put the single chromosomes in each of the four single gametes. Take a picture.
11. Prepare a poster, presentation, or pamphlet using your pictures to describe the steps of meiosis.

11.5 ACTIVITY

Learning the Steps

For this activity you should get a set up index cards or cut 8 1/2" x 11" paper into four squares.

1. Write the name of a phase of meiosis I or meiosis II on the front of the card or paper.
2. On the back of the paper, write what occurs to the various cell structures during that phase.

3. Use these cards to quiz yourself or use them with a partner to quiz each other.
4. After you have covered all the phases, answer the questions for Activity 11.4.

ACTIVITY QUESTIONS

11.4

1. Explain what happens to the chromosomes in metaphase I and metaphase II.
2. How is what happens to the chromosomes during metaphase II similar to what happens to chromosomes during mitosis?
3. How did you simulate crossing-over during the simulation? When during meiosis did this occur?
4. Explain how you could simulate nondisjunction with your chromosomes.

11.5

1. What happens in synapsis and when does it occur during meiosis?
2. How do chromosomes exchange pieces during crossing-over?
3. If a cell had 14 chromosomes, how many unique gamete combinations would be possible? (hint: 2^n)

11.6 REFERENCES

Crow, J. F. (1994). Advantages of sexual reproduction. *Developmental Genetics, 15*(3), 205–13.

List of Organisms by Chromosome. (2017). [On-line] Access date: Aug. 28, 2017. https://en.wikipedia.org/wiki/List_of_organisms_by_chromosome_count#cite_note-104.

Theisen, A., and Shaffer, L. G. (2010). Disorders caused by chromosome abnormalities. *The Application of Clinical Genetics, 3*, 159–74.

FIGURE CREDITS

FIG. 11.1A: Copyright © Brocken Inaglory (CC BY-SA 3.0) at https://commons.wikimedia.org/wiki/File%3ASea_star_regenerating_legs.jpg.

FIG. 11.1B: Copyright © FASTILY (CC BY-SA 3.0) at https://commons.wikimedia.org/wiki/File%3AStolon3800ppx.JPG.

FIG. 11.1C: Copyright © Lifetrance (CC BY-SA 3.0) at https://commons.wikimedia.org/wiki/File:Hydra_oligactis.jpg.

TABLE 11.2: Source: https://en.wikipedia.org/wiki/List_of_organisms_by_chromosome_count#cite_note-104.

FIG. 11.3: Copyright © OpenStax (CC by 4.0) at https://commons.wikimedia.org/wiki/File:0330_Homologous_Pair_of_Chromosomes.jpg.

FIG. 11.5: Copyright © Ali Zifan (CC BY-SA 4.0) at https://commons.wikimedia.org/wiki/File:Meiosis_Stages.svg.

TABLE 11.3: Source: Aaron Theisen and Lisa G Shaffer, "Disorders caused by chromosome abnormalities," The Application of Clinical Genetics, vol. 3. Dove Press, 2010.

FIG. 11.7A: U.S. Department of Energy Human Genome Program, "Down syndrome," https://commons.wikimedia.org/wiki/Category:Down_syndrome#/media/File:21_trisomy_-_Down_syndrome.png. Copyright in the Public Domain.

FIG. 11.7B: Nami-ja, "Klinefelter syndrome," https://commons.wikimedia.org/wiki/File%3AHuman_chromosomesXXY01.png. Copyright in the Public Domain.

Evolution

12.1 INTRODUCTION

What Is Evolution

Evolution is both a scientific **theory** and a scientific **fact**. A theory explains how a natural phenomenon occurs and is supported by a great deal of observations and evidence. The modern synthesis explains how new species arise through processes like natural selection, mutation, genetic drift, gene flow, and sexual selection. A scientific fact is something generally regarded as true because it is consistently observed over and over again. **Evolution** in the age of genetics is defined as a change of allele frequencies over time. This is sometimes referred to as microevolution because it is occurring at the population level. In this context, the term **population** is used to refer to all individuals of the same type or species that live within the same geographic area. Darwin described evolution as "descent with modification", that is, over time, populations change and evolve into new species or even new orders or phyla. This is often referred to as macroevolution.

Early Thoughts About the Diversity of Species

Although he is famous, and rightly so, for the development of his theory of natural selection, there were other thinkers before him who also wanted to make sense of diversity of all the different kinds of organisms. In fact, Aristotle was so intrigued that he wrote a multivolume book on animals called *Historia Animalium* in which he compares the structure of different animals and discusses their behaviors, reproduction, and ecology. Following Aristotle, Theophrastus applied this method to plants in his book called *Historia Plantarum*. Later in the Arab world, there was Abu Uthman Amr Ibn Bahr, mostly known by his nickname al-Jahiz, who built on Aristotle's work by including his own observations and thoughts on animal behavior and ecology. According to some scholars, he was the first to describe food chains in his book *Kitāb al-hayawān*. These three scholars did not attempt to describe a mechanism that led to all the diversity, they were just attempting to organize and describe it so that man could understand it. By the mid-1700s Carolus Linnaeus published his major work *Systema Naturae,* in which he proposed a binomial (two name) naming system and

LEARNING OBJECTIVES

- Understand the historical development of the theory of evolution
- Discuss Charles Darwin and his development of the theory of evolution by natural selection
- Discuss the modern synthesis theory of evolution
- Identify the two main types of speciation
- List the different barriers that lead to reproductive isolation
- Explain the different types of evidence that support the theory of evolution

FIGURE 12.1: **Two giraffes**—These two giraffes are stretching their necks to eat from the tops of trees. According to Lamarck's theory of acquired characteristics, they would pass on their longer necks to their offspring.

a classification system based on structural similarities between groups of organisms. Linnaeus developed his system to help put order to what he considered God's creation and believed that species did not change. Ironically, his classification system supports Darwin's work and modern evolutionary theory. A group of French anatomists working in Paris, including Buffon, St. Hilaire, Cuvier, and Lamarck, were all working on comparative anatomy of plants and animals. They wrote about species changing over time, and Cuvier was one of the first to discuss extinction of species. The term **species** became widely used to refer to a group of organisms capable of breeding and producing viable, fertile offspring with each other but are not capable of producing fertile offspring with individuals belonging to other groups, or species. Although they all added extensive knowledge to what was known of the structures of animals and plants, it was only Lamarck who proposed a hypothesis that tried to explain how—a theory of evolution.

Lamarck proposed that organisms have various needs to fulfill in order to stay alive, for example, accessing food. As an organism used a particular body structure more or less to obtain food, the structure would get larger or smaller. This modified body part would get passed down to the offspring. His most famous example is the giraffe. Giraffes eat leaves off the top of trees. If they continued to stretch their necks to reach up higher and higher in a tree, their necks would get longer and be passed on their offspring. He called it the **theory of inheritance of acquired characteristics**. After enough generations, giraffes would have the incredibly long neck that we see today (Figure 12.1). This was revolutionary at the time because it contradicted the creation story from the Old Testament in the Bible. He had proposed instead a natural explanation of how and why species change over time rather than a supernatural one. It was not well received by many of his colleagues, and we now know that most of what he proposed was incorrect, but it was still groundbreaking for its time.

Darwin and His Trip on the Beagle and Work on Origin of Species

As a child, Charles Darwin was always interested in nature. He spent a lot of time collecting rocks and beetles, and studying plants. When he entered the University of Edinburgh, he originally studied medicine but changed his mind after observing surgery without anesthesia. Darwin's family wanted him to continue his studies, so he decided that a degree in theology and becoming a pastor was a good choice, as it would allow him a career and the ability to continue his love of nature. After completing his BA at Christ's College, Cambridge University, Darwin was presented with an opportunity to travel on the HMS Beagle, which was traveling around the world, mapping English colonies and settlements (Figure 12.2B). Only 22, Darwin convinced his parents that this was a good opportunity for him to travel and collect natural specimens. Darwin left England at the end of December 1831 and wouldn't return for 5 more years.

The HMS Beagle surveyed all along the coast of South America, Australia, and the tip of South Africa. Along the way, Darwin was able to get off the boat and explore inland. He collected fossils of many different animals and plants, and sent them back to England. He also collected living animals, preserved them, and sent those back to England too. He recorded his observations of geological processes, the various people he met, and all his plant and animal collections in a diary, which can be still read today. The collection Darwin is most famous for was the finches found on the Galapagos Islands off the coast of Ecuador. They almost weren't famous because Darwin didn't label the islands the finches came from when he first collected them. Darwin had assumed they were all different species of birds, not different species of finches. Luckily, the captain of the ship had labeled his birds so Darwin's could then be cross-referenced and organized. The experts who identified the birds revealed to Darwin that they were in fact all finches even though they looked very different from one another, especially their beaks.

Darwin returned from his voyage and settled in London. He published the journal that he kept on the journey and began to work on his new theory. He read two influential books at this time, one by his friend Charles Lyell, the geologist. Lyell's *Geology* explains that the Earth is, and always has been, undergoing transformations like erosion, earthquakes, and volcanic eruptions. These changes take place over millions and millions of years and have changed what the Earth looked like, and will continue to change it into the future. The Earth must be very old for all these changes to have occurred. He also read *An Essay on the Principles of Population* by Thomas Malthus, an economist. In his essay, Malthus describes how human populations increase at a rate that outpaces the rate at which resources, like food, can be produced. Eventually, the resources would run out, which could lead to a population collapse. To avoid this collapse, Malthus says that checks on the population like wars, malnutrition, disease, and poverty help keep the human population from outpacing the production of resources.

FIGURE 12.2A: Charles Darwin as a young man—Portrait of Charles Darwin from 1840, about 4 years after he returned from his voyage on the HMS Beagle.

FIGURE 12.2B: HMS Beagle—Painting of the HMS Beagle a few years after Darwin travelled on her around the world.

Over the next 20 years, Darwin would continue to work on his theory to explain the diversity in organisms and the evolution of new species. Darwin worked on his theory almost in secret, with only his wife and close friends knowing what he was working on, because he knew the book would be controversial. In 1858, Darwin received an essay in the mail from Alfred Russel Wallace, a fellow naturalist looking for comments on a theory he had developed. In the essay, Wallace described a theory to explain the diversity of animal and plant species seen on Earth and the mechanism that caused it. It was almost identical to what Darwin had been working on for the last 20 years. Darwin shared the letter with two close colleagues who agreed to present Wallace's essay and selections from Darwin's unpublished work at a meeting of the Linnaean Society. This would allow Darwin to receive credit for developing the theory first, yet still giving credit to Wallace for his work. Darwin's book, *On the Origin of Species,* was published on November 24, 1859.

Darwin's Theory of Natural Selection

In the first chapters of *On the Origin of Species,* Darwin explains that populations must have variations in their traits, and that these traits must be heritable. In both domesticated and natural populations, there are similarities that tie individuals together as the same species, but there are still variations in the color, size, and other traits within the species. Although he doesn't directly use the term **artificial selection,** Darwin discusses how humans domesticated animals and plants by breeding organisms based on traits that were useful or aesthetically pleasing to humans. In this way, humankind has changed these species over time. He then discusses the variations exhibited in wild populations of plants and animals. Referencing Malthus, Darwin discusses that, for most organisms, there are more offspring born than the environment can handle, and that there are limited amounts of resources like food, water, and space. In addition, there are also environmental pressures like climate changes and disease. All of these pressures lead to competition for resources, what Darwin and Wallace called a struggle for existence. Not all organisms in a population will survive this struggle. The ones that do survive do so because they are the most fit to their environment. The set of advantageous traits they inherited that helped them survive and reproduce successfully, are known as **adaptations.** These adaptations are then passed on to their offspring. Those that are less fit do not survive, and their traits are not passed down to the next generation. This is **natural selection,** the differential survival and reproduction of some organisms in a population based on their advantageous traits. Over time, the traits that are advantageous in a population will change as the environment changes. Over long periods of time, these small changes accumulate in a population and eventually a new species evolves. This is **divergent evolution,** when two or more species arise from a common ancestor. The only illustration in Darwin's book shows his depiction of divergent evolution as a tree of life. Each branch depicting the evolution of a new species (Figure 12.3).

Darwin's theory of natural selection was criticized when *Origin of Species* was released. There were, of course, people who objected to his theory based on religious concerns but there were scientific criticisms as well. One criticism was regarding Darwin's emphasis on variations. Critics wanted to know from where did the variations originate. Variation in populations is a key point of Darwin's theory, but Darwin himself couldn't explain their origin. Another criticism of Darwin's theory involves the inheritance of these variations. Initially, Darwin accepted blending as the method of inheritance. Blending inheritance involves the mixing of

FIGURE 12.3: Darwin's tree of life—Darwin's drawing shows how species can branch off of ancestral species. At the end of each branch is a different species.

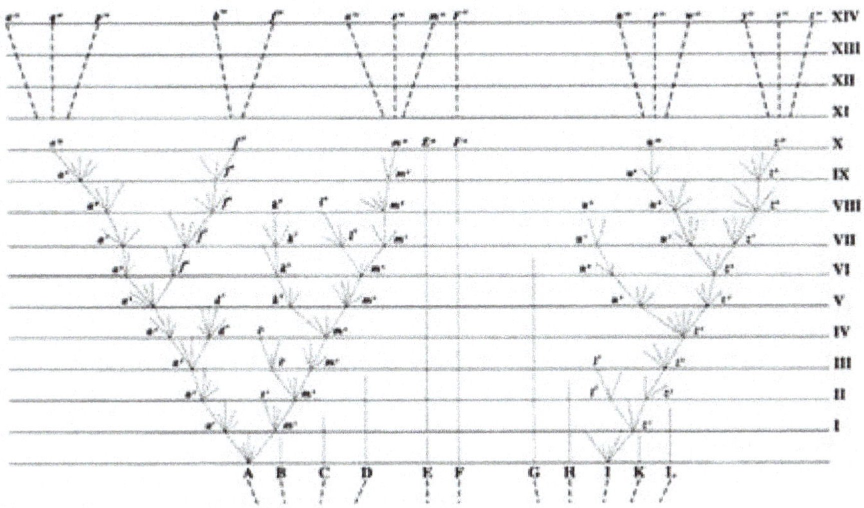

parental traits to produce an intermediate trait in the offspring. For example, if one parent was red and the other was white, the offspring would be pink. This was a commonly accepted mechanism of inheritance when Darwin was working on *Origin*, however, blending inheritance will not work with natural selection. For an advantageous trait to remain advantageous, it cannot be "watered" down in the next generation, which is what occurs in blending inheritance. In later editions of *Origin*, Darwin switched to Lamarck's theory of acquired characteristics. This model of inheritance does work with natural selection because advantageous traits are passed on to offspring and are not "watered" down. At this time, Darwin didn't really understand how inheritance worked, and he certainly didn't know anything about genes and alleles. Around the same time Darwin was working on *Origin of Species*, Gregor Mendel was working with pea plants in Brno and determining the basic laws of genetics. There doesn't seem to be any evidence that either one of them knew what the other was doing. Once Mendel's laws were "rediscovered" in the early 1900s, this prompted a greater understanding of how evolution worked as well. The incorporation of modern genetics into evolutionary theory helped develop a more modern theory of evolution.

Modern Synthesis Theory of Evolution

As the 20th century began, Darwin's theory continued to be modified as more observations and evidence were added to shape it into what is called the Modern Synthesis theory of evolution. The field of genetics was incorporated along with probability and statistics so that predictions of inheritance could occur. It was also determined that individuals do not evolve, only populations of organisms can evolve. Another aspect of the modern theory is the incorporation of other mechanisms besides natural selection that can cause a change in allele frequencies. These other mechanisms include mutations, gene flow, sexual selection, and genetic drift. Introducing new mutations into a population changes the frequency of old alleles, thus changing

FIGURE 12.4A: Two hartebeests fighting—Two male hartebeests fight using their horns for control over their territory and mating access to females, an example of intrasexual selection.

FIGURE 12.4B: Male peacock and female peahen—The large showy feathers of the peacock are used to attract a peahen, who chooses whether to mate with the male or not, in an example of intersexual selection.

the population. **Gene flow** is another mechanism that involves immigration and emigration of members of a population to another population. This mechanism can add new alleles to a population and change the frequency of current alleles. Another mechanism is **sexual selection**, although how much this can change a population varies greatly. Darwin even discusses sexual selection in *Origin of Species*. Generally in populations where there is a large physical difference between males and females of a population, it can mean that sexual selection has played a significant role in causing that change. There are generally two types of sexual selection, intrasexual and intersexual selection. **Intrasexual selection** involves members of the same sex (usually males) competing with one another for a mate. This is sometimes referred to as male-male competition. Males will often have evolved weapons, a larger size, or armor, and control a large group of females in their territory. See Figure 12.4A where two male hartebeests are fighting over territory and females. **Intersexual selection** is often called female choice. Females will choose to mate with males based on some trait or behavior. This can push the evolution of the trait in males in an extreme direction. See Figure 12.4B where a male peacock is showing off his large, beautiful tail in an attempt to woo the gray, drab peahen behind him.

Another mechanism that can cause a change in allele frequencies is **genetic drift**. This mechanism, unlike natural selection, is random. The changes in allele frequencies are not a result of how well adapted some organisms are but rather a random event that permanently removes some alleles from the population. There are two types of genetic drift, founder effect and bottleneck effect. In the **founder effect**, a small group of individuals from a larger population leave and settle in a different area. Once in the new area, they do not become part of an existing population but instead stay isolated. The gene pool that the smaller group initially brought with them will not have the same allele frequencies as the original larger population they left. Over time the alleles in the small population will change in very different ways compared to the original population (Figure 12.5). A well-known example of founder effect is the Old Order Amish of Lancaster County, Pennsylvania. Because of religious reasons, the Amish moved to Pennsylvania and when they arrived they stayed in their own communities. They did not have a very diverse gene pool and unfortunately, some of these earlier settlers brought recessive alleles with them that caused genetic disorders. Now about 300 years later, the frequency of certain genetic disorders is very high in the Amish population compared with the general U.S. population.

Bottleneck effect is the second type of genetic drift and is the result of a large portion of a population dying at or around the same time. This could be the result of a natural disaster like a hurricane or due to human interference like overhunting. The alleles in the original population are randomly removed from the population, and the surviving gene pool doesn't resemble the original one in the number and type of alleles (Figure 12.6). A real-life example occurred in African cheetahs, whose populations passed through two significant bottlenecks about 100,000 and 10,000 years ago. As a result, all surviving cheetahs today are very similar genetically, and many suffer from malformed sperm, which hampers their reproductive success.

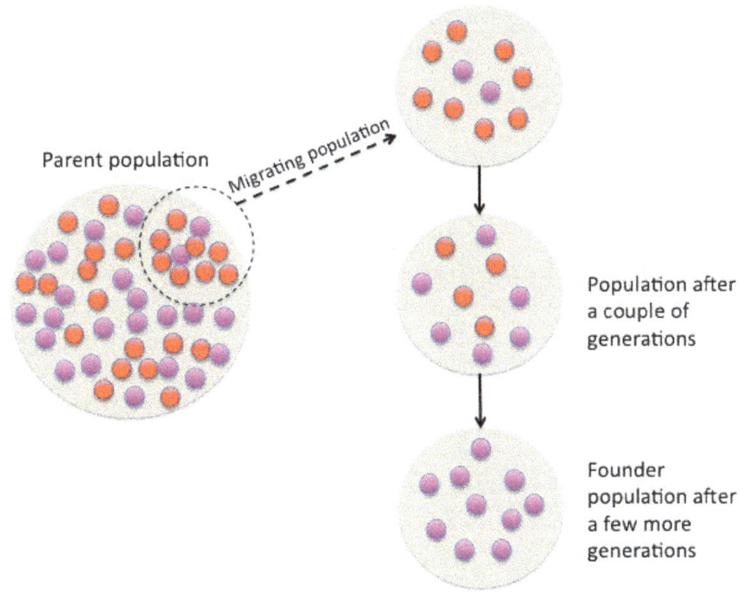

FIGURE 12.5: Illustration of the founder effect—The colored spheres represent two different alleles in a population. Notice that the small population that immigrates has very different allele frequencies than the parent population.

Speciation

How is a species defined? Even Darwin had difficulty defining it in *Origin of Species,* saying that there are fewer differences between members of a species than there are between the members of different species. In the modern synthesis theory of evolution, the most popular definition of a species is the **biological species concept** proposed by Ernst Mayr. This definition states that a species is made of members of a population that can or potentially can interbreed in nature and produce fertile offspring. What separates one species from another is not necessarily how they look, but whether they can reproduce with one another successfully. This definition doesn't work for all disciplines within biology, but it is probably the most well-known and used.

This definition of species means that the process of speciation results in reproductive isolation. **Reproductive isolation** means that two populations cannot interbreed with one another, and there is no gene flow. There are two major types of speciation, allopatric and sympatric, and they differ in how the gene pools are initially separated. The first step of both types of speciation is the isolation of part of a population's gene pool from the larger population. In **allopatric speciation,** this occurs through a geographic barrier like a river or a mountain range, which stops members from mating thus preventing gene flow. Over time, the two populations develop different mutations, and experience natural selection and genetic drift differently. This changes their allele frequencies and, over time, different adaptations accumulate in the

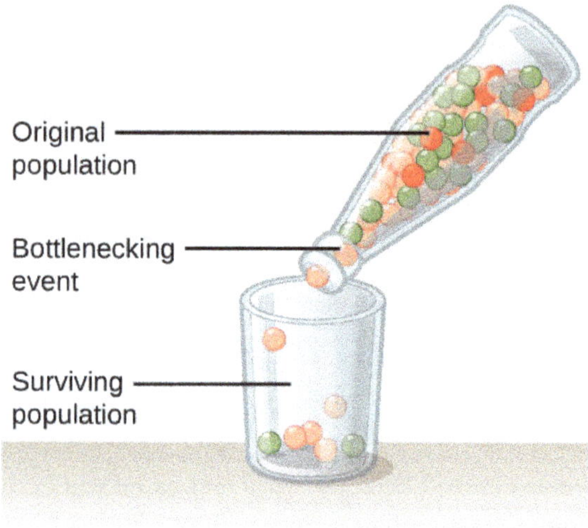

FIGURE 12.6: Illustration of the bottleneck effect—The colored spheres represent alleles. A large portion of the population dies, and only the survivor's allele make it through the bottleneck. The diversity of the surviving gene pool is much less than the original population.

populations. Eventually, if members of these populations were to come in contact again, they will no longer mate. They have become reproductively isolated (Figure 12.7A). Darwin's finches of the Galapagos Islands are a good real-life example. Different islands with different environmental conditions have different species of finches. Their beaks have evolved through mutations and natural selection to be compatible with the types of food available on their island. In **sympatric speciation,** a geographic barrier does not separate the gene pools of the two groups, but instead one group of the population may have started to change their role in their environment, like eating a new type of food or preferring a new type of soil. In plants, oftentimes a problem during meiosis will result in a different number of chromosomes, which isolates the plant. Over time, the two populations develop new mutations, and natural selection and genetic drift work on them differently, changing their allele frequencies. Over time, the two populations become reproductively isolated (Figure 12.7B).

There are many kinds of adaptations that can evolve to prevent reproduction between two species. They are generally divided into two groups, **pre-zygotic**, which prevents fertilization, and **post-zygotic**, which prevents the full development of fertile offspring. Pre-zygotic barriers include behavioral, geographic, gametic, physical, and temporal adaptations. **Behavioral barriers** prevent mating because one of the two sexes doesn't perform the correct courtship ritual, which the other sex doesn't recognize and therefore won't mate with them. An example of this would be birds of paradise from the South Pacific islands. In each species, the males have a specific dance to attract females, which is not recognized by other species of birds of paradise. Two species may not be able to mate because of a **geographic barrier**, which prevents the two species from being in the same place. The snapping shrimp species found in the Pacific Ocean and in the Gulf of Mexico are a good example of this barrier. The shrimp are separated by the Isthmus of Panama and have no way to cross the land. A **gametic barrier** involves blocking fertilization of gametes of two species. Mating has occurred, but the gametes do not chemically recognize each other so fertilization does not take place. An example of this would be

FIGURE 12.7A: Steps of allopatric speciation

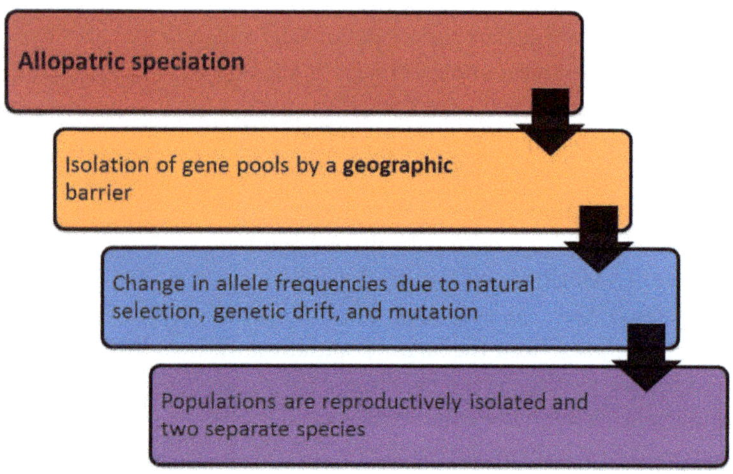

FIGURE 12.7B: Steps of sympatric speciation

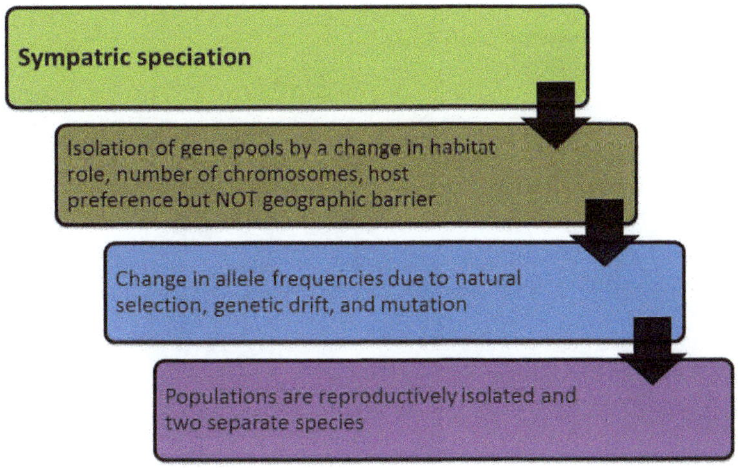

if pollen from an oak tree blew onto the stigma of a daffodil, it would not produce an oak-daffodil hybrid. When the genitalia of two species do not match in size or shape, that is a **physical barrier**. An example of this occurs in the damselfly, an insect closely related to dragonflies. In each species, the male's penis has a specific shape that matches the female's vaginal canal. Different species cannot physically fit together. Another barrier to mating involves a difference in timing of breeding season and is called a **temporal barrier**. Many species will only mate once a year at a very specific time. For example, in red-legged frogs, the breeding season is from January to March, whereas breeding season for yellow-legged frogs is March to May.

Post-zygotic barriers allow fertilization to occur and a zygote to form producing a **hybrid**, or an offspring that has parents of two different species. If the hybrid does not complete development and is miscarried, this is called **hybrid inviability**. Sometimes the hybrid offspring will be born but is infertile. This is **hybrid sterility**. Mules are the result of mating between a female horse and a male donkey, and they are sterile. Therefore mules are hybrids and not considered their own species.

Evidence of the Theory of Evolution

There are many lines of evidence that support the theory of evolution. These can generally be subdivided into three groups.

1. Fossils
2. Homologies, which include anatomical, developmental, molecular, and cellular
3. Biogeography

Fossils are the remains or impressions of a once-living organism. They can be bone that has become mineralized, an impression or cast of an organism in stone, or an organism trapped in tree resin called amber. Not all organisms that die become fossils because they only form under certain conditions. Also many fossils remain undiscovered, so finding a fossil can be very rare.

Fossils form when an organism dies and is usually quickly covered by sand or silt, often at the bottom of a body of water. The soft tissues quickly decompose leaving behind bones or shells. More layers of sediment build up over millions of years on top of the bones or shells, and pressure continues to build. The pressure causes the bone or shell to be replaced by minerals, as can be observed in the mammoth fossil shown in Figure 12.8A. The rock layers of sediment may eventually erode away or shift as a result of earthquakes, eruptions, or other

FIGURE 12.8A: Mammoth fossil—The bones of this mammoth have been replaced by minerals, turning them into fossils.

FIGURE 12.8B: Impression of an ancient maple leaf—This leaf decomposed a long time ago but left an impression in the mud it landed in, producing this fossil.

geological activity, and the fossil may be exposed. Organisms lacking bones like plants can also be fossilized. This occurs when an organism dies and leaves an impression or cast in layers of sediment (Figure 12.8B). Whole small organisms like insects can also be trapped in plant sap that hardens to become amber. In this way the whole organism is preserved (Figure 12.8C).

Fossils can show how extinct and still-living species are related by comparing similar physical features. They can demonstrate how closely related species have changed and diverged over long periods of time. **Transitional fossils** are fossils of organisms that have shared traits between two different but related groups. *Tiktaalik* (Figure 12.9) is considered a transitional fossil because it shows characteristics of both fish and tetrapods (four-legged) land animals. It has a flat head with eyes on top and bones resembling a wrist in its fins, which are indicative of tetrapods, but it also had fins and gills, and lived mostly in the water like fish. The features of *Tiktaalik* help fill in what organisms looked like between the transition from fish to tetrapods. There are other important transitional fossils in the evolution of whales like *Ambulocetus* and in birds like *Archaeopteryx*.

FIGURE 12.8C: An ant trapped in amber—This ant was trapped in tree resin and died. The resin preserved it and prevented it from decaying.

Because fossils are found in layers of sediment, generally the deeper the layer the fossil is found in, the older it is. In some circumstances, this allows for relative dating of a fossil based on its position in the rock and by other fossils found near it in the same layer, see Figure 12.10. **Radiometric dating** of the surrounding rocks where the fossil is found can assign a time range indicating how old the fossil is. Radiometric dating allows scientists to calculate the age of a rock by comparing the amount of a radioactive isotope found in a rock when it was formed to the amount of that isotope present currently in the rock. Carbon-14 dating can only be used on rocks less than 50,000 years old, so it isn't often used to date fossils. Isotopes used for dating of fossils include potassium and argon.

Homologies are another category of evidence to support evolution. They are similarities between organisms that reflect their ancient and current evolutionary relationships. More similarities between two groups indicate that they shared a more recent common ancestor. If there are many differences, then it has been a longer time since they shared a common ancestor. The types of homologies that are examined include anatomical, developmental, and molecular traits.

The forelimbs of tetrapods are a classic example of an anatomical homology. The front limbs of various animal species may have different functions, but looking at the number, layout, and basic shape of each bone, there are many similarities. You can see in Figure 12.11 that the corresponding bones have matching colors. Looking at the phalanges (brown), you can see how the bones have been modified slightly in shape and size, but the general layout is the same. Different mutations over time in the genes that determine bone structure have produced the different shapes and sizes. Natural selection preserved these changes because they provided survival and reproductive advantages to the organisms that had them.

FIGURE 12.9: **Tiktaalik, an important transitional fossil**—In this fossil of Tiktaalik, it is possible to see the flat head with eyes located on the top, which is a tetrapod trait. It is also possible to see the fins, which is a trait of fish.

FIGURE 12.10: **Relative dating of fossils**—In the figure, the dinosaur fossil is found below the 60-million-year-old rock layer. The fossil is about 65–100 million years old.

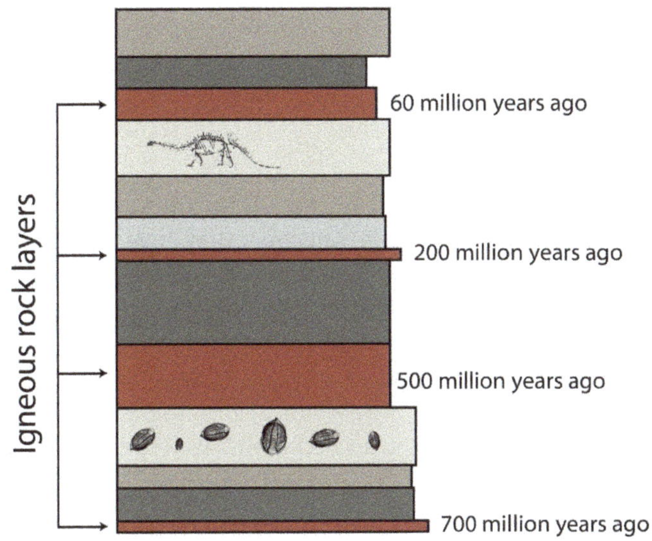

Vestigial structures are structures that may no longer have a function in an organism but did have one in an ancestor or in an extant (still-living) relative. These structures have been modified until they are almost nonexistent and completely lost their function because of mutations. For example, a few species of snakes, pythons, and boas have structures called spurs located near their cloaca (opening used for elimination and mating). They often contain small "floating" femurs that are not attached to a pelvis. They are the remnants of legs, and although they are no longer used for walking, they are still used during mating by males and, in some species, the females

FIGURE 12.11: Comparison of the forelimbs of four tetrapods—Notice that they share the same basic number, size, shape, and layout of bones. The differences are the result of mutations and natural selection.

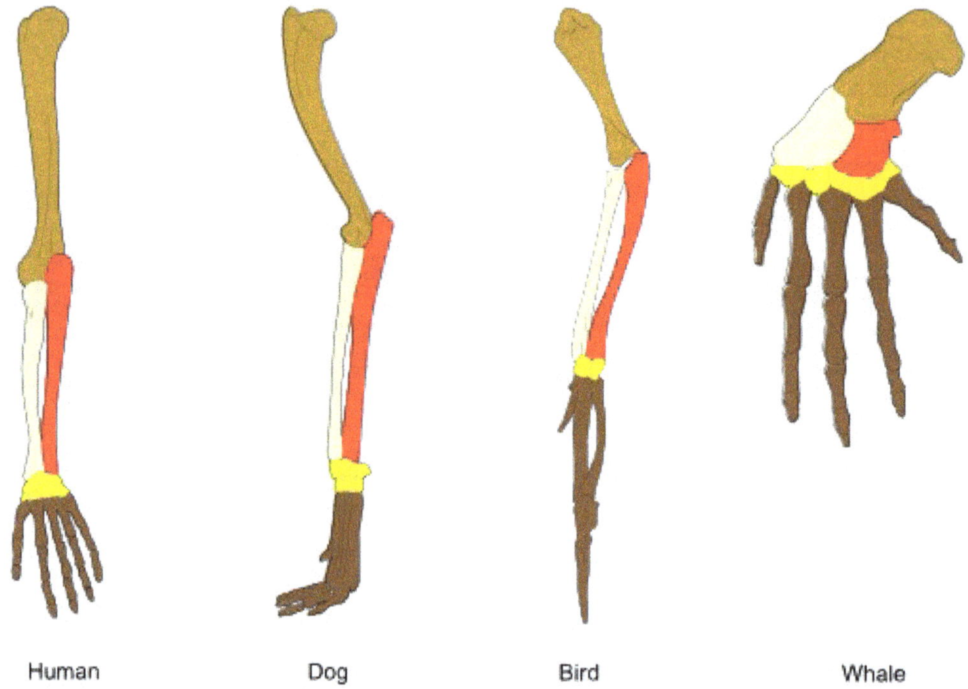

as well. This is evidence that the ancestors of snakes once had legs, but over time because of mutations and natural selection, they have mostly disappeared. Also it shows that snakes are related to other living reptiles that still have legs.

While an embryo develops, the shape and size of certain features will change, and some will even disappear as development continues. Even early on during embryonic development, closely related species share similarities, and it can be difficult to tell them apart. The basic layout for the major anatomical structures is shared, but as development continues, different genes are turned on and off in different species, altering the presence and shape of the structures. In Figure 12.12, it is possible to see that all the embryos have the same basic layout of structures. The more closely related the species are, the more they resemble each other.

Another type of homology involves molecular characteristics like genes and protein sequences. All organisms use DNA as their genetic material, which is made up of same nucleotides with the four nucleotide bases: adenine, thymine, cytosine, and guanine. Also, almost all organisms use the same 20 amino acids to build their proteins. Comparing the nucleotide base sequences of genes and the amino acid sequences in proteins between organisms can indicate how closely related two species are to each other. The more differences there are as a result of mutations, the longer the period of time since the organisms evolved from a common ancestor. When the sequences are more similar, then they split more recently. For example, in plants, evolutionary relationships are examined by comparing sequences of the *rbcL* gene. This gene codes for an enzyme abbreviated RuBisCO, which is used during a step of photosynthesis.

Because all plants are photosynthetic, every plant should have a functioning version of this gene and the enzyme, but because of mutations, they will not be the same. In Table 12.1, there are amino acid sequences for a portion of RuBisCO for three different species of plants. Potato and tobacco only have one difference between their sequences, yet potato and spinach have five. This makes sense because potatoes and tobacco are in the same plant family Solanaceae, whereas spinach is in a completely different family, Amaranthaceae.

Homologies, also known as homologous traits, provide evidence that supports common ancestry between groups because they are structures that are similar although they may have a different function, like a human's arm and a whale's flipper. **Analogies,** on the other hand, are structures that have a similar function but did not evolve from a recent common ancestor. They result from what is called **convergent evolution.** Convergent evolution occurs when organisms evolve similar traits, not because they evolved from a recent common ancestor, but because they live in similar environments. A trait that is advantageous in a particular environment in one group is still advantageous in another group of organisms. A good example of convergent evolution is the pangolin and the armadillo (Figure 12.13). Both have long tongues adapted for eating ants and termites, strong front limbs for digging insect mounds, armored scaled skin, and roll up when threatened.

FIGURE 12.12: **Comparison of four embryos**—Notice that the embryos all share the same basic layout, and many structures are similar in size and shape. A = opposum, B = mouse, C = chicken, D = snapping turtle (HL = hindlimb, FL = forelimb, N and T = specialized brain areas)

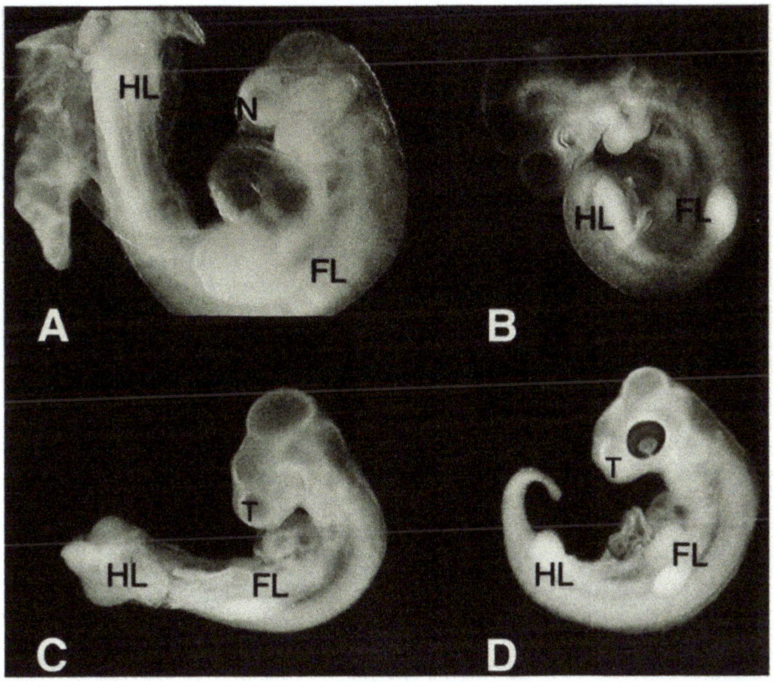

TABLE 12.1: Portion of Amino Acid Sequences for RuBisCO in Three Plant Species

Plant name	Amino acid sequence																
Potato	V	W	K	E	I	V	F	N	V	A	A	M	D	V	L	D	K
tobacco	V	W	K	E	I	V	F	N	V	A	A	V	D	V	L	D	K
spinach	V	W	K	E	I	K	F	E	V	P	A	M	D	T	V	D	K

FIGURE 12.13A: Pangolin

FIGURE 12.13B: Armadillo—Although they share many physical similarities, pangolins and armadillos are not closely related. They evolved similar traits because they live in similar climates.

FIGURE 12.14: Patagonian mara—The Patagonian mara resembles and has the same role in the environment as rabbits, but they are not closely related.

However, pangolins live in Africa and Asia, and armadillos are native to the Americas. Once considered close relatives, anatomical and DNA evidence shows that they are not actually closely related at all but have separately evolved traits that help them survive in their similar environments.

Convergent evolution results in each species having similar adaptations that help them survive in similar habitats, but these adaptations did not evolve from a trait in a common ancestor.

The last category of evidence is **biogeography**. Biogeography examines how species are distributed around the world, and how that is a reflection of their evolutionary past. Biogeography can explain why species that live near each other are more closely related and why some species are not found in habitats that they are well adapted for. It can also give us information about past geological events and ecological change. For example, Darwin noticed that there were no rabbits in South America even though the climate and food sources in some areas could support rabbits. There was, however, an animal called the Patagonian mara (*Dolichotis patagonum*) (Figure 12.14) that occupied the same niche (role) as rabbits would.

This provides evidence for evolution because it helps demonstrate that convergent evolution through natural selection selects similar adaptations in similar environments producing species that resemble each other but aren't closely related.

12.2 SUMMARY

Humans have been wondering about and wanting to classify the organisms around them for centuries. Jean Baptiste Lamarck proposed a theory of evolution of new species, but it was not well accepted and later was found to have some serious problems. Charles Darwin and Alfred Russel Wallace were the first to propose a theory of evolution by natural selection that explained the process of how species change over time and new species evolve. Natural selection works on populations that exhibit heritable variations in their traits. Because resources are limited, competition arises among members of the same species, leading to a struggle for existence. Individuals in a population with advantageous traits will survive and pass on their traits to their offspring. For those that do not survive, their traits are removed from the gene pool. Over time, a new species can evolve.

Because evolution is a theory, it can be modified and updated as new evidence is available. In the beginning of the 20th century, the field of genetics was incorporated into evolution, and the modern synthesis theory was added on to Darwin's work. Mendelian genetics explained how traits were passed down, and statistics and probability could be used to make predictions about the traits of offspring. Mutations are how new variations appear in populations. Also, evolution was described as occurring on the population level and not to individuals. In addition to natural selection, other mechanisms could cause allele frequencies to change in a population causing evolution, including mutations, gene flow, sexual selection, and genetic drift. Also, the biological species concept was introduced, which made reproductive isolation an important determining factor when defining a species. If two organisms are not able to or not potentially able to interbreed and produce fertile offspring, then they are from different species. Biologists were able to study what barriers prevented fertilization and development of hybrid zygotes.

When Darwin's book was published, it was controversial because it contradicted a Biblical explanation of the creation of species. Many lines of evidence support the modern synthesis theory of evolution. Fossils help fill in the evolutionary timeline for various extinct species, showing how structures changed and disappeared over time.

12.3 KNOWLEDGE CHECK

1. Which of the following is required for natural selection to work on a population?
   a. Traits are passed on to the offspring
   b. Traits show variations
   c. There are limited resources and a large population
   d. Some organisms are better competitors and survive and others do not
   e. All of these are required for natural selection

2. Some species of bacteria are now resistant to treatment by antibiotics. The once-effective drugs no long work to control infection. Why do scientists think this happened?
    a. An individual bacteria became resistant to the antibiotics after being exposed to it in the body
    b. The bacteria needed to be resistant to the antibiotics so they could survive after the patient received treatment
    c. The presence of antibiotics in the patient caused the bacteria to be resistant, who were able to pass it on to their offspring
    d. A few bacteria were already resistant to the antibiotics, they survived treatment and passed the trait onto their offspring
    e. The bacteria were created to be resistant to antibiotics

3. The modern synthesis theory of evolution combined Darwin's theory of natural selection with
   a. genetics
   b. chemistry
   c. anatomy and physiology
   d. sociology
   e. physics

4. The biological species concept defines a species as
   a. individuals of a population that are fertile
   b. individuals of a population that resemble each other
   c. members of a population that can interbreed and produce fertile offspring
   d. individuals that can interbreed in isolation
   e. a population that cannot interbreed with another

5. Which of the following would be an example of a fossil?
   a. The thighbone of a Tyrannosaurus rex buried in a shale deposit in North Dakota
   b. The impression of feathers in stone from an ancient bird found in China
   c. A bee encased in amber
   d. Mineralized remains of Tiktaalik
   e. All of these are examples of fossils

6. Which of the following statements is NOT true?
   a. The more differences between the DNA of two species, the more recently they evolved from a common ancestor
   b. All living things on Earth have DNA as their genetic material
   c. The more similarities between the DNA of two species, the more recently they evolved from a common ancestor
   d. Allele frequencies in populations can change over time as a result of mutation, natural selection and genetic drift
   e. Mutations in DNA occur randomly and can accumulate in a population over time

7. Explain why evolution is considered both a scientific theory and a scientific fact.

8. Describe how each of the following can change the allele frequency of a population leading to evolution: mutation, natural selection, gene flow, sexual selection, and genetic drift.

9. Discuss how it is possible for sharks and dolphins to both live in the ocean and resemble one another but not be closely related.

12.4 ACTIVITY

Look Who's Coming for Dinner: Selection by Predation

The purpose of this activity from the HHMI Biointeractive website (biointeractive.org) is to examine how natural selection has influenced the evolution of a species of Anole lizard, *Anolis sagrei.* This species is the only Anole species found on the northern islands of the Bahamas. Another lizard, *Leiocephalus carinatus,* found on the northern islands of the Bahamas, preys on *Anolis sagrei.* The activity involves developing a hypothesis, reading about a real-life experiment with these lizards, analyzing data, and drawing conclusions regarding if and how natural selection has affected the population of *Anolis sagrei.*

Before class, watch the short film "Lizards in an evolutionary tree", which describes different species of Anole lizards living on four different islands in the Caribbean. You can watch it by clicking this link (https://youtu.be/rdZOwyDbyL0) or looking for it on the Biointeractive YouTube channel. Also before class, you should read and answer the questions in Part I: Introduction of the activity "Look who's coming to dinner: Selection by predation" posted by your instructor. In class, you will complete Part II: Hypothesis and experiment, Part III: Data collection and analysis, and Part IV: Conclusions.

12.5 ACTIVITY

The Evolution of Human Skin Color

The purpose of this activity is to analyze and evaluate the role of several important factors that may have shaped the evolution of human skin color during the last 300,000 years. First, watch the TED Talk titled "Skin Color is an Illusion" delivered by Nina Jablonski, which is available using the following link: http://www.ted.com/talks/nina_jablonski_breaks_the_illusion_of_skin_color?language=en. Pay careful attention to this video, and take notes as needed to remember and summarize the core concepts discussed in the video. For the remainder of the activity, follow instructions to work independently or form small groups to complete the indicated sections of the case study titled "The Evolution of Human Skin Color," which is available from the National Center for Case Study Teaching in Science using the following link: http://sciencecases.lib.buffalo.edu/cs/files/skin_pigmentation.pdf. After completing the case study, you should be able to explain the role of each of the following in the evolution of human skin color: foliate, latitude, melanin, natural selection, skin cancer, UV radiation, and vitamin D.

12.6 ACTIVITY

Human Skin Color: Evidence for Selection

This activity provides a slightly different alternative to Activity 12.5 for learning about the evolution of human skin color. The purpose of this activity from the HHMI Biointeractive website (biointeractive.org) is to examine the evidence for the evolution of different skin colors in humans. The activity includes a case study and a film called "The Biology of Skin Color", found at this link https://youtu.be/hFw8mMzH5YA. Your instructor will show the film in class and pause it at certain points to allow for data analysis found in the case study handout. Part I of the activity will examine environmental factors related to human skin color. Parts II and III will look at how natural selection has affected human skin color. Part IV will investigate how human migration and human skin color are often at odds.

12.7 ACTIVITY

Genetic Drift Simulation

Recall that evolution is a change in allele frequencies in a population over time. Several mechanisms can cause allele frequencies to change in a population including natural selection, genetic drift, mutation, gene flow, and sexual selection. Genetic drift causes a change in allele frequencies because of random occurrences and is not based on adaptations in the population. The two different ways that allele frequencies change as a result of genetic drift are the bottleneck effect and the founder effect. The bottleneck effect occurs when a large portion of a population randomly dies as a result of a natural disaster, an epidemic, or hunting. The surviving population's gene pool will be less diverse than the original population's. For example, in the 1800s, Northern elephant seals were hunted almost to extinction for their blubber, reducing their population to about 100 individuals. The surviving population has recovered because of conservation efforts, but the genetic diversity of the population is very different from the pre-1800 population. The founder effect occurs when a small group of individuals moves to a different area away from their original population and then remain isolated in

TABLE 12.2A: Number and Percentages of Alleles of *Darwinius beadus* on Three Different Islands

			Red	Yellow	Purple	Green	Blue	Total
Island A	**Starting population**	# alleles						
		% of alleles						100%
	Population after wildfire	# alleles						24
		% of alleles						100%

TABLE 12.2B: Number and Percentages of Alleles of *Darwinius beadus* on Three Different Islands

			Red	Yellow	Purple	Green	Blue	Total
Island B	**Starting population**	# alleles						
		% of alleles						100%
	Population after tsunami	# alleles						12
		% of alleles						100%

TABLE 12.2C: Number and Percentages of Alleles of *Darwinius beadus* on Three Different Islands

			Red	Yellow	Purple	Green	Blue	Total
island C	Starting population	# alleles						
		% of alleles						100%
	Population after eruption	# alleles						6
		% of alleles						100%

the new location. The genetic diversity of the population is greatly reduced. An example of founder effect in humans involves Afrikaners of South Africa. One or more individuals in the group of Dutch colonists in the 1600s carried the dominant allele that causes Huntington's disease with them from Europe. These white settlers did not integrate with the native African population and stayed isolated. As a result, the modern population of Afrikaners have a much higher frequency of Huntington's disease whereas native South African populations have a very low frequency.

This activity will simulate how genetic drift can change the allele frequencies in a population. The population in this hands-on simulation is an imaginary species of a multicolored bird called *Darwinius beadus* that live on various islands in the South Pacific. Different colored beads represent the population's alleles for feather color in this simulation.

Materials:

Plastic zip-top bag
Paper plate
Five different colors of plastic beads

Procedure:

1. Get a bag containing your population's alleles or gene pool. Each bead represents a different allele for feather color in a population of *Darwinius beadus* living on Island A. Empty your gene pool onto a paper plate.
2. Count how many of each color allele and the total number of alleles found in your gene pool. Enter these data into Table 12.2A for *Island A starting population*. Next, calculate the percentages of each different allele and enter that in Table 12.2A as well.
3. Return the beads back to the bag.
4. Island A is hit by a devastating wildfire, which kills many of the animals on the island. In fact, only 12 *Darwinius beadus* birds survive. Simulate this by closing your eyes and pulling 2 beads at a time until you have a total of 24 beads. These beads represent the 12 pairs of alleles for feather color each individual contributes to the surviving gene pool.
5. Count how many of each color allele remains, calculate the percentages, and record in Table 12.2A in *Island A population after wildfire*. Return the beads back to the bag.

6. Island B also has a population of *Darwinius beadus*, which coincidently has the same allele composition as the starting population on Island A. Enter the starting population data for Island B in Table 12.2B.
7. Island B experiences a massive tsunami that kills many of the animals on the Island except for 6 *Darwinius beadus* birds. Close your eyes and draw another 6 pairs of alleles (12 beads total) to represent the gene pool of the tsunami survivors. Count the number of each color allele. Calculate the percentages and record in your data in Table 12.2B in Island B. Return the beads back to the bag.
8. Island C also has a population of *Darwinius beadus*, which coincidently has the same allele composition as the starting population on Island A. Enter that data in Table 12.2C for Island C starting population.
9. Island C experiences a large volcanic eruption, which destroys most of the island and the animals living on it. Only 3 *Darwinius beadus* birds survive. Draw 3 pairs of alleles (6 beads total) to represent the gene pool of the eruption survivors. Count the number of each color allele. Calculate the percentages and record in your data for Island C (Table 12.2C).

ACTIVITY QUESTIONS

12.4 Look Who's Coming for Dinner: Selection by Predation

1. Follow instructions to work individually or form small groups for completing the activity and answering the indicated questions from the activity.

12.5 The Evolution of Human Skin Color

1. Follow instructions to work individually or form small groups for completing the activity and answering the indicated questions from the activity.

12.6 Human skin color: Evidence for Selection

1. Follow instructions to work individually or form small groups for completing the activity and answering the indicated questions from the activity.

12.7 Genetic Drift Simulation

1. Did the *Darwinius beadus* populations on the different islands evolve? Use your data to support your claim.
2. What type of genetic drift did you simulate? Explain the difference between the type you simulated and the other type of genetic drift.
3. Did you lose any alleles from the post-disaster population? Why did that happen? Propose an explanation for how this allele could "reappear" in the population.

4. Look at your data for the surviving population on Island A. Suppose green alleles are recessive and cause a fatal genetic disorder in *Darwinius beadus*. Over time and many generations, what will happen to the frequency of this genetic disorder in your *Darwinius beadus* population on Island A? Explain using your data.
5. Using this "bead for alleles" model, propose a way to simulate natural selection, gene flow, and mutations.

12.8 REFERENCES

Darwin, C. (1859). *The origin of species by means of natural selection or the preservation of favoured races in the struggle for life*. 1st ed. London: Penguin.

Mayr, E. (1999). *Systematics and the origin of species, from the viewpoint of a zoologist*. Cambridge, MA: First Harvard University Press.

Serafini, A. (1993). *The epic history of biology*. 1st ed. New York, NY: Plenum Press.

Simonetta, A. M. (2003). *Short history of biology: From the origins to the 20th century*. 1st ed. Firenze: Firenze University Press.

The Complete Works of Charles Darwin Online. [On-line] Access date: July, 26 2017. http://darwin-online.org.uk/.

FIGURE CREDITS

FIG. 12.1: 4657743, "Two giraffes," https://pixabay.com/en/giraffe-wildlife-wild-africa-2112414/. Copyright in the Public Domain.

FIG. 12.2A: George Richmond, "Charles Darwin as a young man," https://commons.wikimedia.org/wiki/File:Charles_Darwin_by_G._Richmond.jpg. Copyright in the Public Domain.

FIG. 12.2B: Owen Stanley, "HMS Beagle," https://commons.wikimedia.org/wiki/File:HMSBeagle.jpg. Copyright in the Public Domain.

FIG. 12.3: Charles Darwin, "Darwin's tree of life," On the Origin of Species by Natural Selection. Copyright in the Public Domain.

FIG. 12.4A: Copyright © Filip Lachowski (CC BY-SA 2.0) at https://commons.wikimedia.org/wiki/File:Fighting_Hartebeest.jpg.

FIG. 12.4B: confused_me, "Male peacock and female peahen," https://pixabay.com/en/peacock-bird-plumage-feather-2205201/. Copyright in the Public Domain.

FIG. 12.5: Copyright © Tsaneda (CC by 3.0) at https://commons.wikimedia.org/wiki/File:Founder_effect_Illustration.jpg.

FIG. 12.6: Copyright © OpenStax, Rice University (CC by 4.0) at https://commons.wikimedia.org/wiki/File:Bottleneck_effect_Figure_19_02_03.jpg.

FIG. 12.8A: Copyright © Brett Neilson (CC by 2.0) at https://commons.wikimedia.org/wiki/File:Columbian_mammoth_(Mammuthus_columbi).jpg.

FIG. 12.8B: Copyright © H. Zell (CC BY-SA 3.0) at https://commons.wikimedia.org/wiki/File:Acer_trilobatum_01.JPG.

FIG. 12.8C: Copyright © Anders L. Damgaard (CC BY-SA 4.0) at https://commons.wikimedia.org/wiki/File:Baltic_amber_inclusions_-_Ant_(Hymenoptera,_Formicidae)8.JPG.

FIG. 12.9: Copyright © Camelops (CC BY-SA 4.0) at https://commons.wikimedia.org/wiki/File:FMC08-24_Tiktaalik_rosae.tif.

FIG. 12.10: Copyright © Jillcurie (CC BY-SA 3.0) at https://commons.wikimedia.org/wiki/File:Relative_dating_of_fossils.png.

FIG. 12.11: Copyright © Волков Владислав Петрович (CC BY-SA 4.0) at https://commons.wikimedia.org/wiki/File:Homology_vertebrates-en.svg.

FIG. 12.12: Harvard University, Museum of Comparative Zoology, "Comparison of four embryos," Bulletin of the Museum of Comparative Zoology at Harvard College. Copyright in the Public Domain.

FIG. 12.13A: Copyright © U.S. Fish and Wildlife Service Headquarters (CC by 2.0) at https://commons.wikimedia.org/wiki/File:Manis_temminckii_(29645803646).jpg.

FIG. 12.13B: skeeze, "Pangolins and armadillos are not closely related," https://pixabay.com/en/armadillo-shell-mammal-wildlife-649937/. Copyright in the Public Domain.

FIG. 12.14: diego_torres, "Patagonian mara," https://pixabay.com/en/animal-mara-priroda-952683/. Copyright in the Public Domain.

Immune System

13.1 INTRODUCTION

Virtually every living organism, including bacteria, plants, and animals, have evolved some type of mechanism to protect themselves against foreign invaders, potentially harmful molecules from the environment, and the development of abnormal cells. In bacteria, this defense system is much simpler than in humans, consisting mostly of physical barriers and a set of proteins capable of destroying the genetic material of invading viruses. Plants and invertebrate animals have a more complex immune system, consisting of a variety of physical barriers, chemical barriers, cells, and molecules. Humans and other vertebrate animals have developed the most complex immune systems. This chapter will provide an overview of primitive immune systems, then focus on discussing the main components of the human immune system.

Immune Systems of Bacteria and Plants

As life evolved from rudimentary simple single-celled organisms to produce increasingly complex multicellular beings, the ability to mount a response against pathogens and rogue cells also evolved in complexity. The earliest and simplest type of immunity developed by living organisms is known as the innate immune system. Innate immunity evolved in ancestral organisms over 1 billion years ago, which is far enough back that it was passed down to all plants and animals alive today. Even bacteria and archaea, which are vulnerable to infection by viruses, possess an innate immune system. The innate immune system includes protective physical barriers, such as the cell wall and capsule of a bacterium. Acids and other chemicals that may deter infection are also components of innate immunity.

It has long been known that bacteria possess these simple defense mechanisms, and more recently scientists have discovered that bacteria and archaea also possess a more sophisticated immune system used as a defense mechanism against viral pathogens. This is important because viruses that make it past the physical and chemical barriers of a cell may integrate their genetic material into the DNA of the infected cell, making them extremely difficult to eliminate.

LEARNING OBJECTIVES

- Explain what CRISPR is and what it can be used for
- Identify the main components of the innate immune system and their functions
- Explain how the innate and adaptive immune systems recognize pathogens
- Explain what occurs during inflammation and phagocytosis
- Identify the main components of the adaptive immune system and their functions
- Summarize the difference between innate and adaptive immunity, between cellular and humoral immunity, and between active and passive immunity
- Describe the role of antigen-presenting cells/phagocytes, B cells, CD4 T cells, CD8 T cells, granulocytes, and mast cells in responding to infections
- Identify the roles of antibodies and histamine in fighting infections
- Discuss the role of the immune system in the development of allergies and autoimmunity
- List the main causes and consequences of immunodeficiency
- Identify the function of T regulatory cells
- Identify an example of an opportunistic infection

Of particular interest and significance is an intracellular component of the bacterial immune system known as CRISPR/Cas9. **CRISPR,** which stands for "clustered regularly interspaced palindromic repeats," refers to a set of short nucleotide sequences within the genomes of bacteria and other microorganisms that can be used to guide enzymes to the location of viral genetic material to remove it from the bacterial chromosome. Cas9 is the name of an enzyme that recognizes and cuts the viral genetic material out of the bacterial DNA. Although CRISPR/Cas9 has been useful to bacteria for hundreds of thousands of years, it has recently become extremely valuable to humans as well. Because CRISPR was first characterized in the 1990s, scientists have developed it into one of the most important biotechnologies of the 21st century. CRISPR can be used to remove or edit genomic sequences, facilitating production of genetically modified organisms. The applications of this technology includes the potential production of genetically modified humans. One incredible example of this technology occurred in China in 2016 when a team of scientists successfully used CRISPR to introduce a mutation (CCR5-Δ32) into human embryos. This mutation confers resistance to HIV infection, so the implications of this experiment are that future generations of humans may one day be born with a natural resistance to one of the most devastating pathogens in human history. In 2017, the United States became the second country to report successful modification of human embryos using CRISPR. In this case, the technology was used to repair a genetic mutation called MYBPC3 ΔGAGT, which leads to a type of heart disease called hypertrophic cardiomyopathy. Advancements in CRISPR technology may soon be adapted to fix a wide array of disease causing genetic mutations in human embryos. The potential uses of CRISPR seem exciting and limitless; yet it also poses important bioethical questions. For example, should scientists work to develop ways of removing undesirable genetic variants from human embryos to create genetically modified people with superior characteristics?

Plants are also subject to a variety of infectious agents, including viruses, bacteria, and fungi, as well as damaging animals, and even parasitic plants. The term **pathogen** refers to bacteria, viruses, and any other microorganism that cause disease in another organism. The organism infected by a pathogen is called a host. Similar to bacteria, the first line of defense plants use to protect themselves is composed of physical barriers. Plants cells are protected by sturdy cell walls that may be covered in wax or some type of protective coating. Some plants also secrete sticky and/or poisonous substances to deter predators. Additionally, plants possess aspects of a cell-based innate immune system. However, unlike animals, plants do not have mobile defender cells capable of circulating through their body to search for and respond to pathogens. Plants also lack the ability to mount pathogen-specific adaptive immune responses that could be much more effective than nonspecific innate responses.

Organs of the Immune System

The remainder of this chapter will discuss the immune system of vertebrate animals, specifically humans. The human immune system is composed of chemical defenses and cells, as well as several organs and vessels responsible for producing, maturing, and transporting those chemicals and cells in the body. In humans and other mammals, the organs involved in the immune system include the skin and mucous membranes, bone marrow, the lymphatic system, and organs of the digestive tract.

The skin and mucous membranes help form the first line of defense against foreign invaders. Skin provides a physical barrier that prevents microorganisms from entering the body and also secretes compounds that make it difficult for harmful microorganisms to grow and enter the body. Antimicrobial substances secreted by the skin include acids, oils, and salt. **Mucous membranes** are layers of cells that line the airways, gut, genitourinary tract, and other body cavities, and secrete a thick protective fluid called mucous.

Bone marrow is a fatty, sponge-like material found inside bones responsible for producing most types of immune system cells. Immune cells produced by the bone marrow are immature and not yet fully capable of recognizing pathogens and mounting defenses. These cells then move to other organs and tissues, such as the thymus and spleen, to further develop and become mature. In addition to immune cells, which are also called white blood cells, the bone marrow produces red blood cells, which are responsible for transporting oxygen in the blood, as well as platelets, which help control bleeding. Some serious diseases can result from unhealthy bone marrow, including blood cancers such as leukemia and lymphoma, and certain types of anemia leading to abnormally low levels of functional red blood cells. When less invasive treatments are not effective, treatment of these diseases may require a bone marrow transplant.

The **lymphatic system** is part of the circulatory system that serves as a conduit for the immune system. It consists of lymph nodes, lymph vessels, adenoids and tonsils, the spleen, and the thymus gland. Cells and molecules that are part of the immune system circulate through the lymphatic system within a fluid called lymph. As lymph moves through lymph vessels, it is filtered through a number of small bean-shaped organs called lymph nodes. Lymph nodes are where many immune cells are located and may become enlarged during an infection as immune cells divide and increase in number. Lymph nodes may also trap cancer cells that are moving through the body, and can be the first sites where advanced stages of cancer are detected. During physical examinations, doctors often check for swollen lymph nodes in the armpits, groin, and neck as an indication of infection or cancer.

Two other organs essential to immune system function are the spleen and the thymus. The **spleen**, which is located in the upper left part of the abdomen, acts mainly as a filter for blood. The spleen removes damaged and abnormal red blood cells from circulation and helps to maintain appropriate levels of red blood cells and platelets in the blood. The spleen also serves as a reservoir of several types of white blood cells, including B cells and macrophages, which are activated and released in response to infection. The primary role of the **thymus,** which is a small gland located at the base of the neck behind the sternum, is to control maturation of a type of white blood cell called T cells or T lymphocytes. The thymus promotes development of T cells that can recognize foreign molecules and pathogens, and also helps remove T cells that might improperly attach to molecules found on an organism's own cells.

Cells of the immune system also circulate through a network of other lymphoid tissues that line the mouth and intestines to identify and respond to potentially pathogenic microorganisms. Starting in the head, the first set of lymphoid tissues that foreign invaders entering the body through the nose and mouth encounter includes the adenoids and tonsils. Adenoids and tonsils are masses of soft tissue located between the nose and throat and at either side of the back of the throat. During an infection, these mounds of tissue become enlarged and painful as a result of immune cells and fluid entering the tissue to help clear microorganisms and infected cells. Sometimes, swollen adenoids and tonsils must be surgically removed because they are extremely painful or are interfering with normal breathing.

Strong acid and digestive enzymes in the stomach kill most pathogens that enter the body with food. However, pathogens that do make it past the stomach and into the intestines are identified and targeted by immune cells located primarily in specialized structures in the large intestine called Peyer's patches. The large intestine also contains billions of bacteria known as gut flora, or gut microbiota, which help maintain normal immune system function. The appendix may also have evolved as part of the immune system, perhaps to aid in restoring the natural balance of gut microbes after severe diarrheal illnesses; however, the exact role of the appendix is unclear, and it is not thought to serve an essential function in modern humans.

Innate Immunity

The **immune system** of vertebrates is a complex network of organs, cells, and molecules that protect the body from pathogens and malignant cells, and help clear away foreign substances and debris. **Innate immunity** is the result of a collection of nonspecific defense mechanisms that includes physical barriers, chemical barriers, and cells that respond to infection by pathogens without identifying which specific pathogen is present. Physical barriers that are part of the innate immune system include the skin and the layers of cells that line the airways and intestines. The cells that comprise these barriers form thick, difficult to penetrate layers. These cells often produce or are covered by mucous and chemical barriers, such as acids, salts, and enzymes, that help trap and kill microorganisms. For example, the cells within the skin produce acidic and salty secretions to form an environment inhospitable for potentially pathogenic microbes. Additional chemical barriers with innate immune functions include the strong acid and digestive enzymes found in the stomach, an antimicrobial enzyme found in tears called lysozyme, and the **complement system**, which is a set of proteins found in blood plasma that aid in the destruction of pathogenic microorganisms.

The innate immune system detects pathogens by using cell-based receptors to recognize molecules not present in host cells, and responds to these molecules by releasing defensive chemicals. The molecules recognized by the innate immune system are known as **pathogen-associated molecular patterns** (PAMPs). The innate immune system recognizes a wide variety of PAMPs, including bacterial DNA, viral RNA, and carbohydrates, proteins, and lipids that are only produced by bacteria and fungi. Although innate immunity is effective at detecting the presence of invading pathogens by recognizing PAMPs, it is not sophisticated enough to determine which specific pathogen the PAMPs belong to. This means that innate immunity is nonspecific and not able to refine responses to provide optimal protection against a particular pathogen. Because of this, innate immune responses are not as accurate or efficacious at fighting pathogens than adaptive immune responses, which are specific for each individual pathogen. Additionally, innate immune responses may be wasted by failing to reach a pathogen, or may inadvertently damage nearby host cells. On the other hand, the greatest benefit of the innate immune system is that its physical and chemical barriers provide continual protection against pathogens, and the cells of the innate immune system can respond to invading pathogens almost immediately as they do not require time to recognize and develop a response against each specific pathogen. If not for the quick-acting protection of innate immunity, the majority of organisms would likely succumb to infections and die very early on in their lifetime.

Cells of the Innate Immune System

All cells that are part of the vertebrate immune system are included under the umbrella term **leukocytes**, or white blood cells. This name reflects their lack of red-colored hemoglobin, which is a hallmark of red blood cells. Leukocytes are classified as a number of distinct cell types based on their origin and function. Types of cells that primarily act as part of the innate immune system include dendritic cells, macrophages, granulocytes, and mast cells (Figure 13.1). **Dendritic cells** are found throughout the blood and tissues where they constantly search for and ingest samples of cellular debris and pathogens, then display molecules from what they ingested on their cell surface. Any molecules derived from pathogens, the environment, or cells in the body specifically recognized by the adaptive immune system are known as **antigens**. Therefore, dendritic cells are also known as antigen-presenting cells. Upon encountering a pathogen, dendritic cells become activated and mature, and then move to the spleen or lymph nodes in order to display antigens from the pathogen to lymphocytes. The purpose of this process, which is known as antigen presentation, is to activate robust immune responses against specific pathogens whenever they are encountered. **Neutrophils** and **macrophages** are two other types of antigen-presenting cell with similar roles to dendritic cells. Neutrophils circulate through the bloodstream and respond to infection by engulfing and destroying pathogenic microorganisms. Macrophages may also circulate through the blood, but some macrophages spend their life in one specific organ or tissue. Macrophages are involved in infiltrating cancerous tumors and contributing to the development of fibrosis, which is used to repair damaged connective tissues. Additionally, macrophages are more active than dendritic cells at directly killing pathogens, and dendritic cells are more closely associated with stimulating other cells to help fight infections. As discussed later in this chapter, antigen presentation is one of the key events where the innate and adaptive immune systems intersect. It is the cells of the adaptive immune system that are activated by and respond to antigen presentation.

Another type of innate immune cell, called the **natural killer cell,** circulates through the body like neutrophils and some macrophages but does not engulf pathogens. As an alternative method of helping clear pathogens

FIGURE 13.1: Cells of the immune system—The cells of the innate immune system include dendritic cells, macrophages, mast cells, natural killer cells, and granulocytes. The cells of the adaptive immune system include B cells, CD4+ T (helper) cells, and CD8+ (cytotoxic) T cells.

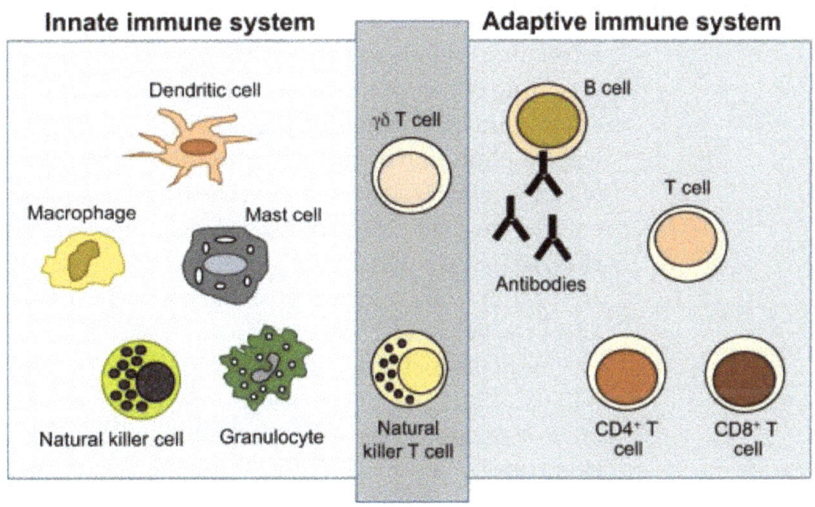

from the body, natural killer cells are able to recognize cells infected by a virus or are cancerous by detecting the presence of abnormal molecules on the surface of these cells, and respond by releasing chemicals that kill the affected cell. The other two types of cells involved in innate immunity that will be discussed here are granulocytes and mast cells. **Granulocytes** received their name because they contain packets of preformed enzymes and proteins called granules, which they release in response to detection of pathogens and tissue damage. The three main types of granulocytes are neutrophils, eosinophils, and basophils. Neutrophils account for 90% of all granulocytes. They are adept at infiltrating tissues and are among the first cells to respond to infections. The two main roles of eosinophils are targeting multicellular parasites and contributing to allergic reactions, whereas basophils are noted mostly for their role in allergies. Both of these cell types release granules containing cytokines, enzymes, and protective chemicals, whereas basophils also produce histamine. **Histamine** is a molecule that modulates the functions of nerves and blood vessels during immune reactions. **Mast cells** are a type of white blood cell found throughout the body, including within the skin. Like granulocytes, mast cells secrete a number of preformed proteins when activated through a process called degranulation. The most notable protein released by mast cells is histamine. Mast cells are best known for their role in allergic responses, yet their normal function of the body is to help fight infections by contributing to inflammation. Finally, mast cells, as well as dendritic cells, neutrophils, and macrophages, are known as phagocytes. **Phagocytes** are immune cells that protect the body by ingesting and destroying pathogens through the process of phagocytosis. It should be noted that the immune system contains far more complexity than is explained in this textbook. The innate immune system alone contains a number of other highly active cell types with very important functions. These include natural killer cells, natural killer T cells, and gamma-delta T cells (Figure 13.1).

Inflammation

If damage occurs to the physical barriers of the body, one of the first responses initiated by the innate immune system is inflammation. **Inflammation** is a complex innate immune response to potentially harmful stimuli that involves immune cells, pro-inflammatory molecules, and blood vessels. Inflammation can occur at many sites throughout the body, including the skin, mucosal surfaces, joints, and within organs and tissues. An example of inflammation involving the skin is illustrated in Figure 13.2. In this common scenario, when a physical barrier such as the skin is broken and skin cells are damaged, those cells release chemical signals to alert the immune system of the possibility that pathogens have entered the body (Figure 13.2). These chemical signals consist of small proteins called **chemokines** that direct the movement of mobile cells of the immune system. Chemokines are part of a larger group of signaling molecules called **cytokines**, which cells release to affect the behavior of other cells.

One of the first types of cells to respond to the chemokines released by damaged skin cells is the mast cell. If mast cells become activated, they degranulate, releasing cytokines and histamine. Histamine released by mast cells quickly binds to receptors located on the cells that line blood vessels and on nerve cells (Figure 13.2). The cells that line blood vessels are affected by histamine in two main ways. First, the blood vessels enlarge, or dilate, allowing more blood to flow into the area of the damage tissue. The second effect of histamine on blood vessels is increased

FIGURE 13.2: Inflammation

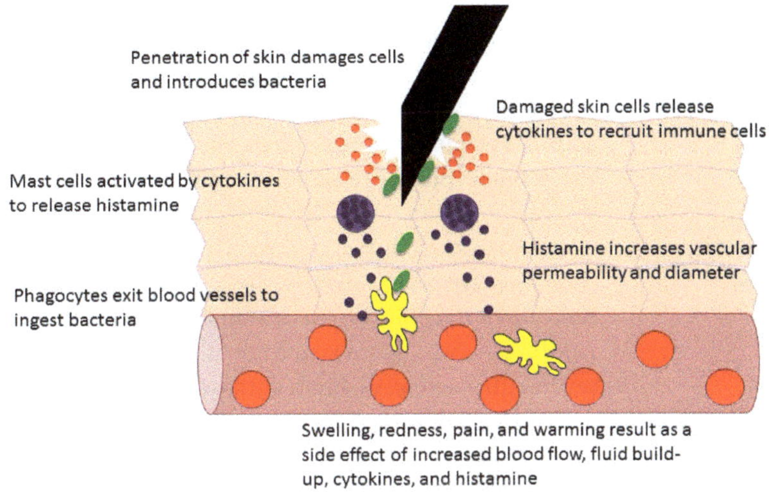

vascular permeability, which means that tiny spaces open up in the blood vessel walls. Together, these changes allow phagocytes to enter the skin and search for potential pathogens. Phagocytes destroy pathogens through the process of phagocytosis, and then engage in antigen presentation to communicate with the adaptive arm of the immune system. During this time, the increased permeability of blood vessels provides excess fluid to wash away debris from damaged cells and dead pathogens. The increase in temperature that results from heightened blood flow reduces the ability of microorganisms to survive and reproduce. This is the same benefit provided by having a fever when ill. As a side effect of the increased permeability, fluid from blood vessels leaks into the site of injury, causing swelling, and the increased blood flow itself causes redness and an increase in body surface temperature. Histamine also binds to nerve cells associated with the skin, airway, and digestive track. Activation of receptors on nerve cells by histamine may cause pain, itching, and even muscle spasms. The cascade of effects initiated by histamine is what produces the characteristic symptoms of inflammation: heat, pain, redness, and swelling. Interestingly, the Roman scholar Celsus first described these classic signs of inflammation in the 1st century AD, long before the underlying mechanisms for these symptoms were understood.

Adaptive Immunity

Adaptive immunity, also called acquired immunity, is composed of a complex network of specialized cells capable of recognizing and eliminating specific pathogens. Because adaptive immunity evolved in our ancestors less than 500 million years ago, it is only found in vertebrate animals. Adaptive immunity provides a second line of defense against pathogens that evade the innate immune system. The main advantage of adaptive immune responses compared with innate immunity is that specific antigens and pathogens can be recognized and targeted, making the adaptive response more efficacious than innate immune responses. The main disadvantage of adaptive immunity is that it may take days to weeks to mount a targeted response against a pathogen's antigens, during which time an infected individual may be very sick and even die.

Cells of the Adaptive Immune System

Cells primarily involved in adaptive immune responses include B cells, CD4+ (helper) T cells, and CD8+ (cytotoxic) T cells (Figure 13.1). Although each of these cells has a distinct role in protecting the body from cancer and pathogens, one commonality shared among them is the ability to recognize and respond to specific antigens. These cells are also referred to as **lymphocytes**, which is the general term of small white blood cells that circulate through the lymphatic system.

About 10% of lymphocytes that circulate in the blood are B cells. **B cells** receive their name because they are produced by the bone marrow. Like dendritic cells and macrophages, B cells are capable of acting as antigen-presenting cells to stimulate the activity of other lymphocytes, such as T cells (Figure 13.3). When antigen-presenting cells display antigens to other cells, they use a type of molecule called major histocompatibility complex (MHC) to hold the antigen. In addition to holding the antigen, the MHC serves as an identifying signal to alert other cells that it is appropriate to respond to the antigen. In theory, this limits the execution of immune responses to only targeting antigens presented by antigen-presenting cells, which prevents unnecessary responses to molecules not associated with potential pathogens.

B cells are best known for their other main function, which is production of antibodies. **Antibodies** are Y-shaped proteins that help the body eliminate pathogens (Figure 13.3). Antibodies, which are also known as immunoglobulins (Ig), are capable of functioning in several different ways. First, antibodies may bind to pathogens or toxic molecules released by pathogens and prevent them from accessing and damaging cells. This is called neutralization. Antibodies may also bind to the surface of bacteria or other pathogens and form a complex with complement proteins from the innate immune system that is able to drill holes into the cell membrane of the pathogen and kill it. This is called complement activation. A third role of antibodies is to bind to pathogens and aid in the process of phagocytosis, making it more effective. This is called opsonization. When B cells first encounter an antigen they recognize, they will be stimulated to proliferate and differentiate into plasma cells (Figure 13.3). It is these plasma cells, which are produced by the thousands, that release antibodies. Plasma cells also release cytokines to activate other cells (Figure 13.3). Once an infection has been cleared, most of the plasma cells will be signaled to undergo apoptosis, yet a small number of them will persist and differentiate into memory B cells. Memory B cells can survive in the body for years and are the main cell type that provides long-term protection against pathogens the body has previously encountered. Immune reactions that are mediated by antibodies, complement, and other molecules found in extracellular fluids, rather than cells directly, are known as **humoral immunity**.

FIGURE 13.3: B cell activation—B cells are activated to produce antibodies after recognizing their cognate antigen and receiving cytokine signals from a helper T cell. Antibodies can then bind to antigens and promote elimination of pathogens by phagocytes or complement-mediated lysis.

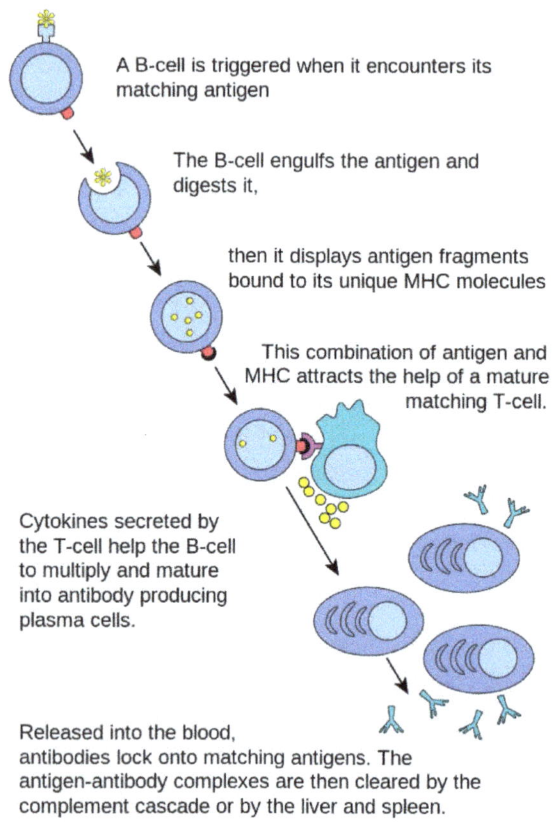

The other main type of lymphocyte involved in adaptive immunity is the T cell. T cells are so-named because they mature in the thymus gland, which is located in the neck. Within the classification of T cell, there are many subtypes of cells with different functions. Two of the most important subtypes of T cells are CD4+ (helper) T cells and CD8+ (cytotoxic) T cells. The name of **CD4+ (helper) T cells** signifies the expression of a cell surface receptor called CD4, and, more importantly the role of helping to recruit and activate other immune cells. The latter is accomplished by secretion of cytokines that attract macrophages and other lymphocytes, and activate lymphocytes to respond to antigens, proliferate, and form memory cells. Although CD4+ (helper) T cells do not directly kill pathogens, they are a crucial part of the overall immune response. The effects of HIV/AIDS illustrate the importance of these cells. Although HIV can infect a number of different cell types, it has evolved to preferentially infect and destroy CD4+ (helper) T cells. Early on in the course of HIV infection, an HIV-positive individual experiences few, if any, noticeable symptoms. However, if left untreated, destruction of CD4+ (helper) T cells leads to a state of immunodeficiency known as AIDS, which is accompanied by susceptibility to opportunistic infections. It is the opportunistic infections, rather than the virus itself, that leads to death of individuals with AIDS. The condition of AIDS is differentiated from the state of being HIV-positive by an HIV+ person experiencing one or more opportunistic infections or having a CD4+ T cell counts of less than 200 cells per microliter. Normal CD4+ T cell counts are greater than 500 cells per microliter.

One of the primary recipients of "help" from CD4+ T cells is the type of lymphocyte known as **CD8+ (cytotoxic) T cells**. As their name suggests, they express a cell surface receptor called CD8, and are cytotoxic, which means they are capable of directly killing pathogens, pathogen-infected kills, or cancer cells. The typical route of CD8+ T cell activation is illustrated in Figure 13.4. First, a dendritic cell, macrophage, or B cell engulfs a pathogen and uses its MHC molecules to present antigens from the pathogen to a naïve T cell (Figure 13.4). During this process, the antigen-presenting cells simultaneously secrete activating cytokines (Figure 13.4). Antigen-presenting cells display molecules from pathogens to cells of the adaptive immune system using specialized structures called major histocompatibility complex (MHC) molecules (Figure 13.4). Antigen recognition combined with cytokine signals causes the T cell to mature (Figure 13.4). Some of the activated T cells will express CD4 and go on to carry out helper T cell functions, whereas others that express CD8 will search for and eliminate pathogens and infected or abnormal cells (Figure 13.4). In contrast to humoral immunity, protection against pathogens conferred by T cells and macrophages is called **cell-mediated immunity**.

FIGURE 13.4: T cell activation

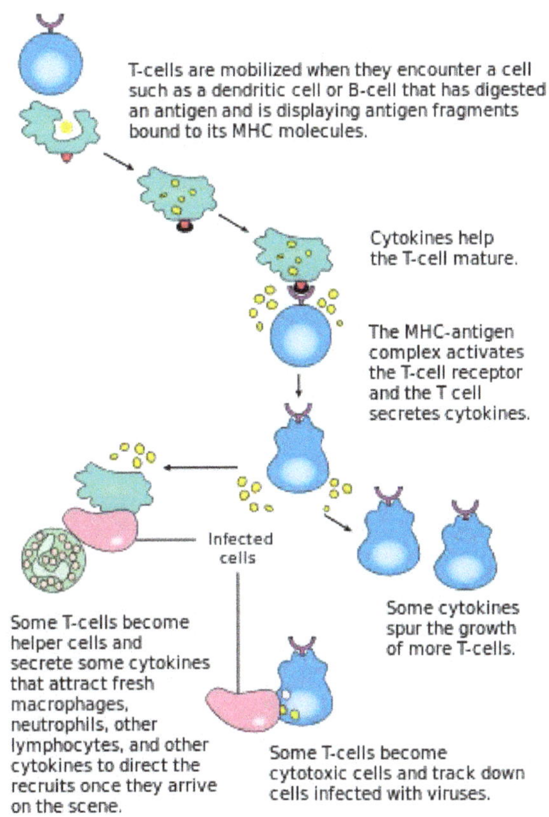

A third type of T cell that is also extremely important to maintaining a properly functioning immune system is the T regulatory cell. This cell type is closely related to the CD4+ (helper T cells) lineage but is characterized by a distinct path of development and role in the body. **T regulatory cells**, also known as Tregs or suppressor T cells, are important regulators of the immune system. Their main function is to prevent immune responses from becoming overly robust or targeting harmless or self-antigens. This is very important because an unnecessarily strong or unregulated immune system can cause tissue damage and even lead to death. Conversely, when T regulatory cells are overactive, protective immune responses to pathogens or cancer may be limited. For example, abnormally high levels of T regulatory cells surrounding cancerous tumors has been linked to poorer prognosis.

Coordination Between the Innate and Adaptive Immune Systems

The innate and adaptive immune systems are described as being different based on their distinct components and mechanisms of action, but these two branches of immunity do not function completely independently of each other. Protection against pathogens requires ongoing coordination among all aspects of the immune system. Examination of the mucosal immunity associated with the mouth and sinuses provides an example of the multifaceted coordination of innate and adaptive immune responses. The physical barriers of the innate immune system provide the first line of defense against pathogens and potentially harmful molecules from the environment (Figure 13.5). Granulocytes, mast cells, phagocytes, and the protective molecules of the innate immune system provide a second line of defense against foreign invaders. Activation of T and B cells by antigen-presenting cells forms a link between innate and adaptive immunity. Cytotoxic T cells (T effector cells) and antibody-producing B cells (plasma cells) help

FIGURE 13.5: Coordination between the innate and adaptive immune systems

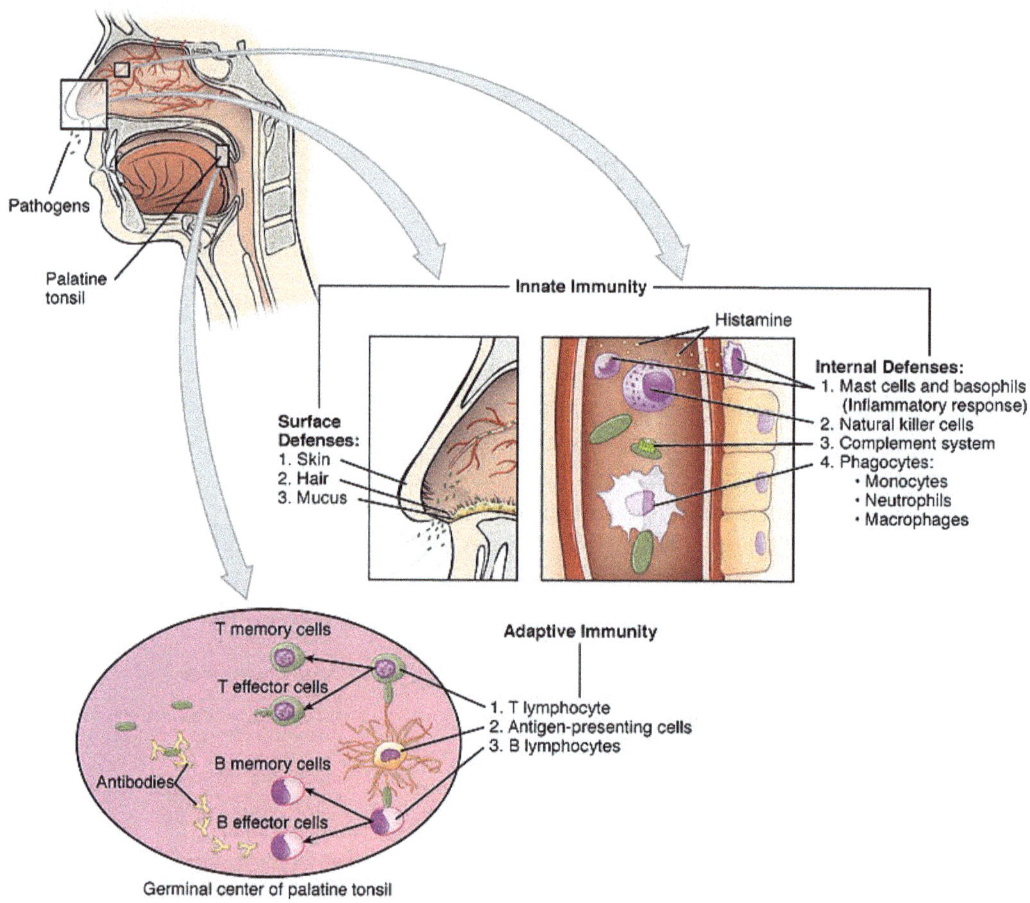

to eliminate pathogens. This is followed by production of memory T and B cells, which provide long-term protection against a previously encountered pathogen. However, future exposure to a pathogen may be thwarted by the innate immune system before it is even encountered by the memory cells. Having multiple lines of protection provides enhanced immunity.

Passive Immunity, Active Immunity, and Vaccination

Active immunity develops when T and B cells are activated by antigen-presenting cells to produce memory cells that can quickly respond to the same antigen if it is encountered again in the future. Active immunity is produced naturally during an infection, often allowing an infected individual to recover. Active immunity is also the product of vaccination. During **vaccination**, a nonpathogenic form of a pathogen or purified antigens from a pathogen is introduced to the immune system to induce active immunity. Most of the time, this protective immunity is in the form of antibody-producing plasma cells. It usually takes 2 to 3 weeks to mount an effective antibody response upon the first exposure to the antigens of a pathogen the body has not been infected with before, during which time an individual may become severely ill, experience irreversible damage, or even be killed by an infection. Vaccinations are effective because they induce active immunity before an infectious pathogen is encountered, allowing the vaccinated individual to benefit from the protection of active immunity almost immediately. As described earlier in this chapter, side effects of normal immune responses may include swelling, pain, fatigue, and flu-like symptoms. Some people mistake these symptoms for an infection; however, because no active pathogens are contained in vaccines, it is not possible to become infected with a pathogen via vaccine. The symptoms caused by vaccination are milder than those caused by infection with the same pathogen, usually resolving within a few days.

Passive immunity, on the other hand, involves transfer of antibodies from one individual to another. The three main ways this occurs are by antibodies crossing the placenta during pregnancy, transfer of antibodies from mother to child during breastfeeding, or through injection of preformed antibodies. Because the antibodies are already fully formed when transferred, passive immunity provides immediate protection. This can be life-saving if antibodies are required to overcome an infection because it could take several weeks to naturally produce antibodies through active immunity. One example of a pathogen that requires treatment with passive immunity is rabies. The rabies virus is so highly virulent that infection is nearly 100% fatal in unvaccinated individuals unless they receive purified anti-rabies immunoglobulin within 24 hours of exposure to the virus. Passive immunity is also used to treat some types of snakebites that deliver quick-acting, potentially fatal venom. The drawback of passive immunity is that it does not provide long-term protection. Because passive immunity does not involve the production or transfer of memory cells capable of producing antibodies over a long period of time, it is only protective for the length of time individual antibody molecules persists in the body, which is only a few weeks at most. After that, an individual treated with passive immunity will again be vulnerable to the same pathogen or toxin, which likely would not be the case if they had been able to produce an active immune response.

Allergy

Up to this point, this chapter has focused on the functions of a normal, healthy immune system. A properly functioning vertebrate immune system is capable of protecting against infection by a wide array of pathogens, as well as limiting the development and spread of abnormal or cancerous cells, and clearing away debris from dead and dying cells. However, sometimes the immune system malfunctions. This may leave the organism abnormally vulnerable to infection or cancer, and may also directly cause damage to the body by attacking an individual's own cells or inappropriately mounting an immune response against harmless molecules. Perhaps the most common type of immune system malfunction is allergy. **Allergies** occur when the adaptive immune system responds to antigens that are not harmful or associated with a pathogen. These are called allergens. Allergens can be found on any molecules that enter the body, including those transmitted through air, food, and the skin. Common allergens include animal dander, bee stings, dust, latex, medications, mold, and pollen. A number of foods are also common allergens. The eight most common food allergens are cow's milk, eggs, fish, peanuts, shellfish, soy, tree nuts, and wheat.

Allergic reactions are initiated by activation of B cells. In most people, B cells will not respond to allergens, which is favorable from an evolutionary standpoint as allergic responses waste the body's resources to generate a needless but potentially harmful response against a harmless antigen. However, in some people, a genetic polymorphism or a lapse in immune system regulation results in B cell recognition of an allergen. The first time the allergen is encountered, a B cell that recognizes it may become activated to proliferate and differentiate into antibody-producing plasma cells (Figure 13.6). Allergic reactions involve production of a specific subtype of antibody known as IgE. Although antibodies are produced from this initial reaction, no noticeable symptoms will be generated. However, some of these IgE antibodies will attach themselves to mast cells, essentially forming allergen receptors on the mast cell surface (Figure 13.6). This transforms the mast cell from a component of the innate immune stem capable only of nonspecific responses to a proxy of the adaptive immune system that will mount a robust response targeting a specific allergen. The second time the same allergen is encountered, it will bind to antibodies on the surface of those mast cells. This will trigger degranulation, releasing chemical mediators of inflammation (Figure 13.6). These include cytokines as well as histamine (Figure 13.6). The characteristic symptoms of allergies result mainly from the release of histamine. Histamine that binds to receptors on blood vessels causes release of fluid from the bloodstream into the affected tissues, resulting in swelling, hives, and/or runny nose or watery eyes. If the allergic reaction is severe enough, dilation of blood vessels or loss of fluid may result in decreased blood pressure, leading to loss of consciousness and even death. Histamine may also activate receptors on nerve cells, causing pain, itching, and smooth muscle contraction. This could affect the gastrointestinal tract, causing nausea, cramps, and vomiting; function of the respiratory system and circulatory system may also be affected, leading to breathing difficulties and increased heart rate.

The reason that some people develop allergies and others do not is not completely understood. Some people have a genetic tendency to develop allergic diseases, which is a condition known as atopy, whereas other people seem to randomly develop allergies without demonstrating any previous indications this would happen. The best way to avoid most allergic reactions is to avoid the antigen that triggers the allergic response. However, many people are allergic to antigens that are difficult to avoid or have allergies they were previously unaware of. Minor allergic reactions can be prevented and treated through use of antihistamines, which bind to histamine receptors and block them from being activated by histamine. More serious allergic reactions may require use of immunosuppressants and medications to normalize blood pressure and heart rate.

Autoimmunity

Normally, an individual's immune system will recognize its own antigens as part of itself and will not attack or mount an immune response to those antigens. This immunological tolerance of self-antigens, or self-tolerance, is an important protective mechanism. Loss of self-tolerance can lead to production of harmful immune responses and a condition known as autoimmunity. **Autoimmunity** results when an individual produces an abnormal immune response that targets antigens on its own cells and tissues. Together with allergies, autoimmune reactions are known as hypersensitivities because they represent abnormally strong immune responses to antigens the immune system usually ignores. However, both T and B cells may be triggered to attack self-antigens because of genetic, environmental, or hormonal factors. Sometimes antigens on pathogens are so similar to host antigens that a cross-reactive response is triggered, turning a protective response into a harmful one. This is called molecular mimicry. Autoimmune responses may be harmless or cause debilitating or fatal disease. Examples of autoimmune diseases include Celiac disease, lupus, type 1 diabetes mellitus, and multiple sclerosis.

Although autoimmune diseases are relatively common, their exact cause is often not known. For example, it is clear that multiple sclerosis results when auto-reactive immune cells destroy the protective insulation of neurons called myelin, resulting in a progressive neurodegenerative disorder characterized by muscle weakness, numbness or tingling, partial paralysis, coordination and vision problems, slurred speech, and fatigue. It is unclear what causes this to happen (Figure 13.7). Studies suggest multiple sclerosis likely results from a combination of genetic and environmental factors, and may be triggered by an unidentified infection, but the precise mechanism of pathogenesis remains to be elucidated.

FIGURE 13.6: Mechanism of allergic reaction—Allergic reactions result when IgE produced by B cells binds to mast cells, causing them to degranulate and release histamine upon exposure to the antigens recognized by those antibodies.

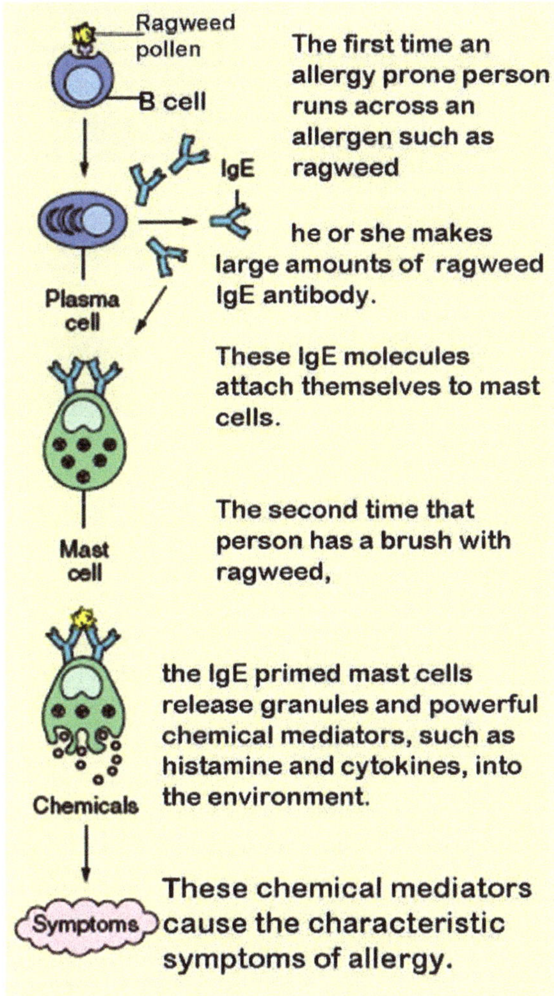

Immunodeficiency

When an individual's immune system is unable to effectively fight infectious diseases and cancer, the body is in a state called **immunodeficiency**. Immunodeficiency may be caused by a number of factors, including chemotherapy, genetic predisposition, malnutrition, pregnancy, very young or old age, and advanced HIV infection. Although aging and malnutrition may result in the failure to maintain and replace existing immune cells and barriers, newborns experience immunodeficiency largely because of the naivety of their adaptive immune system because it has not yet encountered many antigens to trigger production of protective memory

FIGURE 13.7: Multiple sclerosis is an example of an autoimmune disease—Multiple sclerosis is an autoimmune disease caused by immune cells destroying the protective myelin coverings of neurons.

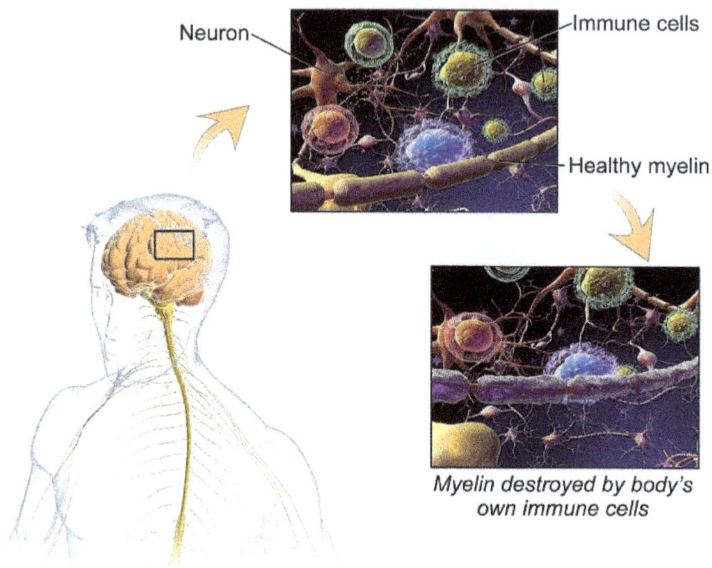

cells. This is one reason that vaccination of children is so important.

Some individuals, however, will never be able to develop a normally functioning immune system as a result of genetic mutations. One example of genetic immunodeficiency is common variable immune deficiency (CVID), which is characterized by dysfunctional B cells unable to make sufficient amounts of antibodies in response to pathogens. Mutations in at least 13 genes have been associated with CVID, yet in 90% of cases, the precise genetic cause of CVID is unknown.

Infections that lead to destruction of immune barriers may also cause immunodeficiency. The presence of immunodeficiency is actually one of the criteria that determine whether someone infected with human immunodeficiency virus (HIV) is classified as HIV-positive or as having acquired immune deficiency syndrome (AIDS). The other factor that results in a diagnosis of AIDS in HIV-positive individuals is the presence of one or more opportunistic infections. An **opportunistic infection** is an infection that occurs more frequently or severely in individuals with weakened immune systems. One example of an opportunistic infection is candidiasis, which is caused by the fungal pathogen *Candida albicans*. Candidiasis may affect the mouth, throat, skin, genitals, and digestive tract, causing itching, pain, soreness, irritation, and vomiting. These symptoms usually only occur in people with weakened immune systems, whereas those with normal healthy immune systems rarely develop candidiasis. Another example of an opportunistic infection is Kaposi's sarcoma, which is a type of skin cancer caused by human herpes virus 8. The most noticeable symptom of Kaposi's sarcoma is development of lesions on the skin, mouth, gastrointestinal tract, and respiratory tract. Because HIV infection itself usually does not cause noticeable symptoms, many HIV-infected individuals who do not receive regular HIV testing learn of their infection status only when they develop an opportunistic infection such as candidiasis or Kaposi's sarcoma.

13.2 SUMMARY

- The immune system consists of physical barriers, cells, and molecules that identify and destroy pathogens, and remove dying and abnormal cells from the body
- The innate system responds to pathogens immediately upon exposure by mounting a nonspecific response

- Inflammation is part of the innate immune response that involves accumulation of fluid and white blood cells at a site of infection or injury accompanied by heat, pain, and swelling
- Adaptive immune responses require days to weeks to develop and involve recognition of specific antigens by T cells and B cells
- The immune system is highly regulated by T regulatory cells, helper T cells, and other mechanisms to prevent unnecessary or harmful reactions
- Vaccines provide protection to pathogens by inducing production of active immunity in the form of a B cell response, leading to a quicker, more robust secondary antibody response if the pathogen is encountered later
- Passive immunity provides immediate but short-term protection against pathogens and toxins through transfer of preformed antibodies, such as through placental transfer, breast milk, and injection of purified immunoglobulin
- Allergies and autoimmunity occur when cells and antibodies of the adaptive immune system react to harmless antigens and produce damaging responses
- Immunodeficiency results when the immune system becomes unable to effectively fight infectious diseases or cancer

13.3 KNOWLEDGE CHECK

1. All of the following are examples of innate immunity, EXCEPT:
   a. antibodies
   b. granulocytes
   c. inflammation
   d. lysozyme
   e. stomach acid

2. Which cells are most closely associated with humoral immunity?
   a. B cells
   b. dendritic cells
   c. mast cells
   d. skin cells
   e. T cells

3. The term for any molecule recognized by an adaptive immune response is:
   a. antigen
   b. autoimmunity
   c. inflammation
   d. PAMP
   e. Pathogen

4. Multiple sclerosis is an example of which of the following:
   a. allergy
   b. autoimmunity
   c. immunodeficiency
   d. innate immunity
   e. passive immunity

5. Which two cell types are most involved in allergic reactions?
   a. B cells and mast cells
   b. B cells and CD4+ T cells
   c. CD4+ T cells and CD8+ T cells
   d. CD8+ T cells and mast cells
   e. mast cells and eosinophils

6. List three cells that are part of the innate immune system, and describe their function.

7. Compare and contrast active and passive immunity.

8. Identify an example of an opportunistic infection, and list several reasons why someone might be susceptible to an opportunistic infection.

13.4 ACTIVITY

CRISPR Analysis and Reflection Essay

The discovery of the double-helix structure of DNA and the development of CRISPR technology are two of the most significant advancements of modern science. The scientific applications of these discoveries have broad social, ethical, and economic implications, and the discovery of each is an important part of the history of science. The purpose of this essay is to demonstrate analytical thinking and written communication skills. In this essay, you will compare and contrast these discoveries, focusing on the findings that led to each discovery, the ethical issues associated with the scientists who contributed to each discovery, and the implications of each discovery to the everyday life of nonscientists. Use the following sources of information when writing your essay, and follow any additional instructions provided.

Bonney, K. M. (2017). Putting the pieces together: The discovery of DNA structure and replication. [On-line] *National Center For Teaching In Science Case Collection*. Buffalo, NY: University of Buffalo. Access date: July 16, 2017. http://sciencecases.lib.buffalo.edu/cs/files/dna_structure_flip.pdf.

Doudna, J. (2015). Genome-editing revolution: My whirlwind year with CRISPR. *Nature, 528*(7583), 469–471.

Lander, E. (2016). The Heroes of CRISPR. *Cell*, *164*, 18–28.

Pollack, A. (2015). Jennifer Doudna, a pioneer who helped simplify genome editing. [On-line] *The New York Times*. Access date: May 11, 2015. www.nytimes.com/2015/05/12/science/jennifer-doudna-crispr-cas9-genetic-engineering.html.

Rothkopf, J. (2016). How one man tried to write women out of CRISPR, the biggest biotech innovation in decades. [On-line] *Jezebel.com*. Access date: January 20, 2016. http://jezebel.com/how-one-man-tried-to-write-women-out-of-crispr-the-big-1753996281.

13.5 ACTIVITY

Sex and Vaccination Case Study

The purpose of this activity is to analyze the legal and ethical implications of a specific case of mandated vaccination. This activity provides an opportunity to demonstrate oral and written communication skills and to connect a scientific topic with its real-world implications. This activity involves working in small groups to complete the case study titled "Sex and Vaccination" published by the National Center For Case Study Teaching In Science (NCCSTS).

"Sex and Vaccination" http://sciencecases.lib.buffalo.edu/cs/files/hpv.pdf

13.6 ACTIVITY

Immunology Virtual Lab

In addition to functioning within the body to protect against infections and disease, the immune system can also be exploited for use in the laboratory. The purpose of this activity is to investigate how antibodies can be used to diagnose an autoimmune disease. This activity involves virtual lab based on a common laboratory technique known as enzyme-linked immunosorbant assay (ELISA). The ELISA uses antibodies and color-changing molecules to detect specific antigens. This activity is provided by the Howard Hughes Medical Institute Biointeractive initiative and can be accessed on their website. Once you input the link below into your web browser, first download the file titled "Worksheet (PDF)" or follow the alternative instructions provided by your instructor to obtain the questions you will answer while completing the lab. Next, click "Start Virtual Lab" at the top of the screen to open the virtual lab activity.

"Immunology Virtual Lab" https://www.hhmi.org/biointeractive/immunology-virtual-lab

ACTIVITY QUESTIONS

13.4

Carefully follow the prompt and any additional instructions provided when writing your essay. Specifically address each of the following objectives in your essay:

- Identify the scientists associated with each discovery
- Describe the main events and findings that lead to each discovery
- Summarize each discovery using nonscientific jargon, including what the discovery was, the applications/uses of the discovery, and the relevance of the discovery to other scientific concepts and to everyday life

- Explain the main ethical issue associated with each discovery; compare and contrast the ethical issues/implications of each discovery
- Evaluate the contributions of the main scientists associated with each discovery; justify what type of credit each scientist should receive for their contributions
- Identify the interactions, claims, and recognition of the scientists involved with each discovery; criticize or defend their actions
- Thoroughly support your arguments with specific evidence from the sources provided and other appropriate sources; at least five appropriate sources should be used and cited
- Demonstrate conciseness, organization, original thought, and proper use of evidence and citations

13.5

Follow instructions to work individually or form small groups for completing the case study and answering the indicated questions.

13.6

Complete all of the questions in the "Worksheet (PDF)" downloaded from the HHMI Biointeractice website unless provided different instructions by your instructor.

13.6 REFERENCES

Baxter, D. (2007). Active and passive immunity, vaccine types, excipients and licensing. *Occupational Medicine, 57*(8), 552–556.

Bischoff, S. C. (2006). Role of mast cells in allergic and non-allergic immune responses: Comparison of human and murine data. *Nature Reviews Immunology, 7*, 93–104.

Bronte, V. and Pittet, M. J. (2013). The spleen in local and systemic regulation of immunity. *Immunity, 39*(5), 806–818.

Ferenbach, D. and Hughes, J. (2008). Macrophages and dendritic cells: what is the difference? *Kidney International, 71*(1), 5–7.

Horvath, P. and Barrangou, R. (2010). CRISPR/Cas, the immune system of bacteria and archaea. *Science, 327*(5962), 167–170.

Jones, J. D. G., and Dangl, J. L. (2006). The plant immune system. *Nature, 444*, 323–329.

Ma, H., et al. (2017). Correction of a pathogenic gene mutation in human embryos. [On-line] *Nature*. Epub ahead of print. doi: 10.1038/nature23305.

Mogensen, T. H. (2009). Pathogen recognition and inflammatory signaling in innate immune defenses. *Clinical Microbiology Reviews, 22*(2), 240–273.

Rosenblum, M. D., Remedios, K. A., and Abbas, A. K. (2015). Mechanisms of human autoimmunity. *The Journal of Clinical Investigation, 125*(6), 2228–2233.

Tam, J. S., and Routes, J. M. (2013). Common variable immunodeficiency. *American Journal of Rhinology & Allergy, 27*(4), 260–265.

Warrington, R., Watson, W., Kim, H. L., Antonetti, F. R. (2011). An introduction to immunology and immunopathology. *Allergy, Asthma & Clinical Immunology, 7*, S1.

FIGURE CREDITS

FIG. 13.1: Copyright © Open-i (CC by 3.0) at https://openi.nlm.nih.gov/detailedresult.php?img=PMC4381333_DMM018036F1&req=4.

FIG. 13.2: Adapted from 2006 Pearson Education, Inc.

FIG. 13.3: Fred the Oyster, "B cell activation," https://commons.wikimedia.org/wiki/File:B_cell_activation.svg. Copyright in the Public Domain.

FIG. 13.4: Hazmat2, "T cell activation," https://commons.wikimedia.org/wiki/File:T_cell_activation.svg. Copyright in the Public Domain.

FIG. 13.5: Copyright © OpenStax College (CC by 3.0) at https://commons.wikimedia.org/wiki/File:2211_Cooperation_Between_Innate_and_Immune_Responses.jpg.

FIG. 13.6: NIH, "Mechanism of allergic reaction," https://commons.wikimedia.org/wiki/File:Mast_cells.jpg. Copyright in the Public Domain.

FIG. 13.7: Copyright © BruceBlaus (CC BY-SA 4.0) at https://commons.wikimedia.org/wiki/File:Multiple_Sclerosis.jpg.

Infectious Diseases

14.1 INTRODUCTION

Viruses, bacteria, and other microorganisms are ubiquitous throughout the environment and within our own bodies. The majority of microorganisms humans encounter do not noticeably affect us, whereas some forms have mutually beneficial relationships with our own cells that help keep us alive. There is also a tiny fraction of microorganisms that sometimes harm or even kill us. Despite their tiny size, these microorganisms have made an outsized effect on human history and continue to affect us every single day. This chapter presents an overview of the main types of microorganisms, discusses examples of some of the infectious diseases they cause, and describes several ways that infectious diseases, and our attempts to prevent them, have shaped human history.

Infections and Disease

An **infection** is defined as the multiplication of harmful microorganisms within a host organism, whereas **disease** refers to a state of health in which the body cannot carry out all its normal functions. Therefore, disorders caused by microorganisms that are spread from one individual to another, either directly or indirectly, are known as **infectious diseases**. Cells or organisms infected by viruses and microorganisms are referred to as hosts. Infectious diseases can affect essentially every cell type in every part of the body, and the effects can range from minor irritation to loss of function to death. Infectious diseases that occur suddenly and last a relatively short amount of time are referred to as **acute diseases**, whereas diseases that take a relatively long time to develop and last for a prolonged period of time are known as **chronic diseases**. For virtually every infectious disease, some period of time passes between the time infection first occurs and the onset of symptoms. This stage of an infectious disease may be referred to as the **latent period** or the incubation period. Some chronic infectious diseases alternate between symptomatic periods of active infection and asymptotic periods of latency. The length of the latent period varies greatly based on the time of infectious disease and characteristics of the infected individual. For example, the virus that causes the common cold has a latent period of 3 to 5 days, whereas

LEARNING OBJECTIVES

- Define the terms infection and disease
- Discuss different modes of transmission for infectious disease-causing pathogens
- Explain the difference between pathogenicity and virulence
- Summarize the defining characteristics of each of the following: viruses, prions, bacteria, protozoans, fungi, and helminthes
- Identify an example of an infectious disease caused by each of the following: viruses, prions, bacteria, protozoans, fungi, and helminthes
- Identify several common shapes of bacteria
- Discuss the differences between Gram-negative and Gram-positive bacteria
- Compare and contrast antigenic drift and antigenic shift
- Compare and contrast bubonic plague, pneumonic plague, and septicemic plague
- Discuss how the plague has influenced human history
- Summarize the history and significance of the smallpox vaccine
- Summarize the origin and history of HIV
- Identify examples of infectious causes of cancer

HIV, if left untreated, has a latent period of 8 to 10 years, on average. The latent period can be considered one of the most dangerous stages of infectious disease. This is because no symptoms are present, but the infected individual may still be contagious and able to spread the infectious agent to others without knowing it.

Infectious diseases may be caused by a wide array of microorganisms, including viruses, bacteria, fungi, and parasites. Any microorganism that causes disease is known as a **pathogen**. Pathogenic microorganisms may enter the body through contact with contaminated air, food, or water; through transfer of blood or other body fluids; or be transmitted by insects or animals. Some infectious diseases are caused by microorganisms that are normally present in the body but usually do not cause disease. These microorganisms are known as resident microflora. Microflora and other pathogens that normally do not cause disease, but occasionally do so by taking advantage of a weakening in a host's immune system, are known as opportunistic pathogens, and the diseases they cause are called **opportunistic diseases**. Some of the conditions that increase the risk of developing an opportunistic disease are advanced age, AIDS, chemotherapy for cancer or immune suppression for organ transplantation, malnutrition, skin damage, and other infections.

Modes of Transmission

Infectious diseases are classified as communicable or non-communicable based on whether they are spread from person to person or transmitted only through contact with animals or inanimate objects in the environment. **Communicable diseases** can be spread between people through exchange of body fluids, such as blood, semen, breast milk, and vaginal secretions, or through physical contact such as kissing, touching of sores, and skin-to-skin sexual contact. Infectious diseases can also be spread through indirect contact. For example, infectious microorganisms can be transmitted between people indirectly if an infected person sneezes on their hand and touches a doorknob or other surface, then another person touches the same surface and later places their hand near their mouth, nose, or eyes. Non-communicable infectious diseases are not spread from person to person but instead may be acquired from contact with microorganisms that reside in the environment. Examples of these diseases are those transmitted through the air, water, food, or by vectors. A **vector** is an organism that does not cause disease itself but spreads infectious diseases by transmitting pathogenic microorganisms. Mosquitos, ticks, fleas, and bats are some of the vectors that most often transmit infectious diseases to humans.

Pathogenicity and Virulence

The magnitude of infectious diseases is often measured in two ways: by the pathogenicity and virulence of the microorganisms that cause them. In simple terms, **pathogenicity** refers to the ability to cause disease. The pathogenicity of a microorganism is a sum of its ability to infect a host, evade the host immune system, persist and replicate within the host, and cause damage to cells or tissues that disrupts normal physiologic functions of the host. **Virulence** measures the severity of disease that infection by a microorganism causes. Highly pathogenic microorganisms establish infection relative quickly and easily, even if only a small number of microorganisms are present. Microorganisms of low pathogenicity are limited in their ability to cause infection and may only be able to do so if a large number of the same type of microorganism is present at one time, or if a breach in the physical barriers or immune system of a potential host increases their susceptibility to infection. Highly virulent microorganisms cause severe disease that may be characterized by extensive tissue damage, debilitation, and even death. Conversely, low virulence produces relatively minor effects and symptoms. Pathogenicity and virulence are not directly linked. For example, HIV spreads easily and causes severe disease (if not successfully treated), so it is characterized by high pathogenicity and high virulence. However, rhinovirus, which causes cold-like symptoms, is easily spread but causes mild disease. Therefore, it has a high pathogenicity but low virulence.

Viral Diseases

Some of the most common and well-known infectious diseases are caused by tiny particles called viruses. A **virus** is a microscopic infective agent that contains genetic material but cannot grow or reproduce independently. Viruses are generally much smaller than bacteria and other microorganisms. The simple structure of viruses is composed of genetic material, which may be either DNA or RNA, surrounded by a protective protein shell called a capsid. Some viruses are enclosed by a membrane made of lipids, proteins, and carbohydrates called an envelope, which is located on the exterior of the capsid (Figure 14.1). Because viruses are not made of cells and lack ribosomes, mitochondria, and all other organelles, they are not capable of producing their own proteins or energy, or carrying out many of the other functions necessary to sustain life on their own. Instead, viruses depend on the cellular machinery of the host cells they infect to survive and replicate. Because of their inability to survive and reproduce independently, many scientists do not consider viruses to be living things. Additionally, unlike all other known living organisms, some viruses do not contain DNA and instead rely exclusively on RNA to store and transmit their hereditary information. One example of an RNA virus is human immunodeficiency virus (HIV). HIV is part of a subgroup of RNA viruses called **retroviruses** that use enzymes to copy their RNA into double-stranded DNA, which is integrated into host DNA during infection. To complete their replication cycle, HIV virus particles also contain the enzyme **reverse transcriptase** for converting its RNA genome into DNA, and the enzyme integrase for inserting the DNA copy of its genome into the DNA of a host cell envelope (Figure 14.1). Once integrated into the host cell, HIV genetic material is capable of being transcribed and translated by the host cell to make the components necessary to build new virus particles. HIV uses an additional enzyme called protease to separate newly translated proteins so they can be assembled into the proper three-dimensional structures. When these particles leave the cell, they often take part of the host cell membrane to use for their envelope. The resulting loss of membrane integrity leads to the death of the infected host cell.

In the United States, the Centers for Disease Control and Prevention (CDC) maintain a list of diseases that are of particular interest to public health officials. These are known as nationally notifiable infectious diseases (Table 14.1). Local, state, national, and international health workers and agencies are strongly encouraged to report all cases of these diseases to the CDC to coordinate a national effort of public health monitoring and protection. In addition to HIV/AIDS, one other common disease caused by a virus that is included in the list of notifiable diseases is influenza (Table 14.1). Influenza virus enters the body because of inhalation of contaminated water droplets or contact with virus-containing respiratory secretions. The virus primarily affects the respiratory system, causing sore throat, coughing, sneezing, fatigue, fever,

FIGURE 14.1: HIV structure—Viruses are made of small, simple structures. HIV is composed of genetic material (RNA) and enzymes needed for replication, a protein capsid, and a lipid envelope containing glycoproteins.

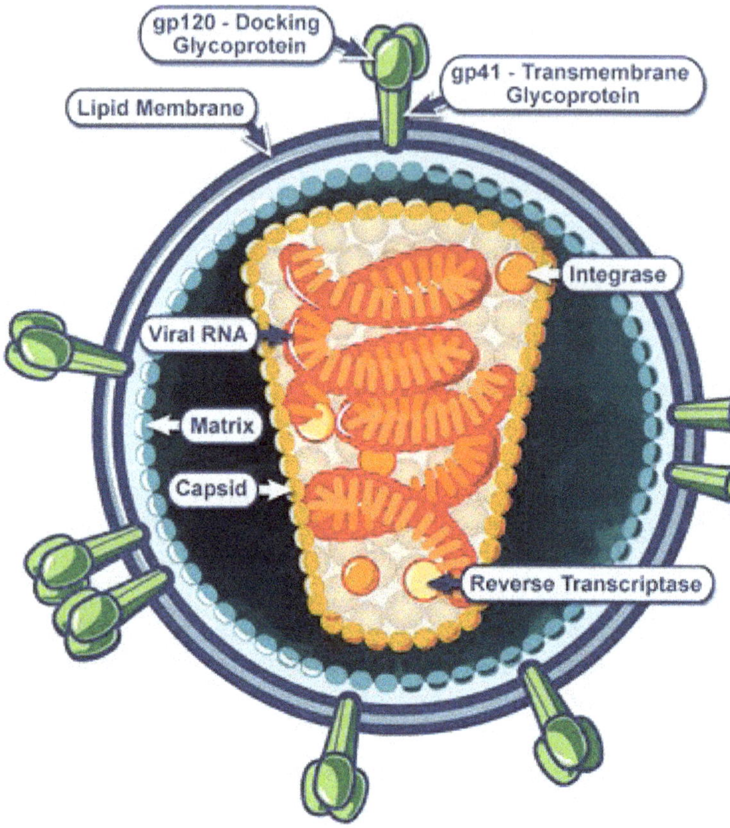

and headache. Influenza infection can be fatal, although fatalities usually only occur in young children, the elderly, or individuals with weakened immune systems. A vaccine for influenza, commonly referred to as the flu shot, is highly effective at preventing infection. However, the influenza virus possesses the ability to change the shape of the antigens that cover its surface as a result of accumulating mutations in the genes that code for two main types of antigens called hemagglutinin and neuraminidase (Figure 14.2).

Because recognition of these antigens by antibodies is a crucial part of protective immunity, this process allows the virus to regain the ability to infect an individual who has mounted a protective immune response to a previously encountered variant of the virus (Figure 14.2). The mechanism by which viruses accumulate genetic mutations that lead to changes in antibody-binding sites is called **antigenic drift**. Antigenic drift is the reason that new seasonal flu outbreaks occur every year and affect people who have previously recovered from the virus (Figure 14.2). Antigenic drift is also the reason that a new flu shot must be produced every year to maintain effectiveness. Although the antibody-producing cells activated by previous flu shots continue to be effective at the strains of influenza virus targeted by those vaccines, they are unable to effectively recognize and eliminate new strains of influenza virus generated by antigenic drift.

Viruses such as influenza are capable of undergoing an even more significant change than antigenic drift, which is called antigenic shift. **Antigenic shift** occurs when two different viruses infect the same organism and exchange segments of genetic material, resulting in the formation a dramatically new strain of the virus. Because antigenic shift changes the morphology of the virus much more than antigenic drift does, antigenic shift is a more effective mechanism for evading the immune system. Occasionally, variants of influenza virus created by antigenic shift produce pandemic outbreaks that are significantly more severe and widespread than the seasonal flu. The deadliest pandemic influenza outbreak occurred in 1918–1919 and was called

FIGURE 14.2: Antigenic drift of influenza virus—The influenza virus evades the immune system by underdoing antigenic drift, causing antibodies to no longer recognize its surface antigens. This is the reason new flu vaccines must be prepared every year to prevent the seasonal flu.

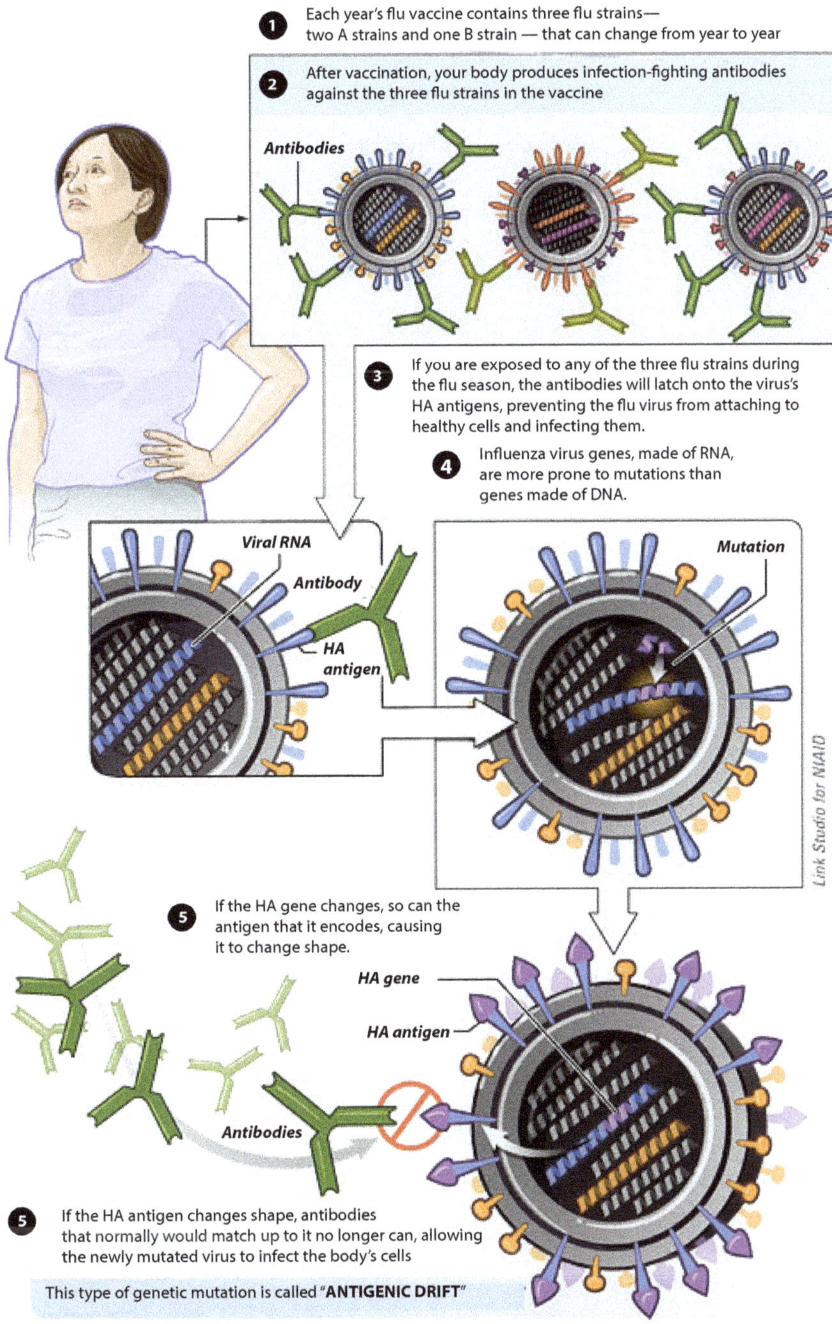

TABLE 14.1: Nationally Notifiable Infectious Diseases—The CDC monitors all cases of certain infectious diseases known as notifiable diseases.

Type of pathogen	Name of notifiable disease
Viral	Arboviral diseases (e.g., West Nile Virus), Dengue, Hepatitis (viral), HIV/AIDS, Influenza, Meales, Mumps, Polio, Rabies, Rubella, Smallpox, Varicella, Viral Hemorrhagic fever (e.g., Ebola), Yellow fever, Zika
Bacterial	Anthrax, Botulism, Brucellosis, Camphylobacteriosis, Chancroid, Chlamydia, Cholera, Diptheria, Ehrlichiosis, Gonorrhea, Legionellosis, Leprosy (Hansen's disease), Leptospirosis, Listeriosis, Lyme disease, Meningococcal disease, Pertussis, Plague, Psittacosis, Q fever, Salmonellosis, Shigellosis, Spotted fever, Syphilis, Tetanus, Tuberculosis, Tularemia, Typhoid fever, Vibriosis
Protozoan	Babesiosis, Giardiasis, Malaria
Fungal	Coccidioidomycosis, Cryptosporidiosis, Cyclosporiasis

the Spanish flu. In less than 2 years, approximately 50 million were killed by the Spanish flu, making it one of the deadliest infectious diseases in history.

Viroids and Prions

Viruses are not the only acellular causes of infectious diseases. Viroids are even smaller and simpler than viruses, consisting only of genetic material with no capsid or envelope. Viroids are known to cause infectious diseases in plants but not in humans or other animals. However, a third type of nonliving particle known as the prion does infect animals and is the cause of several fatal infectious diseases. **Prions** are infective proteins that are not associated with nucleic acids or any other type of macromolecule or structure. Prions result from improper molecular folding during development of the three-dimensional tertiary protein structure.

Diseases caused by prions are rare and have relatively long latency periods, making their transmission difficult to track and study. It may take up to 50 years for symptoms to appear once a person or animal contracts prions. However, once symptoms do appear, these diseases usually progress rapidly, are untreatable, and always fatal. The most famous prion disease is mad cow disease, which, as the name suggests, affects cattle. Human diseases caused by prions include Creutzfeldt-Jakob disease, Kuru, Gerstmann-Straussler-Scheinker Syndrome, and Fatal Familial Insomnia. The first two of these, as well as mad cow disease, are spread through consumption of brains or spinal cords from humans or animals that have been contaminated with prions; the latter two are inherited genetic disorders. All of these are classified as neurodegenerative disorders because they result from damage to the brain and nervous system that progressively worsens, leading to loss of muscle coordination and cognitive function.

Bacterial Diseases

Bacteria are single-celled prokaryotic organisms that lack membrane-bound organelles but contain plasma membranes, ribosomes, cell walls, DNA, and other cellular structures. Bacteria are essentially ubiquitous throughout most environments, as well as inside our own bodies. Rather than living alone, many species of bacteria tend to form communities called biofilms. Biofilms are held together by sticky carbohydrate-rich secretions that allow bacteria to form close associations with each other and attach to surfaces. This behavior promotes survival in the environment, as well as within host organisms. Although most bacteria are harmless, a number of species do cause disease. Biofilms sometimes form on teeth and in airways or intestines, promoting the establishment of infection and making the infection more difficult to eliminate.

To further protect themselves from environmental conditions and from the immune systems of host organisms, bacteria have developed several protective structural features. Many bacteria have flagella that allow them to move through the environment or the body of a host organism in search of nutrients. Some bacteria are surrounded by capsules, which protect them from the host immune system and help them adhere to surfaces. The cell wall of bacteria also plays an important role in promoting its survival and infectiousness. Compared to plants, the cell wall of bacteria is highly complex, consisting of layers of lipids and other molecules juxtaposed with layers of a type of polysaccharide called peptidoglycan. Bacteria are often categorized into two groups based on their cell wall composition and the amount of peptidoglycan they contain. Gram-positive bacteria are surrounded by a thick cell wall that contains many layers of peptidoglycan, causing them to retain a type of dye called Gram stain during a common diagnostic laboratory procedure. Gram-negative bacteria, on the other hand, have a relatively thin cell wall containing less peptidoglycan but are surrounded by another type of protective molecule called lipopolysaccharide. Lipopolysaccharide is also referred to as **endotoxin** because it is an antigen present inside bacteria capable of causing disease on its own. Endotoxin is usually only released when Gram-negative bacteria die. One important implication of this is that Gram-negative bacteria do not need to remain alive to cause damage to the body, because endotoxin may remain intact after the bacteria have been killed. Some bacteria are also capable of releasing a wide range of different harmful molecules into their host or local environment while alive; these molecules are known as **exotoxins**.

To diagnose the presence of Gram-negative or Gram-positive bacteria, a staining technique called Gram stain is used. Use of this technique results in Gram-negative bacteria appearing red or pink and Gram-positive bacteria appearing purple (Figure 14.3). Determining whether an infection is caused by Gram-negative or Gram-positive bacteria is one of the primary factors that determine how bacterial infections will be treated with antibiotics. **Antibiotics** are substances that inhibit the growth of or kill bacteria. For example, the commonly prescribed antibiotic penicillin works by inhibiting one of the final steps in the synthesis of bacterial cell walls, which involves cross-linking peptidoglycan molecules together. Penicillin is highly effective at treating Gram-positive bacteria because of their reliance on peptidoglycan as a crucial component of their cell wall; however, penicillin is not effective at treating Gram-negative bacteria because it cannot penetrate their lipopolysaccharide layer. Other antibiotics are effective at treating Gram-negative bacteria, but their administration must be carefully regulated and monitored to avoid the release of potentially deadly amounts of endotoxin from dying bacteria.

A second way bacteria are classified is by their shape. The four most common shapes of bacteria are rod-shaped (bacillus), round (coccus), comma-shaped (vibrio), and corkscrew-shaped

FIGURE 14.3: Gram staining—After Gram staining, Gram-positive cocci (*Staphylococcus aureus*) appear as purple spheres and Gram-negative bacilli (*Escherichia coli*) appear as pink or red rods.

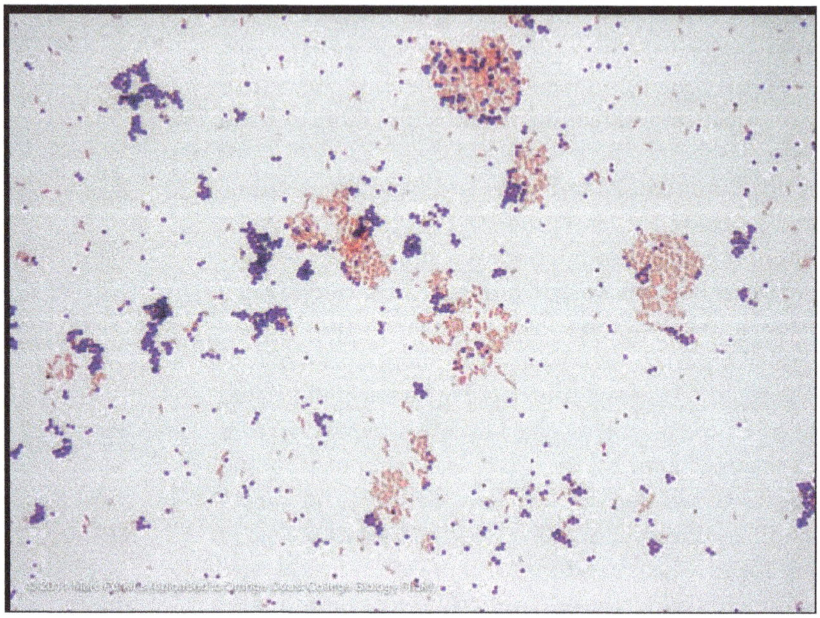

(spirochete). Additionally, some bacteria tend to form chains (strep), grape-like clusters (staph), or pairs (diplo). These shapes are often incorporated into the scientific names of bacterial species. For example, the names of *Vibrio cholera*, which causes cholera, and *Bacillus anthracis*, which causes anthrax, reveal they are comma-shaped and rod-shaped, respectively. Similarly, the name of *Staphylococcus aureus*, which is a common cause of skin infections, indicates it is round and often found in clusters. *Staphylococcus aureus* is Gram-positive so it appears purple after a Gram stain procedure, whereas *Escherichia coli*, a Gram-negative bacillus that is a common cause of food poisoning, looks like pink rods after Gram staining (Figure 14.3).

FIGURE 14.4: Erythema migrans—Erythema migrans is a bull's-eye-like rash that is characteristic of *Borellia burgdorferi* infection, which causes Lyme disease.

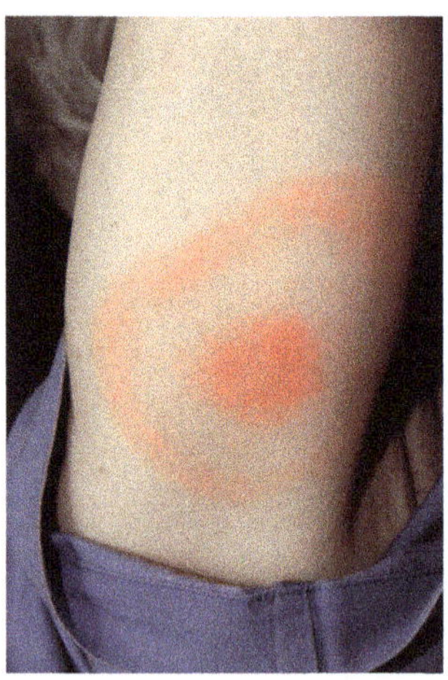

One example of a notifiable infectious disease caused by bacteria is Lyme disease (Table 14.1). Lyme is caused by the bacterium *Borrelia burgdorferi,* which is transmitted to humans by ticks. Lyme disease affects the skin, musculoskeletal system, nervous system, and cardiovascular system. Symptoms including fatigue, joint pain, and memory loss, and can last for up to 6 months or more. The most common symptom, which is often used to diagnose Lyme disease, is a circular red rash called erythema migrans (Figure 14.4). Although Lyme disease has likely existed in North America for thousands of years, and its telltale symptoms have been recorded since at least the early 1800s, the disease was not identified and named until the 1970s. Lyme disease is named after a town in Connecticut in which an outbreak of the disease occurred in a group of children whose parents prompted scientists to study their cases and determine the cause.

Protozoan Diseases

Protozoans are animal-like single-celled eukaryotic organisms. These organisms are related to single-celled plant-like protists, such as algae, but are distinguished by their animal-like behaviors, including motility and predation. Protozoa have evolved three different cellular structures to promote movement and search of prey. Some protozoa have tail-like flagella, similar to those found on some bacteria, which propel them through the environment or the bloodstream of hosts. Other protozoa are covered with hair-like cilia that continuously move like tiny oars to guide the organism around. Finally, some protozoa rely on amoeboid movement to travel from place to place. This is accomplished by pushing out foot-like projections called pseudopodia from the leading edge of the plasma membrane to pull the cell forward, then retracting the lagging edge of the plasma membrane in the same direction.

One protozoan disease that affects a large number of people is malaria (Table 14.1). Malaria is caused by four different species of protozoan parasites: *Plasmodium vivax, Plasmodium malariae, Plasmodium ovale*, and *Plasmodium falciparum*. The most deadly of the four species is *Plasmodium falciparum*. Malaria-causing parasites are transmitted through the bite of the Anopheles mosquito. Once parasites enter the bloodstream, they infect and destroy red blood cells, leading to anemia, and possibly coma, seizures, and death (Figure 14.5). Other common symptoms include chills, fever, sweating, headache, fatigue, nausea, and vomiting. Although individuals who survive a case of malaria experience partial resistance to future infection for several months, this resistance then disappears, leaving the individual vulnerable to repeated infection. Currently there is no widely available vaccine for malaria, but, as of 2017, a promising vaccine candidate is being tested in clinical trials. Fortunately, use of the medications chloroquine and mefloquine can successfully treat many cases of malaria. Prophylactic use of these medications is also effective at preventing malaria, especially when combined with use of insecticide sprays and mosquito netting to block transmission of the parasite by mosquitos.

Despite available treatments and protective measures, malaria affects approximately 200 million people each year, leading to 1–3 million deaths worldwide. Approximately 90% of malaria cases occur in sub-Saharan Africa, where malaria is the fourth leading cause of death in children. Like other pandemic diseases, the extremely high number of deaths and lost productivity caused by malaria has hugely impacted society. One unique effect malaria has had is its unexpected influence on human evolution. Evidence of this effect was revealed through analysis of a genetic mutation called HbS. HbS leads to production of abnormal hemoglobin S instead of the normal form of the protein, hemoglobin A. People who inherit

FIGURE 14.5: Malaria parasites infecting red blood cells—Malaria-causing *Plasmodium falciparum* parasites infect and destroy red blood cells, leading to anemia, fatigue, fever, and even coma and death.

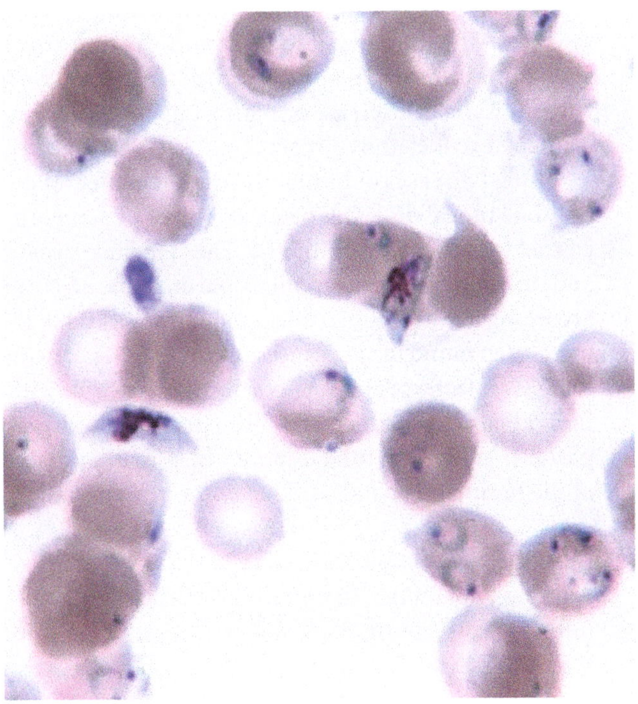

one normal copy of this gene and one mutated copy produce enough normal hemoglobin to maintain normal body functions. However, in individuals who are homozygous for this mutation, production of hemoglobin S causes a disease called sickle cell anemia, which is a potentially fatal condition affecting red blood cells. Instead of being negatively selected by evolution to become exceedingly rare in future generations as might be expected, this mutation is currently found in up to 10% or more of humans whose ancestors lived in certain tropical and subtropical regions. This is unusually high frequency suggests the HbS mutation has been positively selected over time to become more common. Once it was discovered that people who are homozygous for the HbS mutation are resistant to malaria, scientists reasoned it was malaria that provided a strong selective pressure for the HbS mutation. However, the negative selection pressure of sickle cell anemia is also strong in people who are homozygous for the HbS mutation. To compensate for the negative selection pressure of sickle cell anemia, this is the benefit conferred by the heterozygous condition. Individuals heterozygous for HbS also benefit from being resistant to malaria yet do not suffer complications from sickle cell anemia, developing only a mild or asymptomatic version of the condition.

Fungal Diseases

Fungi are a diverse group of eukaryotic organisms that obtain nutrients and energy from other living or dead organisms. Fungi may be single-celled or multicellular, and some species undergo complex life cycles that involve both single-celled and multicellular stages. Although single-celled fungi are microscopic, some multicellular fungi are visible to the unaided eye. In fact, the largest living thing on earth is a fungus of the species *Armillaria solidipes*. The most notable structural feature shared by all fungi is the presence of a cell wall containing a type of carbohydrate called chitin. Interestingly, chitin is not found in the cell walls of bacteria or plants but is one of the primary molecules that form the hard crunchy exoskeleton of insects.

Fungal diseases of humans most often affect the skin or respiratory system. Even though only one fungal disease is listed in Table 14.1 as a notifiable condition, many infectious diseases caused by fungi affect humans. These include aspergillosis, cryptococcosis, histoplasmosis, and pneumocystis pneumonia. One example of a fungal disease that is a notifiable condition, according to the CDC, is coccidioidomycosis (Table 14.1). A fungus called *Coccidioides immitis,* most often found in soil, causes this disease. Like many fungal pathogens, *Coccidioides immitis* is highly infectious but usually only affects people with weakened immune systems. Susceptible individuals can be infected within just a few seconds of being exposed to air containing dust particles laden with fungal spores. In a small minority of cases, infection spreads to the lining of the brain and spinal cord, or to the bones, leading to serious complications.

Helminth Diseases

Although not on the list of notifiable diseases in the United States, many of the most common infectious diseases in the world are caused by yet another type of organism called helminths. **Helminths** are multicellular eukaryotic organisms that range in size from microscopic to more than 10 feet long. Helminths are also known as parasite worms, intestinal worms, nematodes, flatworms, roundworms, and tapeworms. Most helminth infections of humans primarily affects the digestive tract after the parasites enter the body through ingestion of contaminated food or water, or by penetrating the skin. One of the most common helminth diseases, schistosomiasis, affects more than 200 million people per year. Most of these infections occur in poor areas of developing tropical or subtropical countries. The elminthes that cause schistosomiasis usually enter the body through the skin while people are bathing in or wading through fresh water lakes or streams. The parasites move through the bloodstream, causing rash, itchy skin, fever, chills, and muscle pain. Eventually the parasites enter the intestines to lay eggs. This stage of infection may lead to inflammation and malnutrition. The drug Praziquantel can be used to effectively treat and prevent schistosomiasis, though many of the people who are most vulnerable do not have access to this medication. For these people, good hygiene and access to clean water are critical for avoiding the disease.

History of the Plague

Throughout human history, our ancestors have been affected by a variety of infectious diseases. Some infectious disease outbreaks have noticeably shaped human history because of the high number of people killed, the significant economic costs, and the resulting disruption of social fabric. One of the most notable examples of an infectious disease impacting human history is the plague. Plague is a communicable infectious disease caused by a species of bacillus-shaped bacteria called *Yersinia pestis*. *Yersinia pestis* can be transmitted to humans through the bites of fleas and lice that live primarily on rats or humans, or passed directly from person to person through contaminated droplets released into the air. The latter mode of transmission leads to a form of plague known as pneumonic plague, which is a severe and often fatal lung infection characterized by shortness of breath, chest pain, cough, weakness, fever, and headache. Pneumonic plague may also result from *Yersinia pestis* entering the lungs from the bloodstream. Infection of the blood by *Yersinia pestis* is itself considered a distinct form of the plague known as septicemic plague. Septicemic plague is also highly deadly, as it can rapidly lead to hemorrhaging, necrosis of the skin, and shock. Septicemic plague begins when *Yersinia pestis* enter the blood from the bite of a flea or louse. This route of transmission may also cause a third form of the plague called bubonic plague. This is the mildest form of plague, characterized by flu-like symptoms and swollen lymph nodes in the groin, which are called buboes.

Reports of the plague were first recorded in Ancient Greece during the 5th century B.C. Within the next several hundred years, plague was found throughout Europe, Asia, and Africa. The first widespread outbreak of plague began in 541 in Ethiopia. This outbreak is known as the Justinian Plague because it spread through, and devastated, the Byzantine Empire during the rule of Roman Emperor Justinian I. Within 3 years, plague-causing bacteria had spread across the Mediterranean to the bustling port city of Constantinople, following ship routes used for trade. At the peak of the outbreak, the Justinian Plague is estimated to have killed 5,000 to 10,000 people every day in Constantinople alone. Nearly 100 million people may have been killed by the year 546, as the plague spread through Europe, North Africa, and the Middle East. Because the disease affected a large number of people across a large geographic area, this outbreak of plague is classified as a **pandemic**. As a result of the tremendous death toll over a short period of time, the Justinian plague is credited with permanently changing Western society in a significant way. Some suggest the plague precipitated the downfall of the Roman Empire and forced the agrarian system of the empire to be restructured into the feudal system that would later characterize medieval Europe.

FIGURE 14.6: The Black Death—The Black Death was a pandemic of bubonic plague that ravaged Europe and other parts of the world during the 14th century.

After the end of the Justinian plague, smaller outbreaks of the plague occurred throughout the old world for the next 800 years. The next major outbreak of the plague began in 1347 and ravaged most of Europe for the next 5 years, reducing the population of the continent by up to 20%. This pandemic was first referred to as "the pestilence" but later became widely known as the Black Death. The Black Death caused a total of 50 million deaths, half of which occurred in Europe and the other half in Africa and Asia. Like the Justinian Plague before it, the extremely high death toll of the Black Death greatly impacted society. Part of the reason why the plague spread so extensively was that the people of the time did not understand what caused the plague or how it could be prevented. One of the prevailing ideas in the 14th century was that plague, as well as other disease outbreaks, were divine punishments that could only be ameliorated by appeasing God through penance (Figure 14.6). At the height of the plague, many people also feared they would suffer spiritually after their physical death because there were not enough clergy to perform burial rites for all of those who were dying. It was also commonly thought that plague was spread through unpleasant odors called miasmas. To ward away miasmas, some people would wear or burn fragrant herbs and flowers. Inspired by a later outbreak of plague that hit London in 1665–1666, the author Kate Greenaway wrote the following nursery rhyme in her book *Mother Goose*:

Ring, a-ring, o'rosies, (the red blistery pox rash)
A pocket full of posies (fragrant herbs and flowers to ward off "miasmas")
Atishoo, atishoo (sneezing associated with pneumonic plague)
We all fall down. (everybody dies)

The last major pandemic of plague occurred between 1894 and 1959 and primarily affected Asia and Australia. After significant advances in medicine and hygiene during the 1800s and discovery of *Yersinia pestis* in 1894, plague outbreaks become much better managed. Many cases of plague can now be effectively treated using the antibiotic moxifloxacin. However, localized outbreaks of the plague have continued to appear throughout the world up until today. For example, in 2017, more than 2,400 people in Madagascar were struck by the plague within just 4 months and more than 200 of these individuals died. Fortunately, even after hundreds of years of studying the disease, scientists are still learning more about the plague and how to prevent its spread. One major breakthrough came in 2017 when a group of researchers in Norway publish strong evidence suggesting that the Black Death and later plague pandemics in Europe were spread not by rat fleas, as had been widely believed for many years, but instead moved quickly from person to person through the spread of human fleas and lice. This finding sheds new light on what we thought we knew about some of the deadliest pandemics in human history.

Smallpox and Disease Eradication

Smallpox is one of the earliest known diseases to affect mankind. It is thought to have first appeared around 10,000 B.C., in Africa. Smallpox was most often spread through inhalation of contaminated droplets released from the lungs of infected people. This led to development of flu-like symptoms accompanied by a blistering rash, from which many people could recover naturally. However, the mortality rate of smallpox was over 30%, which led to 300–500 million deaths worldwide. Many of those who survived smallpox were permanently disfigured by scars formed from a scabbing rash that often covered most of the upper part of the body, including the face and arms. Thus, smallpox was once known as *variola vera*, which is derived from the Latin words for "spotted pimple." The virus that causes smallpox was named *Variola*. This virus exists in two forms, *Variola major* and *Variola minor*, that are both double-stranded DNA viruses.

As with the plague, smallpox pandemics exerted a tremendous influence on society. Among those who suffered from or were killed by smallpox is a long list of influential people including the Ancient Egyptian pharaoh Ramses V, King Louis XV of France, and 18th century English aristocrat Lady Montague. What is most notable about smallpox, however, is the story of all the people who never contracted the disease. This story begins with the development of the first vaccine, which provided protection against smallpox starting in the 18th century. During the first half of the 18th century, it became common practice for physicians to introduce small amount of material from the smallpox sores on an infected person's body to the skin of an uninfected person. This process was termed variolation after the name of the *Variola* virus that causes smallpox. The majority of people treated in this way developed mild symptoms, then recovered. Importantly, variolation induced an adaptive immune response specific for the *Variola* virus leading to production of memory cells that then conferred protection if the variolated individual was exposed to the virus in the future. However, because of the potential for infection from exposure to active *Variola* virus, a small percentage of people would develop full-blown smallpox after variolation and die.

An improvement to this procedure was made famous, though not invented by, English physician Edward Jenner in 1798. It is widely believed that Jenner's work was based on observations that milkmaids were much less likely to be affected by smallpox than the general population (Figure 14.7). In fact, it is said that milkmaids were known for having perfectly smooth skin at a time when many people carried permanent scars from smallpox lesions. Further investigation revealed that this protection was conferred by exposure to a similar but much less virulent infectious disease called cowpox (Figure 14.7). People who developed cowpox after being unintentionally exposed to infected cows would usually recover from the disease as their immune system fought it off and would simultaneously develop immunity to developing cowpox again in the future. More interestingly, they also seemed to develop immunity to smallpox, as evidenced by their much lower rate of contracting that disease. Recent analysis of historical documents have suggested that Jenner first learned of this phenomenon not by observing milkmaids, but instead by hearing about accounts of farmers who, remarkably, tended not to develop any signs of smallpox when intentionally inoculated with the virus; an observation that had no more plausible explanation than the development of cross-reactive immunity to smallpox after exposure to cowpox on their farms. Whatever his inspiration, Jenner went on to experimentally test the hypothesis that exposure to cowpox leads to protection against smallpox by injecting a small amount of active *Vaccinia* virus, the causative agent of cowpox, into the skin of people who were not immune to smallpox, then exposed them to the *Variola* virus and demonstrated they were protected against smallpox. This process quickly became widely used throughout Europe, then the rest of the world. Over the next hundred years the smallpox vaccine went through several stages of refinement, yet continued to rely on use of active *Vaccinia* virus. This is an important distinction because most vaccines used today do not contain active viruses or live microorganisms but instead rely on the use of purified antigens or inactivated viruses to induce immunity. During this time, the term "vaccine," derived for the Latin word for cow, "vacca," largely replaced the term variolation. Now, **vaccination** is used to refer to the use of any substance to stimulate a protective antibody response specific for any pathogen.

As more and more people were vaccinated against smallpox, the number of new cases dropped dramatically. By 1967, governments and health organizations around the world united together to eliminate smallpox from the globe. This was achieved through widespread vaccination campaigns and diligent surveillance and treatment of infected persons. The last known case of smallpox occurred in 1979. In 1980, the World Health Organization declared that smallpox had become the first, and so far only, human disease to be eradicated. At this point, most countries stopped routinely vaccinating their citizens against the smallpox virus. The United States had already ended routine smallpox vaccination in 1972. No cases of smallpox have been reported since the disease was eradicated, so the risk of being infected is essentially zero. Because of this, most governments and public health agencies agree that the benefit of continuing to vaccinate against smallpox, if there is one, does not outweigh the costs. However, stockpiles of the virus remain stored in laboratories in Russia, South Africa, the United Kingdom, and the United States to be used for research purposes. Some people are concerned that smallpox may be released into the population in the future, either from an unknown natural reservoir, as the result of a mutation in a related virus, or through intentional use of the virus as a weapon. As a measure of caution, the country of China still vaccines citizens against smallpox. The United States

FIGURE 14.7: Origin of smallpox vaccine

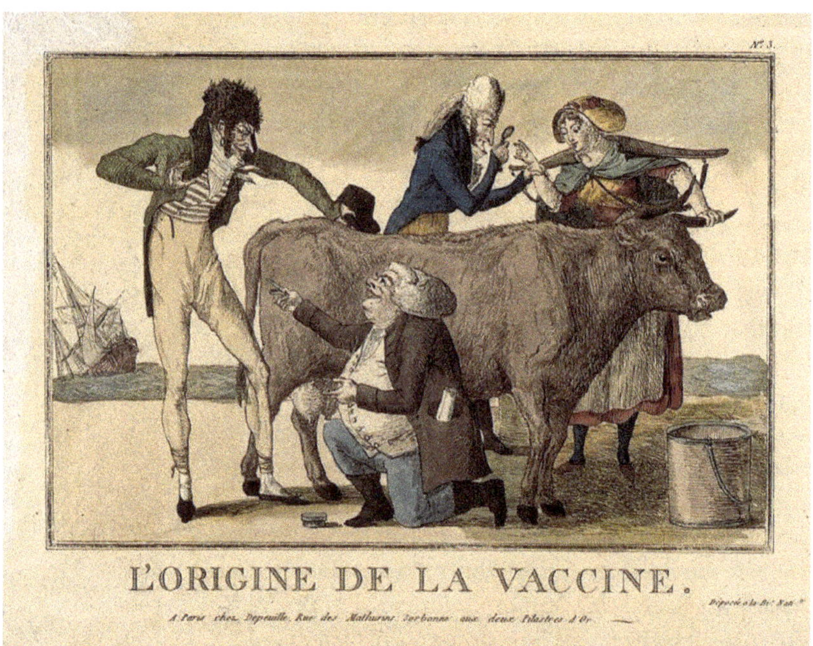

also selectively vaccinates certain military personnel against smallpox, such as those on active duty in the Middle East.

Origin and History of HIV

In June 1981, the United States CDC issued a report describing five cases of *Pneumocystis carinii* pneumonia identified in previously healthy men by the physician Michael Gottlieb working at the University of California, Los Angeles. This finding was noteworthy because *Pneumocystis carinii* pneumonia is an opportunistic infection of the lungs that is closely associated with severe impairment of the immune system, such as that characteristic of people being treated for cancer, organ transplant, or autoimmune disorders. Gottlieb and other physicians and scientists linked these infections with a cluster of other opportunistic diseases, including a type of cancer known as Kaposi's sarcoma, which appeared to follow a pattern of sexual transmission. Because the presence of *Pneumocystis carinii* pneumonia or Kaposi's sarcoma signals lack of proper immune system function, and because the majority of the first cases identified were in homosexual men, this condition was initially referred to as Gay-Related Immune Deficiency (GRID).

The following year, in 1982, the CDC coined the term AIDS to describe this newly identified syndrome, which was then suspected to be spread by both homosexual and heterosexual intimate contact, as well as by transfer of contaminated blood during intravenous drug use and medical procedures. After several years of intense investigation by teams of researchers in the United States and Europe, HIV was identified as the cause of AIDS in 1983. Françoise Barré-Sinoussi, Harald zur Hausen, and Luc Montagnier, whose team had isolated HIV from the T cells of a patient in Paris and identified the virus as the cause of AIDS, were awarded the Nobel Prize in Physiology or Medicine in 2008 for their discovery.

Beginning with burgeoning clusters of infections in major American cities such as Los Angeles, New York City, and San Francisco, reports of HIV and AIDS soon emerged throughout the world, achieving pandemic status by the early 1980s. As surveillance of AIDS intensified and tests to detect HIV became more accurate and widely used, the number of known AIDS-related deaths worldwide surpassed 500,000 by 1990, rose to more than 1,000,000 by 1993, and approached 2,000,000 by 1996. By 1992, more than 10,000,000 people were known to be living with HIV infection globally; this number doubled by 1996, and tripled by 2002. In the early years of the AIDS pandemic, a diagnosis of AIDS or HIV infection was widely considered a death sentence as a result of the absence of any treatment capable of preventing or delaying disease progression effectively. The first anti-HIV drug, known as AZT, was approved for use in the United States in 1986, yet the number of HIV diagnoses and the number of HIV-positive individuals who developed AIDS and died from AIDS-related causes continued to rise year after year for the next decade, even in countries where AZT was readily available. Only in 1996, after the development of highly active antiretroviral therapy (HAART) did the rate of AIDS deaths begin to decrease in the United States and other countries with access to this treatment. In the late 1990s and 2000s, wealthy nations saw AIDS transform from a death sentence to a manageable chronic disease. In 2012, a remarkable treatment known as pre-exposure prophylaxis (PrEP), sold under the brand name Truvada, became the first drug approved for the prevention of HIV. When taken correctly, PrEP can be nearly 100% effective in preventing transmission of HIV between an infected person and uninfected person. However, despite these breakthroughs in HIV treatment, lack of access to medications and an array of other social and political factors have caused infection and death rates to remain high in many at-risk populations, especially in poverty-stricken areas of sub-Saharan Africa. As of 2017, an estimated 76 million people had been infected with HIV since the pandemic began, leading to a total of 35 million deaths from AIDS-related causes. Some estimates place this figure as high as 45 million. Currently, more than 50% of people infected with HIV reside in east and southern Africa, in a region home to only 6.2% of the global population.

After the identification of HIV and AIDS, it quickly become apparent that HIV likely did not originate in Los Angeles or in any of the other cities that had reported the first wave of known infections in the early 1980s. Many scientists then turned their focus to determining where HIV came from and when it first entered the human population. Retrospective analysis of archived samples from people who might have been infected with HIV identified the earliest HIV-containing specimen as a sample taken from a patient in Kinshasa in 1959. Kinshasa, which is the capital of the country now known as the Democratic Republic of the Congo, is one of the most populous cities in Africa and has been an important trade hub since shortly after its founding by Belgian colonialists in 1881 under the name Leopoldville. From the vantage point of today, it should not be surprising that an unknown and undetectable pathogen that spreads through direct human contact might spread across the globe once it reached such a globally connected city as Kinshasa.

Comparative analysis of the genetic material of HIV and similar viruses identified in other animals revealed that HIV most likely entered humans through contact with chimpanzees. For simian immunodeficiency virus (SIV) to gain the ability to spread to humans, it first underwent a genetic mutation in which genetic material from two different strains of SIV combined within a single individual. Several different strains of the viral ancestors of HIV are found in chimpanzees and several species of monkeys. By analyzing the rate and amount of divergence in the genetic code of a number of different samples of HIV and SIV collected at different times and places, scientists have been able to estimate, with a fair degree of confidence, that HIV first entered humans from chimpanzees sometime between 1884 and 1933. However, it is not possible to determine with certainty the exact year that this cross-species transmission first occurred. Genetic analysis of the virus has further revealed that HIV first entered humans in Cameroon, a densely treed Central African country located northwest of the Democratic Republic of the Congo, and that transmission from chimpanzees to humans likely occurred a number of different times, leading to the development of several distinct strains of the virus. In Central Africa in the early 20th century, and even today, wild chimpanzees are sometimes killed by people as a source of food. It is likely that during the hunting and butchering of chimpanzees, transfer of virus-infected blood from prey to hunter lead to the origin of HIV in humans.

Once in humans, HIV probably spread through the population of Central Africa slowly at first, and primarily through heterosexual sexual activity. In Central Africa, the prevalence (proportion of infected individuals in a population) of HIV likely remained between 0.1% and 0.2% for about 50 years before starting to increase around 1970; however, during this time, more concentrated pockets of HIV infection likely existed among high-risk cohorts, such as IV drug users and sex workers. AIDS may have remained a rare but persistent disease of Central Africa for many more decades, if its spread was not hastened by a constellation of environmental, social, political, and economic factors that promoted the transfer of contaminated body fluids to increasing numbers of people. European colonialism in Central

Africa, especially that conducted by the Belgians in the Congo and the Germans and French in Cameroon, was a driving force in propagating conditions that facilitated the spread of HIV through and out of Central Africa. One way this occurred is through the creation of densely populated male-dominated work camps and income inequality that led to the spread of HIV through prostitution networks. Paradoxically, efforts of colonial leaders to safeguard their profits by taking active measures to ensure the health of people living in the African colonies also likely contributed to the spread of HIV. This occurred because large-scale efforts to treat endemic infectious diseases, such African trypanosomiasis, leprosy, malaria, syphilis, and yaws, though ostensibly successful, were conducted without taking precautions to prevent transfer of other infectious agents, such as the then-unknown HIV virus. Medical equipment and blood-containing products were routinely used on multiple patients without being properly sterilized up until the 1980s, not only in Africa, but also throughout the world.

Genetic analysis has further revealed that HIV made its way to the Caribbean during the 1960s. This pathway may have been paved by efforts to bring a large number of health care workers and teachers from Haiti to the Congo to fill skilled positions vacated by Belgian colonists when the country gained independence in 1960. HIV then spread from Haiti to New York City no later than 1974, possibly driven by a combination of sexual tourism and export of contaminated blood products used to treat people with clotting disorders and other blood conditions. Spread of the virus rapidly continued to infect people in several other states including New Jersey, Pennsylvania, Georgia, and California by 1979. During this time, unknown numbers of infected individuals had already transported HIV to Europe, Asia, and throughout Africa, years before Dr. Gottlieb identified a cluster of *Pneumocystis carinii* infections in Los Angeles. We now know that many of the HIV-positive individuals first identified in the early 1980s were infected as much as 10 years earlier but were not diagnosed because they were not exhibiting any symptoms of illness. This is explained by the fact that the majority of people infected with HIV experience a period of viral latency for 3 to 10 years during which time they are infected with the virus, but their immune systems continue to function well enough to prevent development of the opportunistic infections and diseases that are the telltale signs of AIDS.

Neglected and Emerging Diseases

In addition to infectious causes of cancer, another area of the field of infectious diseases often considered to warrant increased attention is that of neglected and emerging diseases. The term "neglected" is used to refer to diseases that receive a disproportionately small amount of funding and attention for research, treatment, and prevention in relation to their effect in the world. For example, the subgroup of neglected diseases known as neglected tropical diseases affect more than 1 billion people worldwide, leading to hundreds of thousands of deaths and tens of billions of dollars in economic costs per year, yet these diseases receive only a small fraction of global public health funding (Table 14.2). This is primarily because of the fact that neglected tropical diseases mostly affect people in developing countries that are not able to attract sufficient foreign aid from richer nations. Neglected tropical diseases may be caused by a variety of different viral, bacterial, protozoan, and multicellular pathogens (Table 14.2). Examples of these include Dengue fever (viral), Leprosy (bacterial), Chagas disease (protozoan), and Dracunculiasis (helminth) (Table 14.2).

TABLE 14.2: Neglected Tropical Diseases

Type of pathogen	Name of neglected tropical disease
Viral	Dengue fever, Rabies
Bacterial	Buruli ulcer, Leprosy (Hansen's disease), Trachoma, Yaws
Protozoan	Chagas disease, Human African trypanosomiasis, Leishmaniasis
Multicellular parasite	Dracunculiasis, Echinococcosis, Foodborne trematodiases, Lymphatic filariasis, Onchocerciasis (River blindness), Schistosomiasis, Soil transmitted helminthiases, Taeniasis/Cysticercosis

Emerging diseases are those whose frequency have sharply increased during the last 20 years or are predicted to experience a dramatic increase in incidence in the future. Even though emerging diseases often pose an impending threat to public health, the governments and organization that oversee public health campaigns face strong pressure to focus resources on diseases that are already widespread and actively affecting people. These diseases represent a vast spectrum of routes of transmission, symptoms, and available preventive measures and treatments. This makes efforts to limit their spread challenging and requires a multipronged approach. For example, one human activity that is highly risky in terms of its potential to spark an outbreak of a new or emerging disease is the hunting of bush meat. **Bush meat** refers to wild animals such as monkeys, reptiles, and birds that are hunted for food, often in tropical forests. This practice is especially common in rural parts of equatorial Africa and in parts of Southeast Asia. Perhaps the most notable consequence of human consumption of bush meat is the HIV/AIDS pandemic, which is through to have originated as a result of the hunting of chimpanzees in central Africa. Despite its dangers, the practice of hunting bush meat persists largely because of lack of affordable alternative food sources and the lack of resources to change current norms.

The tool that is perhaps singularly most important in fighting neglected and emerging diseases is education. Increasing awareness of these diseases not only helps prevent people from contracting infections for which there is often no cure or vaccine available, education also promotes development of new therapies and prevention efforts. Examples of emerging disease include avian influenza, Dengue fever, and Zika virus infection.

Infectious Causes of Cancer

One important and often overlooked aspect of infectious diseases is their contribution to the development of cancer. Globally, one-third of all cancers are attributed to infectious diseases. Of these, 20% of cancers initiated by infectious diseases can be traced to just four pathogens: human papilloma virus (HPV), hepatitis B virus (HBV), hepatitis C virus (HCV), and the bacterium *Helicobacter pylori*. One commonality shared by all four of these pathogens is that they cause a long-term infection leading to production of chronic inflammation. HPV is transmitted through skin-to-skin contact during sexual intercourse. Because it is spread through the skin and not through exchange of body fluids, it can be passed from person to person during sexual activity even when condoms are used for protection. HPV is the primary cause of cervical cancer and has also been linked to cancers of the throat, penis, anus, and vagina. There are approximately 100 different strains of HPV, not all of which cause cancer. However, some strains of HPV that are non-carcinogenic cause the sexually transmitted infection genital warts. The word "papilloma" in the name of HPV means "warts." The name of hepatitis viruses is derived from the fact that they cause inflammation ("itis") of the liver ("hepar"). HBV can be spread through blood, semen, and other body fluids, whereas HCV is only transmitted through blood. One frequent source of transmission for both of these viruses is sharing of contaminated needles by IV drug users, although both can also be transmitted from infected mothers to children during birth, as well as through sexual contact. *Helicobacter pylori* is commonly found in the stomach, where it causes no symptoms in the majority of people. However, up to 80% of incidences of stomach cancer are caused by *Helicobacter pylori* infection.

In the case of all four of these pathogens, as well as the other infectious causes of cancer, there are two primary mechanisms by which infection promotes cancer development. The first of these involves DNA damage induced by viral integration into host DNA during the replication process, and the second involves DNA damage caused by reactive molecules produced during the inflammatory response. Not all DNA damage leads to cancer, but the greater the amount of damage to DNA, the greater chance genes required for proper regulation of the cell cycle will be altered. Because cancer can be initiated by mutations caused by infections even if they are later treated and cleared, the primary method for reducing the incidence of infectious cancers is preventing infection. Effective vaccines are already available for preventing HBV and HPV infections, and candidate vaccines are currently being developed for HCV and *Helicobacter pylori*. The most effective method of preventing cancer caused by infectious agents for which there is not a vaccine is limiting the development of chronic inflammation by prompt and proper treatment of the infection. This requires routine surveillance for these infections, which do not always cause symptoms. In the case of HPV, the virus is may be naturally cleared by the immune system without the infected person being aware they were even infected, yet may still cause irreversible carcinogenic changes prior to that clearance.

14.2 SUMMARY

- Infectious diseases are spread from one individual to another and are caused by a wide array of pathogens including viruses, bacteria, fungi, protozoans, and helminthes
- Pathogenicity measures the ability of a pathogen to cause disease whereas virulence measures the severity of disease that results from infection
- Antigenic drift is the reason new vaccines are needed every year to protect against the seasonal flu, whereas antigenic shift often leads to pandemic outbreaks of disease
- Bacteria are characterized by their shape and by the composition of their cell wall
- The bacterial pathogen *Yersinia pestis* has caused several pandemics of the plague, resulting in a high death toll and significant effect on society throughout history
- Edward Jenner helped popularize the first vaccine, which targeted the virus that causes smallpox, leading to its eradication

- HIV is thought to have first spread from chimpanzees to humans around the turn of the 20th century in Central Africa, and was then spread to the Caribbean, United States, and the rest of the world by the time it was identified in the early 1980s
- Neglected diseases are a threat to public health because they affect a large number of people but receive disproportionately few resources; emerging diseases are a threat because they have recently or are predicted to become significantly more common
- A number of viral and bacterial infections increase the risk of cancer by causing DNA damage and chronic inflammation

14.3 KNOWLEDGE CHECK

1. All microorganisms capable of causing disease are referred to as:
   a. antibiotics
   b. bacteria
   c. pandemics
   d. pathogens
   e. vectors

2. An infectious microorganism that causes disease easily but only causes mild disease has:
   a. low pathogenicity and low virulence
   b. low pathogenicity and high virulence
   c. high pathogenicity and low virulence
   d. high pathogenicity and high virulence
   e. high pathogenicity and no virulence

3. Typhoid fever is caused by a rod-shaped bacterium named *Salmonella typhi* transmitted through ingestion of contaminated food or water. Infection with this bacterium cannot be effectively treated with the antibiotic penicillin. Which of the following best explains this?
   a. penicillin is not effective at treating bacteria, only viruses
   b. penicillin is not effective at treating bacteria containing a large amount of lipopolysaccharide
   c. penicillin is not effective at treating bacteria containing a large amount of peptidoglycan
   d. penicillin is not effective at treating bacteria transmitted through contaminated food or water
   e. penicillin is not effective at treating rod-shaped bacteria

4. The first vaccine protected against which of the following?
   a. AIDS
   b. Influenza
   c. Plague
   d. Smallpox
   e. Typhoid fever

5. What is bubonic plague named for?
   a. the bacterium that causes the disease
   b. the fact that the infection causes neurological problems
   c. the person who first discovered the disease
   d. the swollen lymph nodes in the groin of affected people
   e. the type of immune cell that fights the infection

6. Define the terms "antigenic drift" and "antigenic shift."

7. Identify an example of an infectious disease caused by each of the following: viruses, prions, bacteria, protozoans, fungi, and helminthes.

8. Explain how the plague affected human society and history.

14.4 ACTIVITY

African Illness: A Case of Parasites? Case Study

The purpose of this activity is to investigate the symptoms, diagnosis, and treatment of a tropical disease. This activity provides an opportunity to demonstrate both oral and written communication skills and to connect a scientific topic with its real-world implications. This activity involves working individually or in small groups to complete the case study titled "African Illness: A Case of Parasites?" published by the National Center For Case Study Teaching In Science (NCCSTS).

"African Illness: A Case of Parasites?" http://sciencecases.lib.buffalo.edu/cs/files/safari.pdf

14.5 ACTIVITY

Influenza Case Study

The purpose of this activity is to investigate the symptoms, history, preventive measures, and vaccination associated with influenza. This activity provides an opportunity to demonstrate both oral communication and critical thinking skills and to connect a scientific topic with its real-world implications. This activity presents information in the form of a narrative that can be completed using a role-playing activity or by small or large group members taking turns reading slides that comprise the case study titled "I Don't Need a Flu Shot!" published by the National Center For Case Study Teaching In Science (NCCSTS).

"I Don't Need a Flu Shot!" http://sciencecases.lib.buffalo.edu/cs/collection/detail.asp?case_id=474&id=474

14.6 ACTIVITY

Infectious Disease Writing Project

The effect of infectious diseases often extends beyond the level of the individual to include the society, government, public health organizations, economics, and even culture and histories of the countries that experienced or are threatened by them. For this essay, you will focus on the broad implications of infectious diseases on society. There are two options for structuring this essay, from which you should choose one:

Option A: Choose one country to focus on, and compare how that country has been affected by and responded to two different infectious diseases.

Option B: Choose one infectious disease and compare and contrast how two different countries have been affected by and responded to that infectious disease.

For either option, please choose one infectious disease from the list of CDC Notifiable Diseases (Table 14.1) or the list of neglected diseases recognized by the WHO (Table 14.2). You may also seek preapproval from your instructor to choose a different infectious disease. Be careful to choose an option and topic for which you can adequately address all of the learning objectives provided by your instructor.

ACTIVITY QUESTIONS

14.4

1. Follow instructions to work individually or form small groups for completing the case study and answering the indicated questions.

14.5

1. Follow instructions to work individually or for reading the case study narrative and answering the associated questions.

14.6

1. In your essay:

- Identify the full scientific name of the pathogen that causes the infectious disease(s) you discuss and indicate what type of organism it is (bacterium, fungus, protozoan, or virus)
- Identify the country/infectious disease(s) that your essay will focus on in your introductory paragraph, as well as the time frame of the disease outbreak(s) you will discuss
- Identify the number of people who were affected/killed by the infectious disease in your chosen country/countries as well as worldwide
- Describe how the pathogen that causes the infectious disease is/was/can be diagnosed, treated, and prevented
- The majority of the essay should focus on thoroughly answering the following questions in terms of your chosen pandemic/country/time frame:
    - How did/does the infectious disease affect people socially and culturally?
    - How did/does the infectious disease affect the country economically?
    - How did/does the country respond to the infectious disease in terms of social, cultural, economic, and public health repercussions and interventions?
    - Has the infectious disease resulted in any long-term or permanent changes in the country in terms of social/cultural practices or public policies? If so, describe.
    - Based on published evidence, what is your evaluation of how well the country is prepared to handle an outbreak of this infectious disease? What evidence supports your reasoning?
- Base all statements on properly cited evidence from appropriate sources
- At least five sources of information should be used and cited; at least two primary sources (i.e., original scientific report) should be used; the majority of other sources should be secondary (e.g., description of primary scientific source(s) in a scientific, educational, or news magazine, paper, or website).
- Please carefully review the essay rubric for more information on how the essay will be evaluated for a grade

14.7 REFERENCES

Akpan, N. (2016). America's HIV outbreak started in this city, 10 years before anyone noticed. [On-line] *PBS NewHour.* Access date: October 26, 2016. https://www.pbs.org/newshour/science/america-hiv-outbreak-origins-nyc-gaetan-dugas.

Boylston, A. W. (2018). The myth of the milkmaid. *The New England Journal of Medicine, 378*, 414–415.

Bonney, K. M. (2013). An argument and plan for promoting the teaching and learning of neglected tropical diseases. *Journal of Microbiology and Biology Education, 14*(2), 183–188.

Centers for Disease Control and Prevention. 2017 Nationally Notifiable Conditions. [On-line] Access date: August 4, 2017. https://wwwn.cdc.gov/nndss/conditions/notifiable/2017/.

Dean, K. R., et al. (2018). Human ectoparasites and the spread of plague in Europe during the Second Pandemic. *Proceedings of the National Academy of Sciences of the United States of America*, Early Edition, Jan. 16, 1–6.

De Flora, S., and La Maestra, S. (2015). Epidemiology of cancers of infectious origin and prevention strategies. *Journal of Preventive Medicine and Hygiene, 56*(1), E15–20.

Dickman, M. B., and de Figueriedo, P. (2011). Comparative pathobiology of fungal pathogens of plants and animals. *PloS Pathogens, 7*(12), e1002324.

Diener, T. O., McKinley, M. P., and Prusiner, S. B. (1982). Viroids and Prions. *Proceedings of the National Academy of Sciences of the United States of America, 79*(17), 5220–5224.

Frith, J. (2012). The history of plague – Part 1. The three great pandemics. *Journal of Military and Veterans' Health, 20*(2), 11–16.

Gallo, R. C., and Montagnier, L. (2003). The discovery of HIV as the cause of AIDS. *New England Journal of Medicine, 349* , 2283–2285.

Kovacs, G. G., and Budka, H. (2008). Prion diseases: From protein to cell pathology. *American Journal of Pathology, 172*(3), 555–565.

Luzzatto, L. (2012). Sickle cell anaemia and malaria. *Mediterranean Journal of Hematology and Infectious Disease, 4*(1), e2012065.

Mead, P. S. (2018). Plague in Madagascar — A tragic opportunity for improving public health. *The New England Journal of Medicine, 378*(2), 106–108.

Pepin, J. (2011). *The origin of AIDS*. Cambridge, UK: Cambridge University Press.

Riedel, S. (2005). Edward Jenner and the history of smallpox and vaccination. *Baylor University Medical Center Proceedings, 18*(1), 21–25.

Simon, S. (2016). American Cancer Society updates HPV vaccine recommendations to include males. [On-line] Access date: July 19, 2016. https://www.cancer.org/latest-news/american-cancer-society-updates-hpv-vaccine-recommendations-to-include-males.html.

Timberg, C. and Halperin, D. (2012). Colonialism in Africa helped launch the HIV epidemic a century ago. *The Washington Post*. [On-line] Access date: February 27, 2012. https://www.washingtonpost.com/national/health-science/colonialism-in-africa-helped-launch-the-hiv-epidemic-a-century-ago/2012/02/21/gIQAyJ9aeR_story.html?utm_term=.1ce23556a759.

UNAIDS. (2017). Fact Sheet – World AIDS Day 2017. [On-line] http://www.unaids.org/sites/default/files/media_asset/UNAIDS_FactSheet_en.pdf.

Winslow, C. E. A. (1914). The characterization and classification of bacterial types. *Science, 39*(994), 77–91.

FIGURE CREDITS

FIG. 14.1: Copyright © NIAID (CC by 2.0) at https://commons.wikimedia.org/wiki/File:HI-Virion_Structure_(5080768345).jpg.

FIG. 14.2: Copyright © NIAID (CC by 2.0) at https://commons.wikimedia.org/wiki/File:Antigenic_Drift_of_the_Flu_Virus_(5102830852).jpg.

TABLE 14.1: Source: https://wwwn.cdc.gov/nndss/conditions/notifiable/2017/

FIG. 14.3: Marc Perkins, "Gram staining," https://www.flickr.com/photos/occbio/6946316696/. Copyright © 2012 by Marc Perkins.

FIG. 14.4: James Gathany/CDC, "Erythema migrans," https://commons.wikimedia.org/wiki/File:Erythema_migrans_-_erythematous_rash_in_Lyme_disease_-_PHIL_9875.jpg. Copyright in the Public Domain.

FIG. 14.5: TimVickers, "Malaria parasites infecting red blood cells," https://commons.wikimedia.org/wiki/File:Plasmodium.jpg. Copyright in the Public Domain.

FIG. 14.6: Toggenburg Bible, "The Black Death," https://commons.wikimedia.org/wiki/File:Smallpox_depicted_in_the_Toggenburg_Bible.jpg. Copyright in the Public Domain.

FIG. 14.7: Depeuille, "Origin of smallpox vaccine," https://commons.wikimedia.org/wiki/File:A_physician_inspects_the_growth_of_cowpox_on_a_milking_maid%27_Wellcome_V0011690.jpg. Copyright in the Public Domain.

TABLE 14.2: Source: Kevin Bonney, "An Argument and Plan for Promoting the Teaching and Learning of Neglected Tropical Diseases." Journal of Microbiology and Biology Education, vol. 14, no. 2. CUNY Kingsborough Community College, 2013.

Biotechnology

15.1 INTRODUCTION

Biotechnology is the use of cellular processes, organisms, or whole biological systems to develop products that help humans, including agriculture, industry, and medicine. Biotechnology often, but not always, involves genetic manipulation. Humans have been doing this for thousands of years starting with breeding plants and animals for food and other products. This chapter will look at more recent developments in biotechnology that involve the use of a new set of tools that can be used to manipulate DNA and genes. These tools can then be used to genetically modify organisms, determine the nucleotide base sequence of genes, and even edit genes.

LEARNING OBJECTIVES

- Describe what biotechnology is and what it is used for
- Identify different tools of biotechnology including restriction enzymes, gel electrophoresis, and polymerase chain reaction, and discuss the uses of each
- Discuss the origin, function, and potential uses of CRISPR/Cas9
- Explain how these tools can be used for real-life applications like gene cloning, reproductive cloning, DNA sequencing, and gene editing

Basic Tools and Techniques of Biotechnology

Scientists have developed a whole set of different tools and techniques that can be used to modify molecules or even whole organisms. For example, a gene from one organism can be inserted into another organism enabling them to make a new protein. Because most applications of biotechnology involve using DNA, RNA, or a gene product like a protein, an important technique that helps form the foundation of biotechnology is extraction. The first step in extraction is isolation of the cells of interest. Different cell types are used for different purposes. For example, cells isolated from blood or from swabbing an area of skin could be used for forensic testing in a crime scene investigation or to test for a genetic disorder. Bacterial cells might be cultured for use in creating a genetically modified organism.

DNA is extracted from cells by first breaking open the plasma and nuclear membranes using a solution containing a detergent (Figure 15.1). The detergent breaks down the structure of the membranes by interacting with the hydrophobic ends of phospholipids that make up the membrane. Once the DNA is released, it needs to be separated from the supporting proteins called histones. An enzyme called proteinase that breaks down proteins is added. Salt is also added as it facilitates clumping of the proteins and other cellular components. The mixture is then centrifuged. Centrifugation spins the solution causing denser material to be pulled toward the bottom and materials in solution to remain on top. After centrifugation, the DNA will be in the solution, and the rest of the cellular components

FIGURE 15.1: **DNA extraction**—The steps of DNA extraction involve breaking open the cells of interest and then causing the DNA to precipitate out in an alcohol solution.

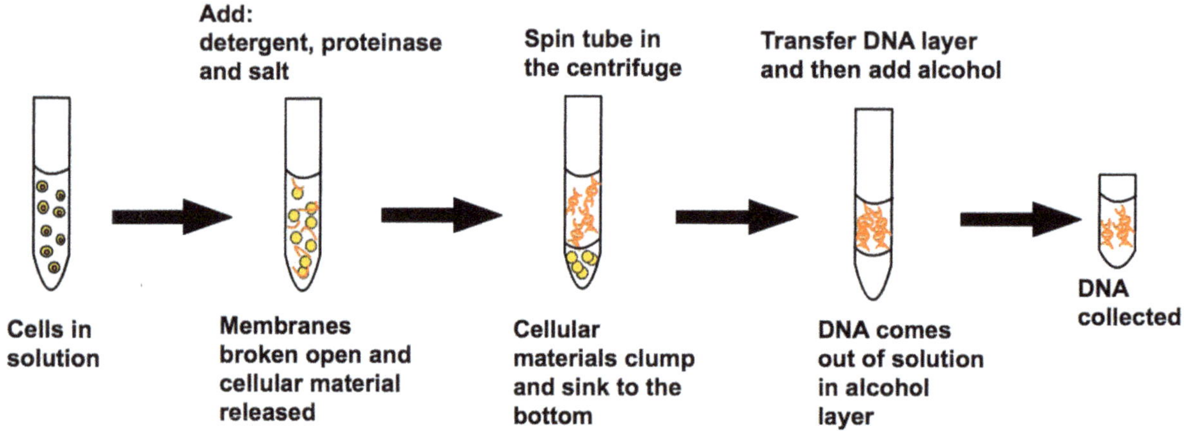

will be in the bottom of the tube. Removing the solution and adding alcohol will cause the DNA to precipitate, or come out of solution, because alcohol is less polar than water (Figure 15.1). This mixture can be centrifuged again so that the DNA will be on the bottom, which makes it easier to purify and isolate the DNA. Then it can be dried and stored or the DNA can be used right away. Depending on the specific application, the DNA may be processed in slightly different ways.

One way that DNA can be processed is through the use of **restriction enzymes**. Bacteria use restriction enzymes to cut out and remove foreign DNA, but now scientists use them to cut DNA from other organisms into smaller pieces. The enzymes are able to recognize specific nucleotide base sequences in the DNA and then break the sugar-phosphate bond in the DNA backbone. For example, the restriction enzyme *Eco*RI cuts between the G and the A in the sequence G/AATTC. On the complementary strand CTTAA/G, it also cuts between the G and the A. Notice that the nucleotide base sequence in one direction is the same on the other strand going the opposite direction. This is called a palindrome. Not all restriction enzymes cut palindromic sequences, but many do. Table 15.1 provides a sample of some of the restriction enzymes and their recognition sites. The names of restriction enzymes come from the name of the bacteria that they were isolated from, the strain, and the order it was found. *Eco*RI was isolated from *Escherichia coli* from strain RY13, and it was the first (I) restriction enzyme isolated from that bacterial species. Scientists can use these to cut DNA in specific places to cut out genes or other portions of DNA that they want to use for analysis.

After the DNA has been cut into smaller pieces, a researcher may need to have many copies of the DNA for analysis. The technique used to amplify or copy DNA is called **polymerase chain reaction (PCR).** It was invented by Kary Mullis who later won the Nobel Prize in Chemistry for this new technique. PCR allows scientists to be able to make millions and millions of copies of a desired DNA sample in a relatively short time frame. This step is necessary because oftentimes the amount of DNA isolated from cells is not sufficient to analyze. The process of PCR mimics cellular DNA replication but in a small reaction tube, so the researcher has control over what

TABLE 15.1: A Sampling of Restriction Enzymes and Their Recognition Sites—The / indicates where the enzyme cuts the sugar-phosphate backbone of DNA.

Restriction Enzyme	Recognition site
BamHI	G/GATCC
HindIII	A/AGCTT
NdeI	CA/TATG
SmaI	CCC/GGG

DNA sequence is copied. The machine that the DNA samples are placed in is called a **thermocycler.** The steps of PCR require the reaction mixture to be heated, cooled, and then heated again repeatedly. The thermocycler can be set for certain temperatures, length of time for each step, and how many times to go through the three-step cycle.

The first step of PCR is to add either restriction digested or undigested DNA to the reaction tube. Also added to the tube is a special type of DNA polymerase. Remember that DNA polymerase is the enzyme that builds new strands of DNA by adding nucleotide bases. This DNA polymerase is called *Taq* polymerase and is named for the bacteria that it was first isolated from, *Thermus aquaticus*, which is a thermophilic (heat-loving) species found in geysers like Old Faithful in Yellowstone Park. This type of polymerase is referred to as thermostable because it can withstand the heating cycles of PCR without denaturing and falling apart. Other DNA polymerases would not be able to stay intact during the heating cycle and would be rendered useless. Also added to the tube is a mixture of A, T, C, and G nucleotide bases. Another ingredient is a small DNA primer. These primers have specific nucleotide sequences complementary to nucleotide sequences on either end of the piece of DNA that you want to copy. You might recall that cellular DNA polymerase can only begin adding new bases in one direction, 5' to 3', and it needs attach to double-stranded DNA to begin. The primers are designed to have specific base sequences so they attach to the left or upstream of the DNA section to be copied as seen in Figure 15.2. After a few cycles, this will eventually allow only the desired sequence of DNA to be copied.

The reaction tube is placed in the thermocycler to begin the process. The first step of PCR involves heating the sample mixture so that the two complementary DNA strands separate (Figure 15.2). This is the denaturing step and is important so that the DNA nucleotide bases are available to be copied. This is followed by the second step, called annealing. During this step, the mixture is cooled slightly, which allows the DNA primers to attach to the single DNA strands (Figure 15.2). The third step is called extending. During this step, the mixture is heated again and the *Taq* polymerase attaches to the primers (Figure 15.2). Using the free nucleotide bases in the mixture, it builds a complementary strand. This three-step process is repeated 10 to 30 times and results in millions and millions of copies of the DNA of interest (Figure 15.2).

Another technique commonly used to analyze DNA, RNA, and proteins is gel electrophoresis. **Gel electrophoresis** is used to separate pieces of DNA, RNA, or proteins based on their size. This section will focus on using DNA. A common type of gel used for this technique is made of agar, which is derived from seaweed and has a consistency like gelatin. Because DNA has a slight negative charge, electricity can be used to push it through the particles of the gel. The samples are loaded into small wells at one end of the gel, which is then placed in a box containing a buffer solution that can conduct electricity (Figure 15.3). The DNA samples have oftentimes been cut with restriction enzymes or processed in a way that there will be DNA pieces of multiple lengths. There is also usually a dye added so that it is possible to see the DNA moving through the gel. Each end of the gel box is attached to a power source. The

FIGURE 15.2: Steps of polymerase chain reaction

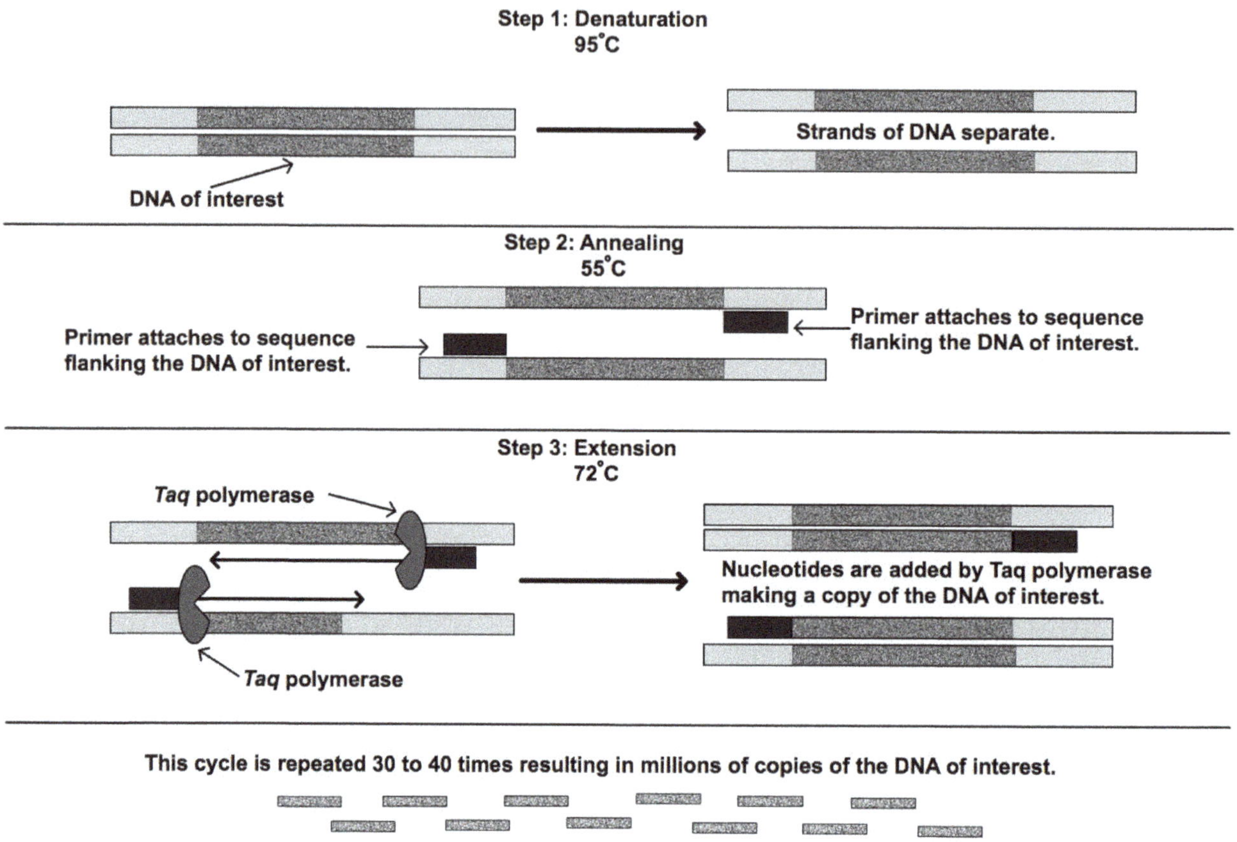

end where the samples are loaded has a negative charge, and the opposite end has a positive charge when the power is turned on. DNA molecules have a slight negative charge and are repelled by negative charges. When the power is turned on, the DNA moves away from the negative end toward the positive end, with the longer pieces moving more slowly. When the samples are near the end of the gel, the power is turned off and the molecules stop moving. The shorter pieces will be closer to the bottom of the gel, and the longer pieces will be closer to the end where they were loaded.

The scientist needs to record where the different pieces of DNA have landed and unfortunately a gel is very temporary. A common way to preserve the locations is to use a technique called **Southern blotting**. A membrane made of a special type of paper or nylon is laid on top of the gel and weighted down to help the DNA in the gel transfer to the membrane. The membrane is removed and exposed to radioactive DNA probes. The probes are short pieces of complementary DNA that match the DNA on the membrane. When exposed to the DNA in the membrane, the probes bond and "stick" to the membrane. Because they are radioactive, the membrane can be placed on photographic film to expose it. Wherever there is a probe, it will expose the film, and the pattern of the DNA pieces will now be visible and permanent. The

FIGURE 15.3A: **Gel electrophoresis set up**—The DNA samples are loaded into small wells in the gel at one end. When the power is turned on, the negatively charged DNA is pushed through the gel toward the positive end of the apparatus.

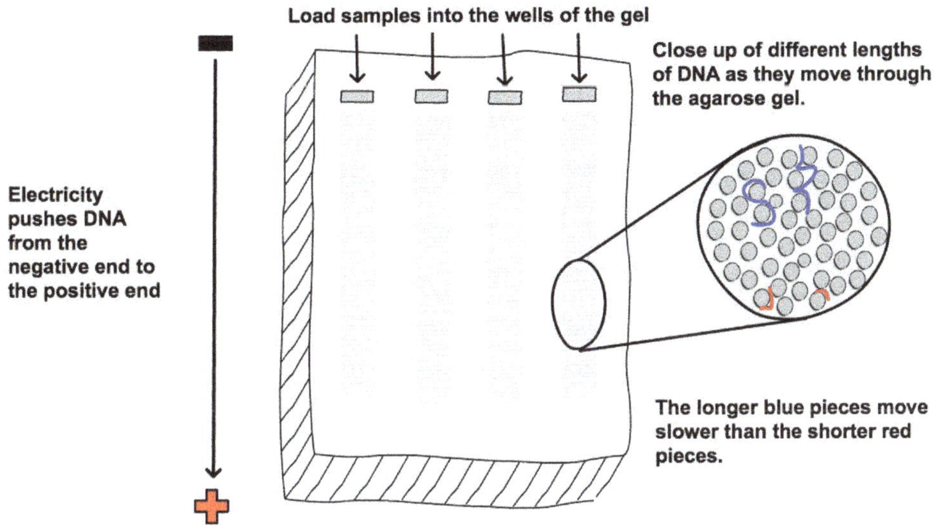

pattern can then be analyzed. Figure 15.4 shows a Southern blot of a comparison of two species of bacteria.

A **genetically modified organism** or GMO is an organism that has had its genetic composition changed to contain DNA sequences that it does not normally have. In the past, this was achieved by selective breeding of organisms to produce offspring with desirable traits. Humans have been using this method of primitive biotechnology to enhance crop yields since the Neolithic Revolution that occurred around 10,000 BC. The effect of thousands of years of selective breeding is so strong that many modern crops, such as corn, broccoli, and bananas, bear little resemblance to their wild ancestors. During the twentieth century, some scientists began to experiment with mutation breeding, which uses chemicals or radiation to induce genetic variations in plants. Although this process sometimes produced desirable variants, mutation breeding was unpredictable, untargeted, and inefficient. Most recently, production of desirable genetic variants has been achieved using **recombinant DNA technology** (rDNA), which is the insertion of DNA from one organism into another. When most people think of GMOs, this is the kind that they think of. One of the most common methods is to use bacteria, often *E. coli*, to clone or make exact copies of the gene of interest. *E. coli* has a chromosome made of DNA and also has a circular molecule of DNA called a **plasmid**. The bacterial plasmid can be removed from the bacterium and then be cut by a restriction enzyme to open up. Foreign genes can be mixed with the plasmids and will bond

FIGURE 15.3B: **Results of gel electrophoresis**—After the power is turned off, the DNA pieces stop moving. The longer pieces of DNA move more slowly through the gel, and the shorter pieces move faster.

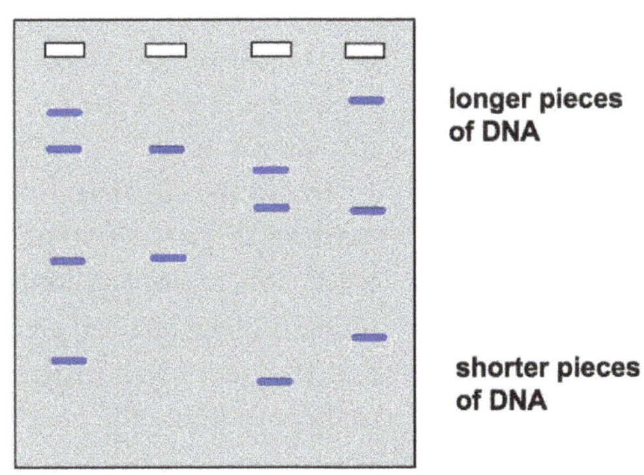

FIGURE 15.4: Southern blot of two bacterial species—Each band represents a piece of DNA of varying sizes. Longer pieces move more slowly and are found closer to the top, while shorter pieces are found closer to the bottom. The bp stands for base pairs and is a unit of measure for the length of DNA.

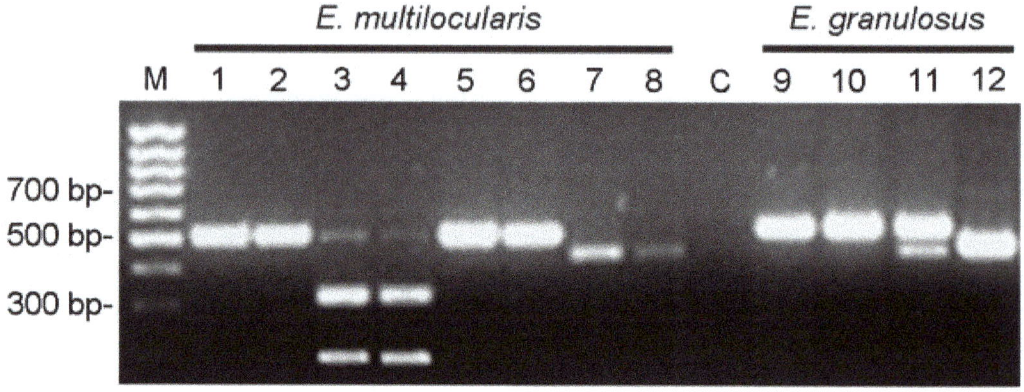

to the complementary bases on the open ends of the plasmid. Another enzyme, DNA ligase, seals the strands to close up the plasmid. This is called **recombinant DNA** because it has DNA from two different organisms. The recombinant plasmid is then inserted back into a bacterium. All the bacteria with the foreign gene are cultured, and every time they reproduce, the offspring receives the recombinant plasmid. The gene itself can then be collected from the bacteria and inserted into another organism. The process of making multiple identical copies of a gene in this way is known as molecular cloning, or **gene cloning** (Figure 15.5). This is not to be confused with another fascinating type of biotechnology called **reproductive cloning**, which is a method of creating a multicellular organism that is genetically identical to an existing individual. Reproductive cloning involves removing the nucleus of an egg and replacing it with the nucleus of a donor cell, then triggering the egg to divide and develop into an embryo.

In plants, the insertion of DNA from another organism is often done using the gene gun method (Figure 15.6). A "bullet" made of gold particles is covered with the cloned gene. The plant cells to be modified are grown in a culture. The "bullet" is then shot at the cultured cells using high-pressured gas in a vacuum (Figure 15.6). Once inside the cell, the gold dissipates and the genes will insert themselves into chromosomes. Now the cell has the genetic information to make the gene product. Another technique used to insert genes into cells is called the bacterial DNA transfer method. To transform plant cells, the bacteria *Agrobacterium tumefaciens* is often used. This bacterium naturally infects plant cells causing tumors. *A. tumefaciens* has a plasmid called a Ti plasmid, which can have the DNA of interest inserted into it, using a method like the one used in gene cloning described earlier. The *Agrobacterium* with Ti plasmids carrying the gene of interest, now called T DNA, are placed with plant cells that are in culture. The T DNA moves into a plant cell and then inserts into a plant chromosome. The plant cells are stimulated to grow and divide producing a plant whose cells have been genetically modified.

FIGURE 15.5: Adapted from *Campbell Biology in Focus*, "Figure 13.24: An overview of gene cloning and some uses of cloned genes."

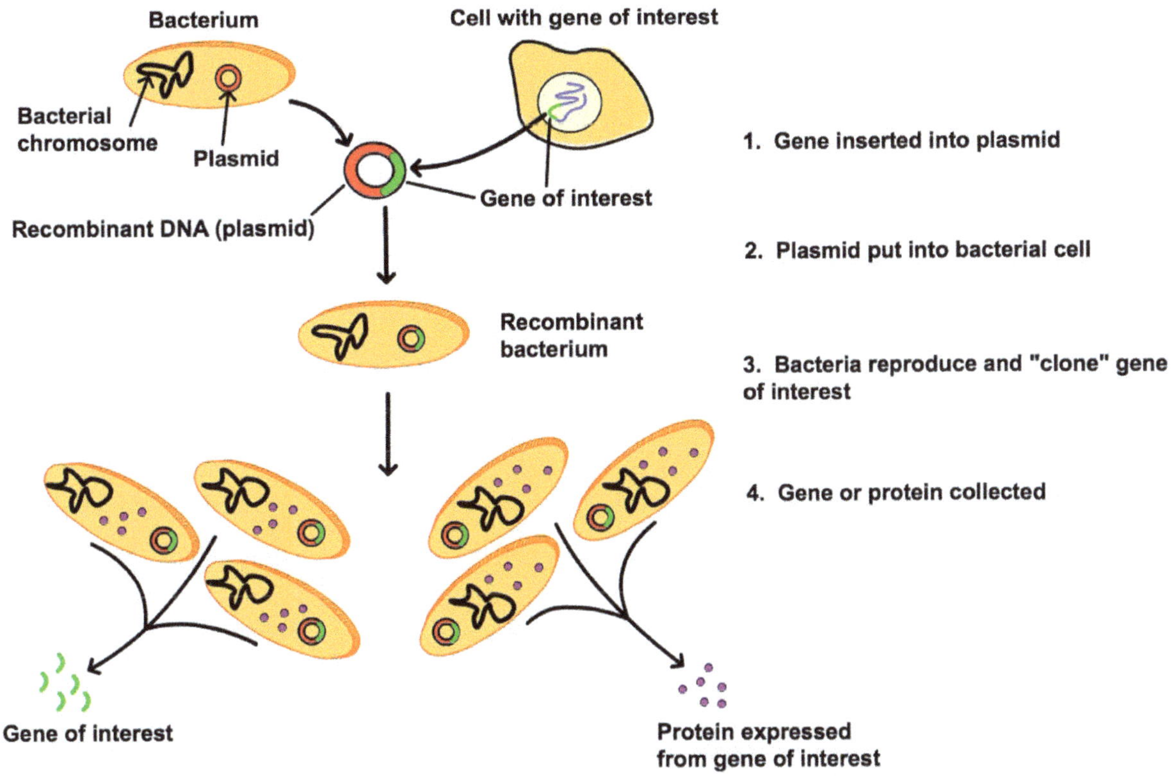

Real-Life Application: DNA Sequencing

Another application of biotechnology is **DNA sequencing**, which is determining the nucleotide base sequence of a gene or other types of DNA. Because DNA strands are complementary, it is possible to exploit this to determine nucleotide base sequences. The Sanger method, which is also called the chain-termination method, was the first method developed in 1975 by Frederick Sanger. Sanger won his second Nobel Prize in Chemistry, which he shared with Peter Gilbert, for developing this sequencing technique. His first Nobel Prize, also in Chemistry, was won in 1958 for determining the amino acid sequence of the insulin protein. Sanger's technique utilizes **dideoxynucleotides**, which are nucleotide bases that are almost identical to regular nucleotides except they have a hydrogen atom instead of a hydroxyl group on the number 3 carbon of deoxyribose. Figure 15.7 shows the difference between an adenine nucleotide and an adenine dideoxynucleotide. This means, in simple terms, that a dideoxynucleotide will always be the last nucleotide in a chain, and no more nucleotide bases will bond with it to elongate the chain. The hydrogen atom cannot share electrons with another atom, which stops elongation.

Sanger's sequencing method utilizes dideoxynucleotides or didNTPs to stop DNA replication in a test tube at specific points. By being able to stop DNA replication at certain points, it is possible to figure out each nucleotide base in a sequence one by one. The setup of the experiment has four separate test tubes, one for each type of nucleotide base in DNA: adenine, thymine,

FIGURE 15.6: The gene gun method of inserting DNA—When the gene gun is turned on, the helium gas forces the gold "bullet" to enter the plant cells, transferring the DNA.

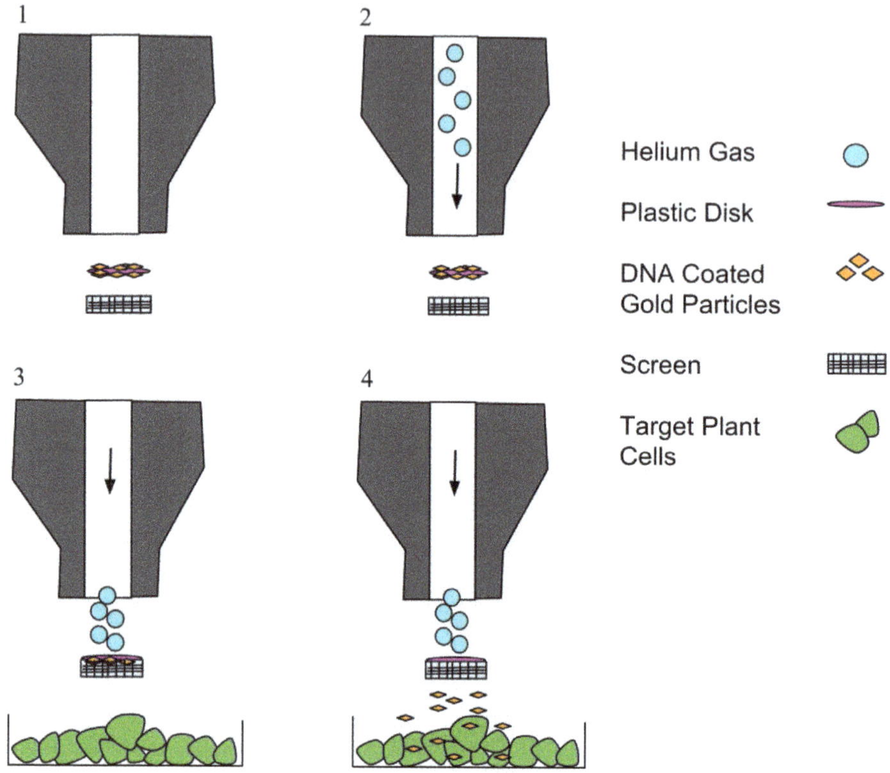

FIGURE 15.7A: Dideoxynucleotide of adenine

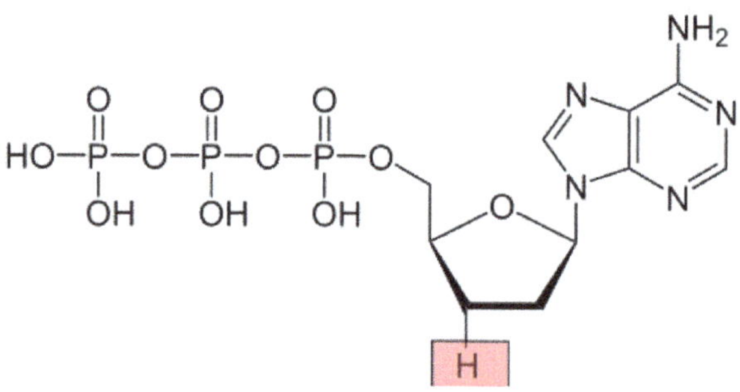

FIGURE 15.7B: Normal nucleotide—Structure of an adenine dideoxynucleotide and normal nucleotide. Highlighted in red is the structural difference between the two.

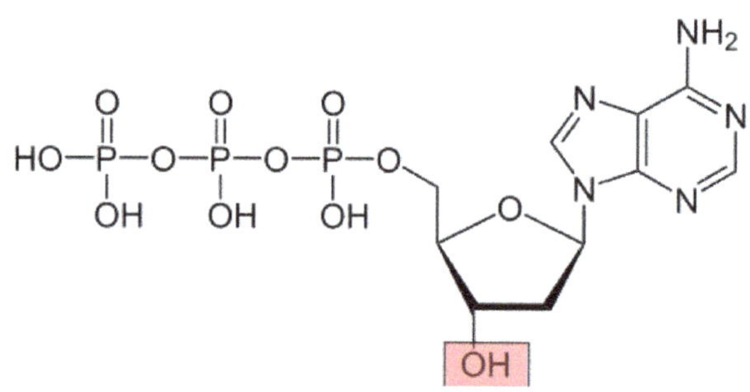

cytosine, and guanine. Each tube receives the DNA template to be sequenced, a supply of all four regular nucleotides, a short DNA primer, and DNA polymerase. Some of the regular nucleotides are radioactively labeled. Then each tube receives one of the dideoxynucleotides, didATP, didTTP, didCTP, or didGTP. The reaction tubes are warmed and cooled to allow the

DNA strands to separate, primers to attach, and DNA polymerase to begin adding new bases. Using complementary base pairing rules, A with T and C with G, the new strand of DNA is built until a dideoxynucleotide is added and elongation stops. The reaction will result in each tube containing many lengths of DNA that all end with the same nucleotide base. There should be enough pieces to account for every nucleotide base in the sequence (See Figure 15.8).

Figure 15.8 shows what the content of each reaction tube should have, but it is not possible to see this in real life. The DNA pieces can be separated from one another using gel electrophoresis. Each lane of the gel is loaded with a sample from one reaction tube. The power is turned on, and the pieces of DNA migrate down the gel, the shorter pieces traveling faster and reaching the bottom of the gel first. A Southern blot is done, and the gel can be read from the bottom to the top. Look at Figure 15.9 for an example of a Southern blot of a Sanger sequencing gel and how to read it.

Sanger developed this method in the 1970s and was limited in the length of DNA sequences he could determine. Since then, sequencing has become an important tool in biology, and the technique has been improved mainly through the use of computers. The **Human Genome Project** began in 1990 with the goal to map and sequence every gene in humans. It was a global, publicly funded project, which had the participation of many different research institutions and universities. James Watson was the original head of the project in the U.S. and was later replaced by Francis Collins. It used a variation of Sanger's technique called **clone-by-clone sequencing**.

FIGURE 15.8: Sanger method of DNA sequencing—Each of the reaction tubes gets the same reaction components except they each get a different dideoxynucleotide (didNTP). The tube on the left has didATP added so all the pieces end with an A nucleotide base. The pieces of DNA are of differing lengths and account for every place in the sequence that an A occurs.

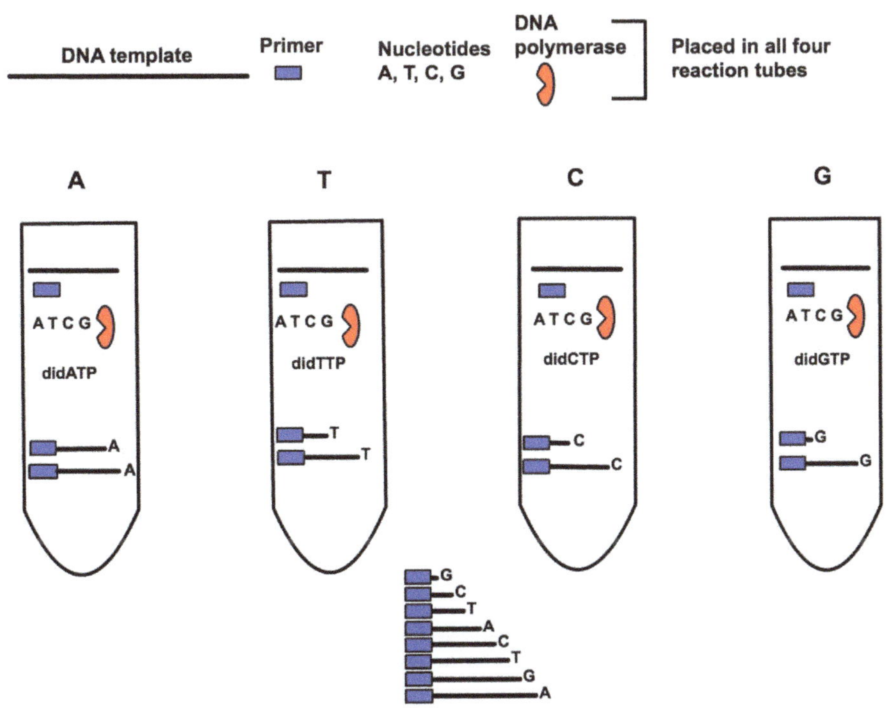

FIGURE 15.9: Sanger sequencing gel—You read the gel starting from the bottom, which is the shortest piece. Keep reading the sequence one band at a time moving toward the top of the gel. The sequence read of the gel is actually the complement of the original DNA sequence. The original DNA sequence can be determined using base pairs, A-T, G-C.

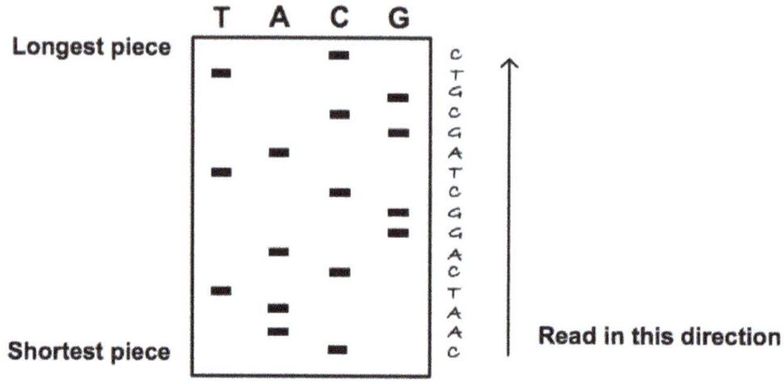

Sequence from the gel CAATCAGGCTAGCGTC

Original sample sequence GTTAGTCCGATCGCAG

In this method, a large genome is broken into many small pieces. The pieces are mapped to the chromosome they came from so that, later, the map can be used to put all the pieces of the genome back together. These pieces are inserted into bacterial plasmids and cultured so the bacteria create multiple copies of each piece. The small pieces are sequenced and then using the map the pieces are put together. In 1998, J. Craig Venter established a private company called Celera Genomics, which also had the goal to sequence the human genome. Scientists at the company had developed a different technique that they called **whole-genome shotgun sequencing**. This method is quicker than the clone-by-clone method because it doesn't build a map and skips the cloning process. Instead, the whole genome is broken into many, many small pieces all at once. Each of the small pieces is sequenced and then computers are used to find overlapping sequences to assemble longer and longer pieces of the genome. Although it is quicker, it can be less accurate initially as it may be difficult to find overlapping sequences to fill in gaps. In fact, the Celera team ended up using the public project's map to help with the final assembly of the whole genome. It was also more expensive initially as the company needed to buy a lot of new sequencing equipment. The final version of the human genome was released in 2003. Besides the human genome, other organism's genomes have also been sequenced, like the dog, corn, and our closest relative, the chimpanzee. Newer techniques of sequencing have made the process faster and cheaper. Where The Human Genome Project took over a decade to finish and cost millions of dollars, new methods can potentially do the same thing in less than a day for around $1,000.

An increase in genomic research has led to the production of enormous amounts of data, which need to be stored and analyzed. This can only be accomplished with the help of computers. This combination of large biological datasets and computer hardware and software is called **bioinformatics**. Because this data is stored in computers, it can be made searchable. A great deal of DNA sequences have been uploaded to databases, which are accessible to the public. One popular database is called BLAST, or Basic Local Alignment Search Tool, which allows users with an Internet connection to enter a nucleotide or amino acid sequence and find matches. By comparing sequences between organisms, it can be possible for researchers to establish and analyze evolutionary relationships. Genomic information can also be used in medicine. The Cancer Genome Atlas has the goal of finding mutations associated with 10 types of cancer. Gene sequences from normal and cancerous cells are compared from many patients to build a database of different mutations in genes and other parts of the genome. Researchers hope to use this information to develop cancer detection tests and drugs to treat patients.

Real-Life Application: Gene Editing

In the last few years, a new technique has emerged in molecular biology called **CRISPR/Cas9 system**. This system allows specific genes in a cell to be targeted and edited to be made nonfunctional or to repair mutations. Cas9 is an enzyme naturally found in bacteria. In bacteria, it is used to remove foreign DNA inserted by **bacteriophages**, which are viruses that infect bacteria. This is similar to what restriction enzymes do. The CRISPR in the system stands for Clustered Regularly Interspaced Short Palindromic Repeats. What this refers to are short DNA sequences in the bacterial genome that are palindromic, like AAAATTTTAAAA, and that have short "spacer" DNA sequences between them that are not repetitive. What's most interesting about CRISPR is that it is very similar to an immune system as it allows the bacteria to "remember" specific phages if it is infected again. When a phage infects a bacterial cell the first time and the cell survives, some of the phage DNA is integrated into the bacterial genome in the CRISPR region. If the same phage infects the same bacterium again, it actually causes transcription of the CRISPR DNA to occur, making RNA. Because the RNA contains a complementary sequence of the foreign DNA, it can find it and bind to it and is called guide RNA. Before it does that, it first binds to Cas9 enzyme and helps direct it to where the viral DNA is located. The Cas9 enzyme can now cut out the phage DNA thus ending the life cycle of the phage.

Scientists can utilize this system by designing an artificial guide RNA complementary to the gene sequence they want to target. The guide RNA and Cas9 are allowed to bind to each other and then they are introduced into a cell. Once in the cell, the guide RNA finds the target gene using its complementary nucleotide base sequence, and the Cas9 cuts the target gene. The damage to the DNA triggers the cell's DNA repair mechanisms. If there is no undamaged version of the gene for the repair enzymes to use as a template, they will add or delete random nucleotides. This can cause a large enough change in the gene sequence to result in a nonfunctional gene. Once the gene has been rendered nonfunctional, researchers can figure out what the gene actually does in the organism. Researchers have discovered that if the gene of interest has a mutation in it that causes a disease, it is possible to use CRISPR/Cas9 to repair the mutation. In this scenario, a normal gene with no mutation is introduced. When the repair enzymes begin to fix the broken DNA, they will use the normal gene as the template this time and add the correct nucleotide bases. The mutation that caused the disease will have been edited and corrected. Scientists have been able to do this in lab mice successfully and recently in humans. Scientists in China and the U.S. were recently able to successfully modify genes that caused a heritable heart problem in embryos. The system has also been used to successfully edit genes that code for receptors on T cells. The loss of these receptors can prevent infection by HIV. None of these procedures are available to patients currently, but it is hoped with further testing they will provide a new option for patients.

15.2 SUMMARY

Humans have engaged in simple forms of biotechnology for thousands of years by cultivating crops and domesticating animals. In the last few decades, however, biotechnology has changed to allow for modifications at the molecular level. Scientists have developed and use a variety of tools, like restriction enzymes, gel electrophoresis, and PCR to change the DNA of various organisms. These tools are used to produce genetically modified organisms through recombinant DNA technology like pest-resistant corn and bacteria that can produce human insulin. The human genome and other organism's genomes were determined using DNA clone-by-clone sequencing and whole genome shotgun sequencing. Newer methods of sequencing are helping researchers learn more about the relationships between organisms and genetic diseases. The CRISPR/Cas9 is a new tool available to researchers that will hopefully be successful in helping correct genetic mutations that cause disease in humans.

15.3 KNOWLEDGE CHECK

1. Which of the following is true about restriction enzymes?
   a. Restriction enzymes can cut DNA in any place
   b. A restriction enzyme will always recognize the same DNA base sequence and cut the DNA wherever that sequence is located
   c. Restriction enzymes were originally isolated from bacteria
   d. Bacteria use restriction enzymes to insert foreign DNA
   e. B and C are correct

2. Which of the following list the steps of polymerase chain reaction in the correct order?
   a. Annealing, extension, denaturation
   b. Denaturation, extension, annealing
   c. Denaturation, annealing, extension
   d. Annealing, denaturation, extension
   e. Extension, annealing, denaturation

3. What is the function of gel electrophoresis?
   a. To separate pieces of DNA based on their length
   b. To separate pieces of DNA based on their base sequence
   c. To make millions of copies of a piece of DNA
   d. To stop the elongation of a piece of DNA
   e. To join together two pieces of DNA

4. Why are dideoxynucleotides used in DNA sequencing?
   a. They will bond faster to other nucleotide bases than regular nucleotides
   b. Their structure ensures that they will stop elongation of a DNA molecule
   c. They are radioactive and can be used for tracing
   d. They are denser and will sink to the bottom during centrifugation
   e. None of these choices is correct

5. How does Cas9 know where to cut DNA?
   a. It recognizes phage sequences in the DNA
   b. It has repeats of nucleotide bases
   c. It is guided by RNA that has complementary bases to the DNA
   d. It randomly cuts the DNA
   e. It triggers transcription of complementary RNA

6. Why does the plasma membrane of a cell need to be broken down before DNA extraction can be completed?

7. Explain why it is necessary to use a special DNA polymerase in polymerase chain reaction.

8. Describe one advantage and one disadvantage of using the whole-genome shotgun sequencing method over the clone-by-clone sequencing method when determining the human genome.

15.4 ACTIVITY

Golden Rice

This case study looks at the genetically modified organism called golden rice. Golden rice is rice that has had the gene to produce β carotene added to its genome. The body produces vitamin A using beta-carotene. Vitamin A is helpful for proper immune system function, to prevent blindness, and ensure proper growth and development. It is called golden rice because of the slight yellow color of the rice. The hope was that golden rice could help with vitamin A deficiencies in developing countries where there are limited food supplies. This activity will allow students to debate why or why not golden rice should be grown in developing countries. Students will work individually or in small groups. They will be assigned a side, either pro or con, to research and then will conduct an intimate debate with class members as outlined in the case study "Golden rice: An intimate debate" published by National Center for Case Study Teaching in Science (NCCSTS).

Golden rice: An intimate debate http://sciencecases.lib.buffalo.edu/cs/files/golden_rice_debate.pdf

FIGURE 15.10A: ★ gel— Choose one of these three gels to start with and record its sequence.

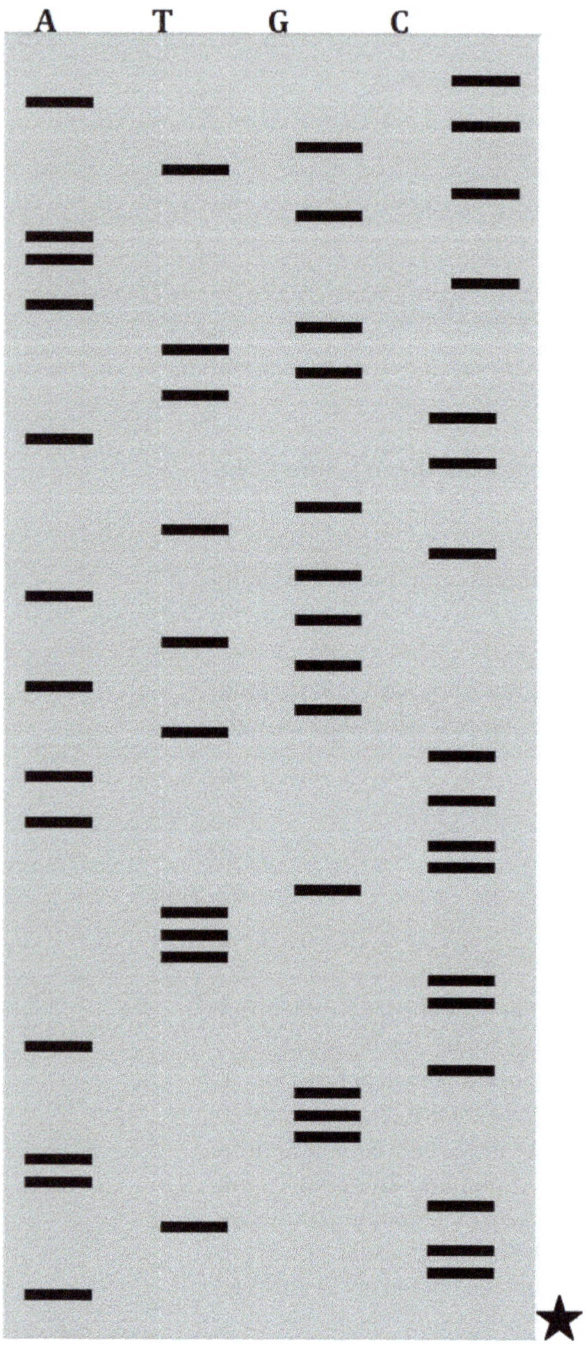

15.5 ACTIVITY

Gene Sequencing

During this activity students will work individually or in small groups to read three partial gene sequences from a simulated Southern blot. The three different partial gene sequences will be assembled into one complete gene sequence by working with other groups in the class. After assembling the whole gene sequence, students will use an online search tool called BLAST to find out what the name of the organism is and what gene was sequenced.

1. Your group should pick one of the Southern blots provided or your instructor will give you one (Figure 15.10A–C). Each blot has a different shape, so you can keep track. Make a note of which symbol you have.
2. Read the gene sequence from the gel starting at the bottom and working up the gel. It may be helpful to use a straightedge to help you keep your place. One person in the group should read off the nucleotide bases in the sequence while another records the bases, either on paper or in a text document on a computer.
3. Find another group with a different gene sequence with a different symbol. Compare the two sequences and see if you can find any overlapping sequences on the ends. If you find overlapping sequences this will tell you what order the two sequences are to each other. Add this sequence to your original sequence. If there are not overlapping sequences, find a group with the third symbol. Add the new sequence to your original sequence. Remember to only include the overlapping nucleotide base sequence once.
4. Find a group with the third sequence. Find the overlapping sequences to determine its order compared to the other two sequences. Record this sequence with the other two to construct a full sequence.
5. If you have not done this already, type the sequence into a text document. You can use lower case or uppercase letters but be careful not to leave out or change any letters.

FIGURE 15.10B: ✚ gel

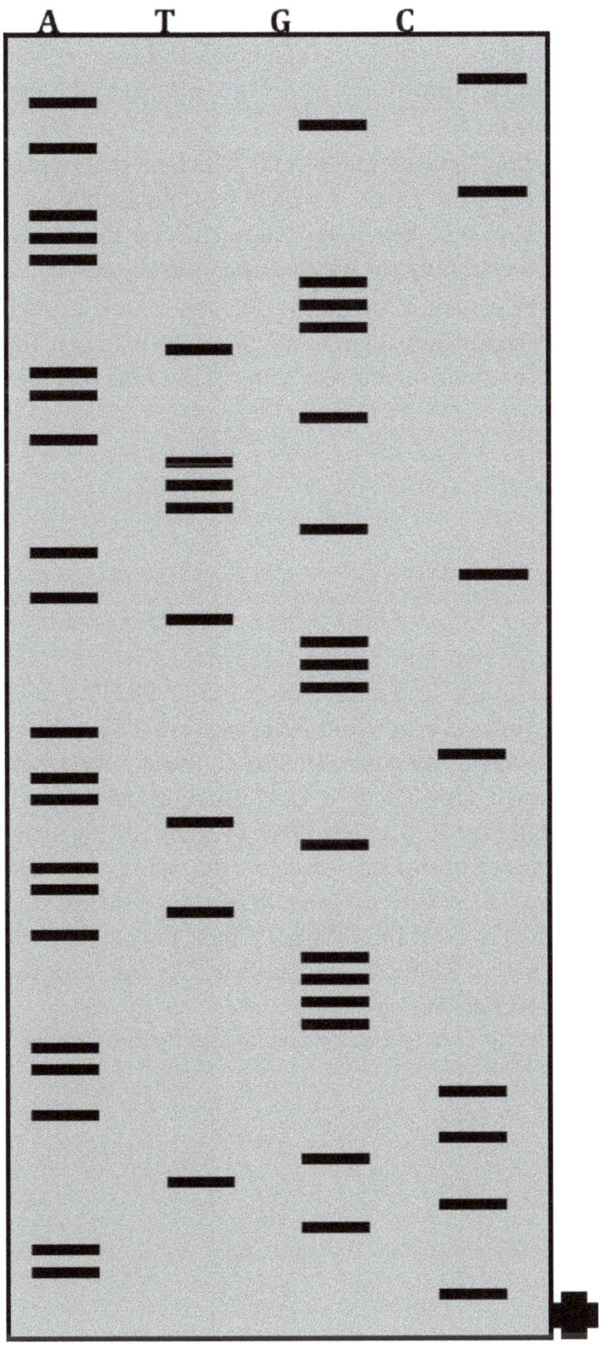

FIGURE 15.10C: ◆ gel

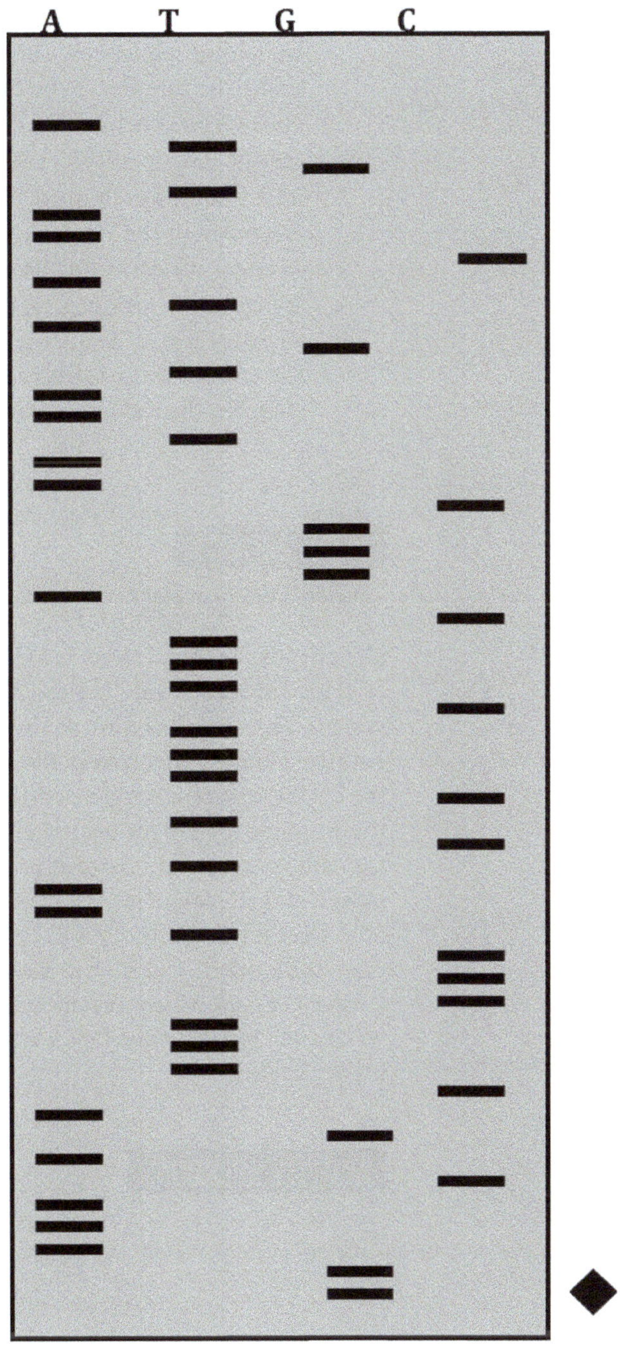

6. The tool you will use to search for your gene sequence is called BLAST. The letters stand for Basic Local Alignment Search Tool, and it allows users to input any nucleotide base or amino acid sequence into the tool and it will search its database for matching sequences. The tool is free to use, is open to the public, and is accessible using the Internet. Type this URL into your browser search window to go to the site: (https://blast.ncbi.nlm.nih.gov/Blast.cgi).
7. On the homepage you will see the option "Nucleotide BLAST". Click on that option and a new page to open. You should copy and paste or type in your nucleotide base sequence in the "Enter Query Sequence" box. You will need to click on the button marked "BLAST" for the tool to begin searching the database for your sequence.
8. It may take a minute to get results, so be patient. When the results appear, scroll down to where it says "Sequences producing significant alignments" to see the matches. The list will show the scientific name of the organism and the name of the gene that best matches the sequence searched.

15.6 – ACTIVITY

CRISPR/Cas9 Case Study

The purpose of this activity is to learn about the gene editing technique of the CRISPR/Cas9 system. This case study reviews the basic technique and applications of the CRISPR/Cas9 system. This case has four parts. In Part I, students watch and reflect upon a TED talk by Jennifer Doudna that covers the basics of the CRISPR system. In Part II, the application of the CRISPR system for gene editing is examined. How CRISPR/Cas9 functions as an adaptive immune system for bacteria is discussed in Part III. Lastly, in Part IV, gene editing using CRISPR is explored in more depth. Also examined in this case study are the strengths, weaknesses, and ethical issues associated with using CRISPR to edit genes in humans. The learning objectives for this activity will be accomplished by working in small groups to complete the case study titled "Cut it Out! Editing DNA with CRISPR/Cas9" published by the National Center For Case Study Teaching In Science (NCCSTS).

Cut it Out! Editing DNA with CRISPR/Cas9 http://sciencecases.lib.buffalo.edu/cs/files/crispr.pdf

ACTIVITY QUESTIONS

15.4

1. At the end of the activity, students should individually write a one-minute paper, which is a summary of the opposing side's argument and the evidence to support it.

15.5

1. What are the scientific name of the organism and the name of the gene that matches your sequence?
2. Why did you write down your gene sequence starting at the bottom of the gel and not at the top of the gel?
3. Is the sequence you recorded from the gel that of the original gene in the reaction or the complement of the gene? Does it matter? Why or why not?

15.6

1. Follow instructions to work individually or form small groups for completing the case study and answering the indicated questions from the case study.

15.7 REFERENCES

Collins, F. S., Green, E. D., Guttmacher, A. E., and Guyer, M. S. (2003). A vision for the future of genomics research. *Nature, 422*(6934), 835–47.

Comparing DNA sequences to understand evolutionary relationships with BLAST. (2012). *AP® Biology investigative labs: An inquiry-based approach teacher's manual, T76*. New York, NY: The College Board.

Hughes, S. S. (2011). *Genentech : The beginnings of biotech.* [On-line] Chicago, IL: University of Chicago Press. Access date: August 22, 2017. ProQuest Ebook Central.

Ma, H., Marti-Gutierrez, N., Park, S. W., Wu, J., Lee, Y., Suzuki, K., Koski, A., et al. (2017). Correction of a pathogenic gene mutation in human embryos. *Nature*, 548(7668), 413–9.

Sanger, F., Nicklen, S., and Coulson, A. R. (1977). DNA sequencing with chain-terminating inhibitors. *Proceedings of the National Academy of Sciences of the United States of America, 74*(12), 5463–7.

Sternberg, S. H., and Doudna, J. A. (2017). Expanding the biologist's toolkit with CRISPR/Cas9. *Molecular Cell, 58*(4), 568-74.

Venter, J. C., Adams, M. D., Myers, E. W., Li, P. W., Mural, R. J., Sutton, G. G., Smith, H. O., et al. (2001). The sequence of the human genome. *Science, 291*(5507), 1304–51.

FIGURE CREDITS

FIG. 15.4: Epp Moks et al./CDC, "Southern blot of two bacterial species," https://commons.wikimedia.org/wiki/File:05-0339_b.jpg. Copyright in the Public Domain.

FIG. 15.5: Adapted from Lisa A. Urry, Michael L. Cain, Steven A. Wasserman, Peter V. Minorsky, and Jane B. Reece, "Figure 13.24: An overview of gene cloning and some uses of cloned genes," Campbell Biology in Focus. 2nd ed. Copyright © 2017 by Pearson Education, Inc.

FIG. 15.6: Copyright © RachelBrooks15 (CC BY-SA 4.0) at https://commons.wikimedia.org/wiki/File:Gene_Gun.svg.

FIG. 15.7A: Adapted from "Normal nucleotide," https://commons.wikimedia.org/wiki/File:Desoxyadenosintriphosphat_protoniert.svg. Copyright in the Public Domain.

FIG. 15.7B: Adapted from "Dideoxynucleotide," https://commons.wikimedia.org/wiki/File:Desoxyadenosintriphosphat_protoniert.svg. Copyright in the Public Domain.

Glossary

#

5' cap modified guanine nucleotides added to 5' end of RNA during RNA processing

A

acid any substance that causes an increase in the H+ concentration of a solution

active immunity production of memory T or B cells induced by infection or vaccination

acute disease state of poor health that occurs suddenly and lasts a relatively short amount of time

adaptation heritable trait that makes an organism better suited to its environment and increases its fitness

adaptive immunity network of specialized cells capable of recognizing and eliminating specific pathogens

adenosine triphosphate (ATP) molecule known the "energy currency" of the cell

allele alternative version of a gene responsible for production of different observable traits

allergy abnormal response of the adaptive immune system to antigens that are not harmful

allopatric speciation type of speciation where the isolating barrier is geographic, for example a river or mountain range

amino acid monomer of a polypeptide that contains an amino group and a carboxyl group

analogy presence of similar structures on organisms that do not share a common ancestor

anaphase third phase of mitosis during which chromosomes separate and move toward opposite poles of the cell

aneuploidy the condition of having the incorrect number of chromosomes

antibiotic substance that inhibits the growth of or kills bacteria

anion negatively charged ion

antibody Y-shaped protein produced by B cells that helps the body eliminate pathogens

antiparallel property of DNA double helix; the two strands are always the same distance apart but are oriented in opposite directions

anticodon nucleotide base sequence located on a transfer RNA that is complementary to a codon on the mRNA

antigen molecule recognized by the adaptive immune system

antigenic drift accumulation of genetic mutations in a virus that leads to changes in the shapes of antibody-binding sites

antigenic shift when two different viruses infect the same organism and exchange segments of genetic material, resulting in the formation of a dramatically new strain of the virus

apoptosis programmed cell death that naturally eliminates aging, defective, infected, or unwanted cells

archaea any prokaryotic organisms that are not classified as bacteria because of their chemical composition, genetic code, and other characteristics

artificial selection a process where humans select the most practical or pleasing traits in organisms through breeding

asexual reproduction a type of reproduction which results in offspring that are genetically identical to the parent

A site first of three binding sites in ribosomes; location where the anticodon of a tRNA first binds to the corresponding codon in the mRNA

atomic mass mass of an atom, expressed in atomic mass units, is approximately equivalent to the combined number of protons and neutrons contained in its core

atomic number number of protons in an atom

atom smallest particle of an element that retains the properties of that element

autoimmunity abnormal immune response that targets an individual's own cells or tissues

autosome chromosome that is not a sex chromosome

autotroph organism that obtains food and energy without eating other organisms or consuming organic molecules produced by other organisms

B

bacteria single-celled prokaryotic organisms that lack membrane-bound organelles, but contain plasma membranes, ribosomes, cell walls, DNA, and other cellular structures

bacteriophage virus that infects bacteria

base any substance that causes an decrease in the H+ concentration of a solution

B cell lymphocytes produced by the bone marrow; release antibodies during adaptive immune responses

behavioral barrier barrier that prevents mating because one of the two sexes does not perform the correct courtship ritual, which the other sex doesn't recognize

benign tumor noncancerous mass of abnormal cells that is not capable of infiltrating nearby tissue or spreading to other parts of the body

binary data information that results from categorizing items into one of two mutually exclusive groups

binary fission method of asexual reproduction in prokaryotes in which the cell grows in size, duplicates its chromosome, then divides into two identical cells

biogeography field of study that looks at the distribution of different organisms around the world and how that reflects their evolutionary past

bioinformatics field of study that combines large biological datasets with computer software and hardware

biological species concept a species is made of members of a population that can or potentially can interbreed in nature and produce fertile offspring

biology the branch of science that focuses on the study of living things

biotechnology use of organisms or whole biological systems to develop products that help humans, including agriculture, industry and medicine

bone marrow fatty sponge like material inside bones that produces immature red and white blood cells and platelets

bottleneck effect type of genetic drift that is the result of a large portion of a population dying at or around the same time

bush meat wild animals hunted for food

C

Calorie the amount of energy needed to raise the temperature of one gram of water by one degree Celsius

capsid protective protein shell surrounding a virus

capsule sticky layer of polysaccharide that surround the cell wall of some prokaryotes; used for protection and adherence

carbohydrate macromolecule made of one or more sugar subunits

carcinogen substance or exposure that can lead to the development of cancer

carrier heterozygous individual who does not exhibit a disorder or disease but carries one copy of a recessive mutation that causes the disorder or disease

cation positively charged ion

causation when one factor, action, or event directly produces an observable change in other factor, action, or event

CD4+ T cell type of lymphocyte that helps activate other immune cells by secreting cytokines

CD8+ T cell type of lymphocyte involved in direct killing of pathogens and abnormal or infected cells

cell fundamental unit of life capable of carrying out all of the functions necessary for a living thing to survive

cell cycle ordered series of events in the life cycle of a cell, including interphase (G1, S, G2 phases) and mitosis

cell-mediated immunity protection against pathogens that is conferred by T cells and macrophages

cellular respiration the process of breaking down carbohydrates, lipids, and other molecular fuels into a form of energy that cells can use

cell wall protective structure found outside the plasma membrane of a variety of different cells, including those of bacteria, fungi, and plants.

centriole cellular structure made of tubulin involved in mitosis and meiosis

centromere the site of attachment for sister chromatids

centrosome structure formed by pairs of centrioles which organize mitotic spindle fibers that during mitosis.

chemokine type of cytokine that directs movements of cells to a particular area

chiasmata an area where two homologous chromosomes touch each other and exchange pieces during crossing-over

chloroplast organelle that captures energy from the sun to produce organic molecules through a process called photosynthesis

chromatin complex DNA, histones, and other associated proteins that condense to form chromosomes

chromosome structure composed of DNA and associated proteins

chronic disease state of poor health that lasts for a prolonged period of time

cilia small hairlike projections on the exterior of some cells to aid motility

clone-by-clone sequencing a method of gene sequencing that involves dividing the genome into small pieces that are cloned in bacteria before being sequenced; initially used in The Human Genome Project but was mostly replaced by whole genome shotgun sequencing

codominance mode of inheritance that occurs when both alleles for a trait are simultaneously expressed in heterozygotes

codon three nucleotide base code found in mRNA that corresponds to an amino acid

cohesion linking together of similar molecules, such as through formation of hydrogen bonds

communicable disease disease that can be spread between people through contact or exchange of body fluids

complementary property of DNA that determines which nucleotides form base pairs together

complement system set of proteins in blood plasma that help clear microbes as part of the innate immune system

complete dominance pattern of inheritance observed when individuals who are homozygous dominant and heterozygous have the same phenotype

continuous data information that can be represented by any numerical value within an infinite or defined interval

control group groups in which the independent variable is not applied or changed

control variable factor kept equal and left unchanged through an experiment

convergent evolution occurs when organisms evolve similar traits, not because they evolved from a recent common ancestor but because they live in similar environments

correlation a relationship between two observable phenomena that can be used to make predications about how those phenomena behave in relation to each other

covalent bond bond formed when two or more atoms share pairs of electrons

CRISPR/Cas9 system "clustered regularly interspaced palindromic repeats;" and enzymes used for editing DNA using a complementary RNA sequence as a guide

crossing-over the process during meiosis when homologous chromosomes exchange pieces resulting in new combinations of alleles. See also recombination

cyclin-dependent kinases (CDKs) enzymes that speed up various chemical processes important for cell division

cyclin protein that accumulates throughout different phases of the cell cycle until they reach a level that signals the cell to advance to the next phase

cytokine signaling molecule that cells release to affect the behavior of other cells

cytokinesis separation of the cytoplasm and other cellular contents into two distinct cells that occurs in the last stage of eukaryotic cell division

cytoplasm jelly-like substance known as cytosol that contains all of the cells organelles; found inside the plasma membrane

D

deductive reasoning type of logical thinking in which specific results are predicted based on known information and general principles

deletion genetic mutation involving improper removal of one or more nucleotides

denaturation process during which proteins lose their natural quaternary, tertiary, or secondary structure because of disruptive changes in external conditions

dendritic cell cell found throughout blood and tissues that ingests cellular debris and pathogens and displays antigens on its cell surface

density-dependent inhibition mechanism of cell cycle regulation that prevents overcrowding by stopping cell division in cells that are closely touching neighboring cells

deoxyribonucleic acid (DNA) nucleic acid containing adenine, cytosine, guanine, and thymine; usually exists as a double-helix molecule

dependent variable factor that is observed and measured during an experiment

dideoxynucleotides nucleotide bases that are almost identical to regular nucleotides except they have a Hydrogen atom instead of a hydroxyl group on the number three Carbon of deoxyribose.

diffusion tendency for molecules to spread out evenly into the available space

dihybrid cross mating of individuals who are heterozygous for two Mendelian traits

diploid having a complete set of chromosomes, sometimes written as 2n

discrete data information that can only be represented by a finite set of values

disease state of health in which the body cannot carry out all its normal functions

divergent evolution when two or more species arise from a common ancestor

DNA ligase enzyme that links nucleotides together during DNA replication and repair

DNA polymerase enzyme that catalyzes synthesis of DNA

DNA replication copying of one DNA molecule to make a second identical DNA molecule

DNA sequencing biotechnology technique that involves determining the nucleotide base sequence of a gene or other types of DNA

dominant allele allele that is completely expressed in heterozygotes

double helix structure of DNA formed by two intertwined antiparallel strands

duplication when a segment of a chromosome is repeated

E

electron negatively charged, low-mass subatomic particle

electronegativity strength of an atom's tendency to attract electrons

electron shell space in which electrons are grouped around the nucleus of an atom

elements distinct substances that cannot be chemically converted or broken down into other substances

endosymbiont theory explanation for how complex eukaryotic cells are thought to have evolved from symbiotic relationships with prokaryotic cells

endotoxin lipopolysaccharide; toxic molecule found in the cell membrane of Gram-positive bacteria

enzyme molecule that accelerates the rate of a chemical reaction without being consumed by the chemical reaction

epigenetic inheritance transmission of traits from parents to offspring that does not rely on the genetic code

epistasis mode of inheritance that occurs when expression of one gene alters or interferes with the expression of one or more other genes

E site last of three binding sites in ribosomes; location where the uncharged tRNA dissociates from the mRNA

essential amino acid one of nine amino acids that cannot be synthesized in the human body and must be consumed as part of the diet

eukaryote organism composed of cells containing a membrane-bound nucleus and organelles

evolution a change of allele frequencies over time

exocytosis process that actively transports molecules out of the cell

exon sequence of RNA transcript that remains after RNA processing

experimental group group in which changes to the independent variable are tested

exotoxin harmful molecule released by bacteria into their host or local environment

F

F1 generation first filial generation; second generation in Mendel's experiments produced from mating of individuals in the P generation

F2 generation second filial generation; third generation in Mendel's experiments produced from mating of individuals in the F1 generation

facilitated diffusion passive movement of molecules across the plasma membrane through a transport protein, and down their concentration gradient

fact something that is generally regarded as true, based on repeated observations and experimental evidence

fermentation the breakdown of glucose or other small carbohydrates to releases ATP as well as byproducts including carbon dioxide and various acids or alcohols

fertilization the process of combining two gametes, like an egg and sperm, to create a zygote

flagellum tail-like structure used to propel a cell through the environment

fluid mosaic model characterization of the plasma membrane as a movable collection of different types of molecules drifting in a fluid phospholipid bilayer

fossil the remains or impressions of a once living organism (i.e., it can be bone that has become mineralized, an impression or cast of an organism in stone or an organism trapped in tree resin called amber)

founder effect a type of genetic drift where a small group of individuals of a population leave and move to a different area and stay isolated

frameshift mutation a type of mutation where nucleotide bases are either inserted or deleted, which results in a shift in the reading frame of the gene

fungi type of single-celled or multi-cellular eukaryotic organism that obtain nutrients and energy from other living or dead organisms

G

G0 phase non-dividing, quiescent phase in which cells that have exited the cycle cell exist

G1 phase first stage of interphase during which cells grow and prepare for DNA replication

G2 phase third stage of interphase during which cells grow, duplicate organelles, and store energy

gamete haploid cell that are the result of meiotic cell division and used in sexual reproduction

gametic barrier when gametes of two species meet but do not chemically recognize each other and fertilization does not occur

gel electrophoresis technique used to separate pieces of DNA, RNA or proteins based on their size

gene cloning making multiple, identical copies of a gene

gene expression process in which the information in a gene is used to make a gene product, most often a protein

gene flow mechanism of evolutionary change that involves the immigration and emigration of members of a population into another population

gene sequence of DNA that functions as the basic unit of heredity; usually encodes at least one polypeptide

genetically modified organism (GMO) an organism that has had its genetic composition changed to contain DNA sequences that it does not normally have

genetic drift a mechanism of evolutionary change that permanently removes some alleles from the population as a result of a random event

genetic recombination mode of inheritance that results in offspring expressing combinations of alleles that differ from those found in either parent

genome an organism's complete set of genetic material, including genes and noncoding sequences

genomic imprinting type of epigenetic modification in which the expression of certain genes is dependent upon which parent the gene was inherited from

genotype set of alleles that an individual possesses for their genes

geographic barrier reproductive barrier that prevents mating because the two species are not found in the same place

glycolysis the breakdown of the carbohydrate glucose into two smaller molecules called pyruvate, releasing two molecules of ATP

glycosidic linkage covalent bond between two monosaccharides

Golgi apparatus part of the endomembrane system that packages and sorts molecules produced by the endoplasmic reticulum.

granulocyte innate immune cells that release packets of pre-formed enzymes and proteins in response to pathogens and tissue damage

growth factor protein or lipid hormone that stimulates cellular growth, differentiation, and replication

H

haploid a cell that contains only one of each chromosome, often written as n

helicase enzyme that unwinds DNA double-helix molecules

helminth multicellular eukaryotic organisms commonly referred to as parasitic worms

heterotroph organism that obtains food and energy from eating other organisms or by consuming organic molecules produced by other organisms

heterozygous genotype composed of two different alleles for a given gene

histamine molecule primarily released by mast cells that modulates the functions of nerves and blood vessels during immune reactions

histone positively charged protein around which negatively-charged DNA is wrapped to form chromatin

homeostasis actively maintaining stable internal conditions to support the processes of life

homologous chromosomes chromosomes that are similar in size and genetic information

homology type of evidence to support evolution, which is a similarity between organisms that reflect their ancient and current evolutionary relationships

homozygous genotype composed of two copies of the same allele for a given gene

hormone chemical messenger produced by a specialized gland that helps control a wide range of bodily functions

Human Genome Project a global, publicly funded project with the goal to map and sequence every gene in humans

humoral immunity immune reactions mediated by antibodies and other molecules found in extracellular fluids

hybrid an organism with parents of two different species

hybrid inviability reproductive barrier that results when a hybrid does not complete development and is miscarried

hybrid sterility reproductive barrier that results when a hybrid is born but is not fertile

hybridization mating of two true-breeding individuals with different phenotypes

hydrogen bond bond that results from an attraction between a hydrogen atom in one molecule that has a slight positive charge and a slightly negatively charged atom in a nearby molecule

hydrophilic describes molecules that readily interact with water

hydrophobic describes molecules that do not readily interact with water

hypertonic environment in which the concentration of water is higher inside of a cell than outside

hypothesis proposed explanation for an observation that can be tested

hypotonic environment in which the concentration of water is higher outside of a cell than inside

I

immune system network of organs, cells, and molecules that protects organisms from pathogens and malignant cells

immunodeficiency condition where the immune system is unable to effectively fight infectious diseases and cancer

incomplete dominance form of inheritance in which the dominant allele only partially masks the recessive allele, resulting in expression of an intermediate phenotype in heterozygotes

independent variable condition or item deliberately changed or manipulated during the experiment

inductive reasoning type of logical thinking that generates conclusions by forming connections between multiple related observations

infection the multiplication of harmful microorganisms within a host organism

infectious disease state in which the body cannot carry out all its normal functions because of the presence of pathogens that can be spread from one individual to another

inflammation innate immune response to harmful stimuli that involves immune cells, pro-inflammatory molecules, and blood vessels

innate immunity collection of nonspecific defense mechanisms that includes physical barriers, chemical barriers, and cells that respond to infection

insertion genetic mutation involving improper addition of one or more nucleotides

interphase non-mitotic phase of the cell cycle during which cells grow, replicate their DNA, and carry out their necessary functions

intersexual selection evolutionary mechanism where females will choose to mate with males based on some trait or behavior; also referred to as female choice

intrasexual selection evolutionary mechanism where members of the same sex (usually males) competing with one another for a mate; sometimes referred to as male-male competition

intron sequence of RNA transcript that does not code for part of a polypeptide and is removed during RNA processing

inversion occurs when a fragment of a chromosome breaks off and reattaches at the same place but in the reverse orientation

ion atom that has a charge as a result of exiting an ionic bond

ionic bond bond that forms when one atom gives or takes electrons from another atom

isotonic environment in which the concentration of water is the same on both sides of a plasma membrane

isotope form of an element that contains a different number of neutrons than protons

K

karyotype a picture of chromosomes, often used for diagnosis of chromosomal abnormalities

kinetochore complex of proteins that forms at connection of mitotic spindle fibers and the centromere

L

lagging strand strand of DNA synthesized in fragments during DNA replication because of its 3' to 5' orientation

latent period stage of an infectious disease that occurs between the time infection first occurs and the onset of symptoms

law a description of a natural phenomena based on observations and evidence

law of dominance part of Mendel's model of inheritance that states one version of a gene completely masks the other version of the gene when both alleles are present

law of independent assortment part of Mendel's model of inheritance that states alleles for different genes assort independently during meiosis

law of segregation part of Mendel's model of inheritance that states parents have two copies of each gene, but can only pass on one copy to each offspring; therefore the pair of alleles must separate (segregate) from each other during gamete formation

leading strand strand of DNA synthesized continuously in the 5' to 3' direction during DNA replication

leukocyte colorless cell of the immune system found in the blood and body fluids

linked genes genes that are inherited together more often than independent assortment would predict because they are located close together on the same chromosome

lipid type of hydrophobic molecule consisting mostly of carbon and hydrogen

locus specific location of a gene on a chromosome

lymphatic system part of the circulatory system that consists of lymph nodes, lymph vessels, the spleen, tonsils, and the thymus gland

lymphocyte small white blood cells that circulate through the lymphatic system

lysosome organelle that contains an acidic solution of digestive enzymes capable of breaking down large molecules

M

macromolecule large molecular formed by joining together smaller molecules

macrophage phagocytic cells involved in fighting infections, eliminating cancerous cells, and aiding tissue repair

malignant tumor cancerous group of abnormally proliferating cells capable of damaging nearby tissues, and possibly spreading to other parts of the body

mast cell immune cells that secrete histamine and other proteins through a process called degranulation during allergic reactions and in response to tissue damage

meiosis a type of cell division occurring in organisms that undergo sexual reproduction that results in production of haploid gametes

melanin dark pigment found in skin, hair and eyes

messenger RNA (mRNA) type of RNA produced by transcription of a DNA template and used in translation during polypeptide synthesis

metaphase second phase of mitosis during which chromosome pairs line up at the center of the cell

metaphase plate imaginary structure located in the center of a cell along which chromosomes appear to line up during metaphase

metastasis spreading of a malignant tumor from one site in the body to another

mismatch repair mechanism for recognizing and repairing genetic mutations

missense mutation genetic mutation that results in change from one amino acid to a different amino acid during translation

mitochondrion site of cellular respiration, which is the process that produces energy by breaking down carbohydrates, lipids, and other molecular fuels

mitosis division of chromosomes that occurs during the process cell division in eukaryotic somatic cells

mitotic spindle fiber assemblage of microtubule proteins that separate chromosome pairs during mitosis and meiosis

monohybrid cross mating of two individuals who are each heterozygous for one Mendelian trait

monosaccharide simplest form of carbohydrate (single subunit) with the general molecular formula of CH_2O

monosomy occurs when a zygote only contains one copy of a chromosome

mucous membrane layer of cells that line the respiratory tract, digestive tract, genitourinary tract, and other body cavities, and produce a thick protective liquid called mucous

multiplication rule states that the probability of two events occurring can be determined by multiplying the probability of each individual event

mutation physical or chemical agent that causes mutations in DNA

N

natural killer cell immune cell capable of recognizing and killing cancerous and virally-infected cells

natural selection evolutionary mechanism characterized by the differential survival and reproduction of some organisms in a population based on their advantageous traits

neutron neutrally charged, subatomic particle found in the atom's nucleus

neutrophil immune cell that circulates through the bloodstream and phagocytizes pathogens

nominal data information that can be placed into discrete categories, but not numerically measured or ordered

nondisjunction occurs when one pair of chromosomes does not separate properly during meiosis I

nonpolar covalent bond bond formed when two or more atoms share electrons equally

nonsense mutation genetic mutation that results in change from an amino acid-encoding codon to a stop codon

nucleic acid polymer made of nucleotide monomers

nucleoid region of prokaryotic cell where genetic material is concentrated; not surrounded by a membrane

nucleotide monomer of nucleic acids

nucleotide excision repair removal of damaged DNA followed by replacement with undamaged DNA synthesized using the complementary strand as a guide

nucleotide-pair substitution genetic mutation involving the change of one nucleotide base pair for a different nucleotide base pair

nucleus membrane-bound organelle that contains genomic DNA in eukaryotic cells

O

Okazaki fragment short segment of DNA synthesized during replication of the lagging strand

oncogene gene that promotes development of cancer when it becomes improperly activated or overactive

opportunistic infection infection that occurs more frequently or severely in individuals with weakened immune systems

ordinal data type of information for which variables have a natural or implied order, but the distances between values cannot be known or reliably measured

organelle cellular structure with a specialized function

organic composed of carbon-based molecules

osmosis diffusion of water across a selectively permeable membrane

P

pandemic infectious disease that affects a large number of people across a large geographic area

passive immunity transfer of preformed antibodies from one individual to another

pathogen a microorganism that causes disease

pathogen associated molecular patterns molecules on infectious microorganisms that are recognized by the innate immune system

pathogenicity the ability of a microorganism or virus to cause disease

peer review critical evaluation of academic or scientific work by knowledgeable professionals in the same field

P generation parental generation; first, true-breeding generation in Mendel's experiments

peptide bond covalent bond that forms between the nitrogen-containing "amino" group of one amino acid and the carbon-based "carboxyl" group of an adjoining amino acid

peroxisome organelle that help detoxify drugs and poisons, and to break down large molecules

phagocyte immune cells that protect the body by ingesting and destroying pathogens through the process of phagocytosis

phagocytosis process of "cellular eating" in which large particles are taken into a cell

phenotype observable physical and biochemical characteristics determined by the genetic makeup

phosphodiester bond covalent bond that joins subunits of nucleic acids

physical barrier reproductive barrier that prevents mating because the genitalia of the two sexes are not compatible in size or shape

pinocytosis process of "cellular drinking" in which large particles dissolved in liquid are taken into a cell

plasma membrane structure that separates a cell from its external environment; comprised primarily of two layers of phospholipids

plasmid a circular molecule of DNA found in most bacteria that is separate from the bacterial chromosome

pleiotropy the production of two or more independent characteristics as a result of a single gene

point mutation genetic mutation that affects a single nucleotide pair

polar covalent bond bond formed when two or more atoms share electrons unequally

polar molecule molecule with an uneven distribution of charge

poly A tail sequence of 50 to 250 adenine nucleotides added to the 3' end of an RNA transcript during RNA processing

polygenic inheritance mode of inheritance in which the phenotype for a single trait is determined by the additive effects of two or more genes

polymerase chain reaction (PCR) biotechnology technique that is used to amplify or make millions of copies of a DNA sample

polyploidy the condition of having additional complete sets of chromosomes, like 4n

population group of individuals of the same species that live within the same geographic area

post-zygotic barrier reproductive barriers that occur after fertilization

pre-zygotic barrier reproductive barriers that prevent fertilization

primary structure level of protein structure formed consisting of amino acids joined by peptide bonds

primer short nucleotide sequence that forms the starting point for DNA replication

prion infective proteins that are not associated with nucleic acids or any other type of macromolecule or structure

prokaryote organism made of a single cell that does not contain a nucleus

promoter nucleotide sequence associated with a gene that is the site of RNA polymerase binding

prophase first phase of mitosis in which nuclear membrane dissolves, chromosomes condense, and mitosis spindle fibers form

protein polymer made of amino acids linked by peptide bonds; folded into three-dimensional conformation

proton positively charged, subatomic particle found in the atom's nucleus

protozoan animal-like single-celled eukaryotic organism

P site second of three binding sites in ribosomes; location where the peptide bonds form between amino acids in the newly forming polypeptide

pseudoscience claims that are purported to be scientific and but are not based evidence collected using the scientific method

Punnett square diagram used to predict the allele compositions of offspring resulting from a cross or mating

Q

qualitative data information that can neither be measured numerically nor ordered along a scale

quantitative data information that can be measured numerically

quaternary structure level of protein structure formed by association of two or more polypeptides

R

radiometric dating process by which the age of a rock can be calculated by comparing the amount of a radioactive isotope found in the rock when it was formed to the amount of that isotope present currently in the rock

recessive allele allele that is completely masked in heterozygotes

recombinant DNA technology biotechnology technique that involves combining two genes or pieces of DNA from different organisms; may involve inserting DNA from one organism into another

recombination the process during meiosis when homologous chromosomes exchange pieces resulting in new combinations of alleles. See also crossing-over

replication fork Y-shaped structure that forms where the DNA double-helix separates during DNA replication

reproductive cloning method of creating a multicellular organism that is genetically identical to an existing individual

reproductive isolation occurs when two populations cannot interbreed with one another and there is no gene flow

restriction enzyme enzyme isolated from bacteria that can recognize specific DNA nucleotide base sequences and cut DNA

retrovirus type of virus that uses enzymes to reverse transcribe its RNA into DNA before integrating into host DNA

reverse transcriptase enzyme used by retroviruses to produce DNA from RNA

ribonucleic acid (RNA) nucleic acid containing adenine, cytosine, guanine, and uracil

ribosomal RNA (rRNA) nucleic acid that forms the structure of ribosomal subunits

ribosome cellular structure responsible for protein synthesis in all eukaryotic and prokaryotic cells

RNA polymerase enzyme that catalyzes synthesis of RNA

RNA processing modification of newly transcribed RNA; involves removal of introns and addition of 5' cap and poly-A tail

rough endoplasmic reticulum network of membranes that connects nucleus with the rest of the endomembrane system; has ribosomes attached

S

salt compound formed by an ionic bond

sample size number of observations or individuals included in an experimental group

saturated fat fat that contains the greatest number of hydrogen atoms possible, given the number of carbon atoms it contains

science the intellectual and practical process of systematically studying the natural world through observation and experimentation

scientific method iterative process that consists of making observations, asking questions, proposing hypotheses, conducting experiments, analyzing data, and refining this approach until conclusions can be drawn and shared with others

scholarly source article written by scientists or other trained professionals who are experts in their fields

secondary structure alpha helix and beta pleated sheet structures formed by hydrogen bonding between amino acids in a polypeptide

sex chromosome chromosome responsible for determining the genetic sex of an individual

sex-linked trait phenotype that results from expression of genes located on the sex chromosomes

sexual reproduction occurs when sex cells from two parents are combined to produce a new unique offspring that is different from both parents

sexual selection evolutionary mechanism that results in noticeable differences between males and females of a species; see also intrasexual selection and intersexual selection

silent mutation nucleotide substitution that does not result in a change of the amino acid sequence during translation

single-strand binding proteins proteins that bind to each strand of DNA to stabilize the replication fork during DNA replication

sister chromatids identical chromosomes joined together at their centromeres

smooth endoplasmic reticulum part of the network of membranes that connects nucleus with the rest of the endomembrane system that does not have ribosomes attached

somatic cell diploid cells that are not gametes

solute substance dissolved in a solvent

solution homogenous mixture of two or more substances formed by dissolving one or more solutes in a solvent

solvent substance capable of dissolving another substance

Southern blotting biotechnology technique that involves producing a photographic record of the DNA banding pattern after gel electrophoresis

S phase second phase of interphase during which DNA replication occurs

species a group of organism capable of breeding and producing viable, fertile offspring with each other but are not capable of producing fertile offspring with individuals belonging to other groups, or species

spleen organ that filters blood, maintains levels of red blood cells and platelets, and is involved in distributing activated white blood cells to the bloodstream

sympatric speciation a type of speciation that occurs one two one ancestral species develops into two distinct species while a geographic barrier does not separate the gene pools of the two groups; may occur when one group of the population may have started to change their role in their environment

synapsis process of homologous chromosomes pairing up together during meiosis I

T

telomerase enzyme that elongates telomere sequences

telomere repetitive sequence at the end of a DNA molecule that protects against loss of genetic code during successive rounds of DNA replication

telophase last phase of mitosis during which mitotic spindle fibers dissolve and a nuclear membrane forms around each daughter nucleus

template strand strand of DNA involved in transcription

temporal barrier a reproductive barrier that prevents mating because of different timing of breeding seasons

tertiary structure three-dimensional shape of a entire polypeptide formed by interactions between amino acid side chains

test cross mating of an individual of unknown genotype that expresses a dominant phenotype with an individual known to be homozygous recessive for that trait

theory broad explanation for why some aspect of the natural world occurs the way it does that is well substantiated by a large amount of evidence

theory of acquired characteristics evolutionary theory proposed by Lamarck that states that traits acquired during the lifetime of an organism that help it survive can be passed on to its offspring changing the population

thermocycler the machine that is used to perform polymerase chain reaction, which can be programmed to heat and cool the samples

thymus organ that is the site of T cell maturation

topoisomerase enzyme that facilitates the unwinding of the DNA double helix during DNA replication by reducing physical strain on the molecule

trans fat type of lipid that contains one or more double bonds, and are thus unsaturated fats, but has a three-dimensional structure that resembles a saturated fat

transcription synthesis of RNA from a DNA template

transfer RNA (tRNA) RNA molecule involved in translation; contains anticodon on one end and amino acid on the other end

transitional fossil fossil that has shared traits between two different but related groups

translation production of polypeptide from mRNA sequence

translocation occurs when part of a chromosome breaks off and attaches to a different, non-homologous chromosome

T regulatory cell type of lymphocyte that prevents immune responses from becoming overly robust or harmful

trisomy when a cell or individual contains three copies of the same chromosome

true-breeding refers to organisms that always pass on the same phenotype for given characteristics to their offspring after self-fertilization

tumor suppressor gene gene that produces products that normally limit progression through the cell cycle, decreasing the risk of cancer development

U

unsaturated fat fat that contains one or more double bonds

V

vaccination use of a substance to stimulate a protective antibody response specific for a pathogen

vacuole membrane-bound storage organelle derived from the endomembrane system

valence electrons electrons in an atom's outer electron shell

variable any condition or item that can be changed or controlled during an experiment

vector an organism that does not cause disease itself, but spreads infectious diseases by transmitting pathogenic microorganisms

vestigial structure structures that may no longer have a function in an organism but did have one in an ancestor or in a still living relative

virulence the severity of disease caused by infection with a microorganism or virus

virus a microscopic infective agent that contains genetic material but cannot grow or reproduce independently

W

whole-genome shotgun sequencing DNA sequencing technique that involves breaking the whole genome into many small pieces all at once, sequencing them and then reconstructing the sequence; used later for The Human Genome Project

X

x-ray crystallography use of x-ray diffraction to analyze crystalized molecules

x-ray diffraction scattering of x-rays by molecules; used to determine molecular structures during x-ray crystallography

Z

zygote diploid cell formed by the fertilized of a haploid egg by a haploid sperm

www.ingramcontent.com/pod-product-compliance
Ingram Content Group UK Ltd.
Pitfield, Milton Keynes, MK11 3LW, UK
UKHW052008041026
15590UKWH00008B/2209

9 781516 508952